Oxford Avian Biology Series

Oxford Avian Biology Series

Series Editor: Professor T.R. Birkhead FRS (University of Sheffield, UK)

A new series of exciting, innovative contributions from the top names in avian biology. Topics have been carefully selected for their wider relevance to both students and researchers in the fields of ecology and evolution.

Volume 1: Avian Invasions
Tim M. Blackburn, Julie L. Lockwood, Phillip Cassey

Volume 2: Interspecific Competition in Birds
André A. Dhondt

Volume 3: The Sensory Ecology of Birds
Graham R. Martin

The Sensory Ecology of **Birds**

Graham R. Martin
School of Biosciences, University of Birmingham

OXFORD
UNIVERSITY PRESS

OXFORD
UNIVERSITY PRESS

Great Clarendon Street, Oxford, OX2 6DP,
United Kingdom

Oxford University Press is a department of the University of Oxford. It furthers the University's objective of excellence in research, scholarship, and education by publishing worldwide. Oxford is a registered trade mark of Oxford University Press in the UK and in certain other countries

First Edition published in 2017

Published in the United States of America by Oxford University Press
198 Madison Avenue, New York, NY 10016, United States of America

British Library Cataloguing in Publication Data
Data available

Library of Congress Control Number: 2016951955

ISBN 978–0–19–969453–2 (hbk.)
ISBN 978–0–19–969454–9 (pbk.)

DOI 10.1093/acprof:oso/9780199694532.001.0001

Dedicated to the memory of Marie-Anne Martin

Preface

It is just 100 years since the publication of the book which started the area of enquiry that has become known as Sensory Ecology. That book was *The Fundus Oculi of Birds*, written by the Canadian-born clinician, Casey Albert Wood. At the time of its publication Wood was Professor of Ophthalmology at the University of Illinois. Although he was a clinician, he had a passion for birds and he managed to combine his two interests in this book. A recent exhibition, 'The Bird Man of McGill', was promoted by the rare books and special collections section of McGill University. It presented information on Wood's life and his many interests. Wood gave his personal archive and collections to the University and now the exhibition can be readily browsed in digital form at http://digital.library.mcgill.ca/caseywood/.

Casey Wood employed a comparative approach to the study of birds' eyes. Using relatively simple techniques, he examined and described the eyes of a wide range of bird species held in various zoos. He viewed the eyes of live birds through a hand lens or an ophthalmoscope, and he also examined whole retinas of excised eyes using a microscope. Wood recorded in drawings, and in paintings executed by Arthur Head, some of the diversity in structures found in the retinas of birds. The book was of a large format and took as its inspiration some drawings published 20 years earlier by J.R Slonaker in a paper titled, 'A comparative study of the areas of acute vision in vertebrates' in the *Journal of Morphology*. Remarkably illustrations from both Slonaker and Wood still have value to anyone wishing to start describing the diversity of birds' eyes and also for anyone wishing to provide a framework that accounts for this diversity in both functional and ecological terms.

Wood's final conclusion in his book was, 'In future no report upon a particular avian species can be held complete that ignores the visual apparatus'. It is very clear that this conclusion still holds. Indeed all of the detailed understanding gained in the past century on the physiology, anatomy, function, and evolution of vision in birds has reinforced many times over the wisdom of Wood's conclusion. Yet many people still ask very general questions about 'What can birds see?' They often ask this because they have a general understanding that the sensory world of birds is in some way different from ours. They often ask in the hope that there will be just a few simple differences between our vision and the vision of birds, simple facts that can be readily learnt and perhaps be of value when faced with an applied question, such as how to prevent birds flying into obstacles, or getting trapped in fishing nets.

Today, we not only know about the diversity of vision among bird species, but we also have potent ideas that can be used to account for this diversity. We are also now gaining a clearer understanding of the capacities and diversity of the other senses of birds, of the trade-offs within a particular sensory capacity, and of the trade-offs and complementarity between different sensory information in the control of particular behaviours. For example, we now have ideas for understanding how hearing can complement vision in the execution of particular tasks, or touch sensitivity can complement vision in others.

One hundred years on from Wood's book is a good time to bring together much of what is known about the senses of birds. Much of this information can now be interpreted in the broad framework of what has become known as sensory ecology. From one perspective this whole book is an argument in support of Wood's final conclusion, but we can now broaden his conclusion and state that, 'No report upon a particular avian species can be held complete that ignores its sensory systems'. This captures the fact that birds are far more than animals guided by vision, while vision may often be the dominant sense, many senses come into play at any one instant. Furthermore, particular behaviours may, in fact, be controlled by information which at best can be described as sparse. In some important instances, birds conduct key behaviours when their senses provide what seems to be a paucity of information.

This book initially sets some broad questions and provides a brief historical perspective on how senses in non-humans have been understood. These are followed by some general surveys of the sensory capacities of birds. In the latter half of the book, this information is used to answer questions on the information that birds have available to guide the execution of key tasks. Finally, the question of why birds are often unable to meet novel challenges posed by humans is addressed. Throughout the book, many old assumptions about the sensory worlds of birds are challenged by the research findings which have been published since Woods' seminal work. There is much here to interest anyone who has more than a passing interest in birds, in sensory systems, or in behaviour. I also hope there is much that will challenge reader's assumptions about the sensory information that they use to control their own behaviours.

Acknowledgements

Individual acknowledgements of the people who have helped to complete this book are all but impossible. So many people have contributed in so many ways and over many years, to the ideas presented here. Colleagues, research collaborators, and students have all provoked my ideas, framed questions for me to ponder, and provided support and companionship throughout my career. They have all helped to give me the motivation and shored up my confidence when it came to putting this book together.

The most tangible help came from my late wife, Marie-Anne, who initially encouraged me to put it all down on paper in one place, and she encouraged me to carry on its completion even though she knew she would never see the result. She would have been my proofreader and indexer since those were her professional skills and I am sure the text will have ragged edges which she would never have allowed slip through.

Other specific acknowledgements are due to people who have allowed me to reuse and edit their illustrations and photographs to help clarify and enlighten the text. I hope I have acknowledged appropriately all of their specific contributions in the figure legends. A small number of people set me on the path which was to lead eventually to this book about avian sensory ecology. Ian Gordon, Bill Muntz, and John Lythgoe were all there encouraging me when I first started asking questions about birds' sensory capacities and the information that they provided. What they started off has continued to fascinate me for nearly 50 years. Sadly none of them is alive to see this book and provide the feedback and comments which they generously offered so many years ago.

Contents

1 Birds' Eye Views 1

1.1 Metaphor and Reality 1
1.2 Many Birds, Many Views 3
1.3 The Tuning of Senses 6
1.4 Epicurus, Sextus, and the Sceptics 9
1.5 Sensory Ecology 11

2 Vision 19

2.1 The Emergence of Vision 19
2.2 What Eyes Do 21
2.3 Optimal Eyes 22
2.4 A Fundamental Trade-off in Vision 24
2.5 The Primacy of Vision in Birds 27
2.6 Sources of Variation in Camera Eyes 28
2.6.1 *The Basic Functional Components* 28
2.6.2 *Optical Systems of Camera Eyes* 31
2.6.3 *The Image-analysing System* 33
Colour vision and sensitivity in the spectrum 34
Photoreceptors and visual pigments 36
2.7 Variation of Image Analysis 40
2.7.1 *Variations in the Distributions of Receptors in a Single Eye* 43
2.8 Variation of Optical Structure 49
2.9 Variation of Visual Fields 55
2.10 Comparing Doves and Shearwaters: An Example of the Visual Ecology of Optical and Retinal Structures 57
2.11 Measures of Spatial Resolution 58
2.11.1 *Acuity* 58
2.11.2 *Contrast Sensitivity* 61
2.12 Conclusion: Vision in Birds 62

3 Hearing and Olfaction 65

3.1 Hearing 66

3.1.1 *The Sound Stimulus* 68

3.1.2 *The Hearing System of Birds* 69

3.1.3 *Hearing Sensitivity* 70

3.1.4 *Locating Sounds* 73

Determination of sound direction 73

Determination of sound distance (sound ranging) 77

3.1.5 *Echolocation (active SONAR)* 78

3.1.6 *Conclusion: Hearing in Birds* 82

3.2 Olfaction 83

3.2.1 *Organization of the Olfactory Systems of Birds* 84

3.2.2 *The Importance of Olfaction and Olfactory Bulb Size* 86

3.2.3 *Olfactory Information and Foraging for Specific Items* 86

3.2.4 *Detection of Foraging Locations using Olfaction* 87

3.2.5 *Body Odours and Semiochemicals* 89

Odour-based recognition of species 89

Odour-based recognition of individuals 90

Odour-based recognition of individual quality and mate choice 90

3.2.6 *Odours and Nests* 91

3.2.7 *Conclusion: Smell in Birds* 92

4 Touch, Taste, and Magnetoreception 94

4.1 Touch or Somatic Sensitivity 94

4.1.1 *Somatic Sensitivities* 96

Mechanoreception 96

Herbst corpuscles 96

Grandry corpuscles 97

Thermo-sensitive receptors 97

Cutaneous nociceptors 98

4.1.2 *Bill Tip Organs* 98

Bill tip organs in waterfowl 98

Bill tip organs in parrots 99

Bill tip organs in shorebirds, kiwi, and ibises 101

4.1.3 *Conclusion: Somatic Sensitivity in Birds* 103

4.2 Taste 103

4.2.1 *Taste Buds* 104

4.2.2 *Taste Genes and Taste Receptors* 104

4.2.3 *Relative Numbers of Taste Receptors* 105

4.2.4 *Taste Categories in Birds* 105

Sweet 105

Umami 106

Bitter 106

Calcium 106

Salt 106

Sour 107
Fat 107
4.2.5 *Taste and Foraging in Shorebirds* 107
4.2.6 *Conclusion: Taste in Birds* 108
4.3 Magnetoreception 108
4.3.1 *Animals that Detect the Geomagnetic Field* 109
4.3.2 *Magnetic Compass Mechanisms* 110
4.3.3 *Detection of the Geomagnetic Field* 110
The magnetite model of magnetic field detection 110
The radical pair model 111
4.4 Conclusion: Magnetoreception in Birds 112

5 From Senses to Sensory Ecology **114**

5.1 Making Sense of the Diversity of Bird Senses 114

6 Birds in the Dark: Complementary and Partial Information **117**

6.1 The Problem of Night-time 119
6.2 Absolute Visual Sensitivity and the Challenges of the Nocturnal Environment 125
6.3 Visual Sensitivity in Context: In and Out of the Woods 126
6.4 Nocturnal Birds 127
6.5 The Owls' Solutions to Nocturnality 128
6.6 The Oilbirds' Solution to Nocturnality 132
6.7 The Kiwi's Solution to Nocturnality 134
6.8 Nocturnality in Other Birds 141
6.8.1 *Nocturnal Parrots* 141
6.8.2 *Nightjars, Frogmouths, and Potoos* 145
Nightjars 145
Frogmouths and Potoos 149
6.9 Occasional Nocturnality 151
6.9.1 *Occasional Nocturnal Foraging* 151
6.9.2 *Nocturnal Migration* 156
6.9.3 *Night Attendance at Nests* 158
6.10 Conclusion: Birds in the Dark—Complementary and Partial Information 159

7 Birds Underwater: A Paucity of Information **161**

7.1 The Underwater Foragers 162
7.2 Optical Challenges of Foraging Underwater 163
7.3 Light Levels and Spectral Challenges of Foraging Underwater 166

7.4 The Challenge of Rapidly Changing Light Levels 168
7.5 Aquatic Foraging and Nocturnal Foraging 169
7.6 Tactile Information and Underwater Foraging 170
7.7 Solutions to Underwater Foraging 172
7.7.1 *Cormorants* 174
7.7.2 *Penguins* 178
7.7.3 *Auks* 180
7.8 Birds Underwater: A Paucity of Information 181

8 What Drives Bird Senses? 183

8.1 Visual Ecology, Trade-offs, and 'Just-so Stories' 185
8.2 Which Tasks Drive the Evolution of Sensory Systems in Birds? 186
8.2.1 *Key Tasks and Perceptual Challenges Faced by Birds* 187
Flight 187
Foraging 188
Predator detection 189
Reproduction 189
8.3 Competing Tasks and Competing Information 190
8.4 General Characteristics of the Visual Fields of Birds 191
8.5 Functional Interpretations of the Visual Fields of Birds 194
8.6 The Key Functions of Bird Visual Fields 194
8.6.1 *Control of Bill Position in Foraging* 194
8.6.2 *Panoramic Vision* 195
8.6.3 *Differences in Visual Fields between Closely Related Species* 196
8.6.4 *The Perceptual Demands of Bill Control versus Predator Detection* 196
8.7 What Is the Function of Binocular Vision in Birds? 198
8.7.1 *Binocular Vision in Birds* 198
Binocular field width 200
Vertical extent of binocular fields 201
Abolishing binocular vision 202
Binocular field widths, nocturnality, and predation 204
8.8 Visual Fields, Eye Size, and Imaging the Sun 207
8.9 Summary: The Key Drivers of Visual Fields in Birds 209
8.10 Binocular Vision, Optic Flow-fields, and Contralateral Vision 210
8.11 Summary: The Drivers of Visual Fields and their Fine Tuning 212
8.12 What Drives Colour Vision in Birds? 213
8.13 What Drives Bird Senses? 215

9 The Sensory Ecology of Collisions and Entrapment 217

9.1 Why Do Flying Birds Collide with Static Objects? 219

9.2 Information Available to Flying Birds 220
9.2.1 *Colour Vision* 220
9.2.2 *Spatial Resolution* 220
9.2.3 *Relative Depth, Distance, and Time-to-contact* 221
9.2.4 *Fields of View* 222
9.3 Comparing Bird and Human Views of their Worlds 222
9.4 The Functions of Lateral Vision in Birds 223
9.5 When Birds are Flying in Open Airspace, What are they Doing? 224
9.5.1 *Looking but Failing to See* 224
9.5.2 *Can Flying Birds Adjust their Rate of Gain of Visual Information?* 225
9.5.3 *Are Flying Birds Always Looking Ahead?* 226
9.6 The Sensory Ecology of Collisions 227
9.7 A Sensory Ecology Perspective on Collision Mitigation 229
9.7.1 *Collisions with Static Hazards* 229
Visual factors 231
Perceptual factors 232
Diverting and distracting birds 232
Tailored rather than general solutions 233
9.7.2 *Collisions with Moving Hazards: Aircraft and Wheeled Vehicles* 233
Manipulating the environment 233
Manipulating the hazard 234
Are vehicles perceived as predators? 234
9.8 Entrapment: The Problem of Gillnets and Diving Birds 235
9.9 Gillnet Bycatch Bird Species 235
9.10 The Role of Vision in Seabird Gillnet Bycatch Species 236
9.11 Distracting Birds from Nets 237
9.12 A Sensory Ecology Solution to Gillnet Bycatch 237
9.12.1 *Making Nets Conspicuous* 237
Lighting nets 238
Warning birds of the presence of nets 238
9.13 Warning Panels 239
9.13.1 *Patterns on Warning Panels* 239
9.13.2 *The Size of Warning Panels* 239
9.13.3 *The Colour of Warning Panels* 240
9.13.4 *The Detection Distances of Warning Panels* 240
9.13.5 *How Many Panels?* 240
9.13.6 *Would Warning Panels Be Effective?* 241
9.14 Collisions with Glass 241
9.14.1 *Mitigation Measures* 241
Patterns on glass surfaces 242
UV patterns on glass 242
9.15 The Sensory Ecology of Collisions and Entrapment: Conclusions 243

10 Postscript: Conclusions, Implications, and Comment **244**

Appendix 1 247
Appendix 2 253
References 259
Index 287

1
Birds' Eye Views

1.1 Metaphor and Reality

The bird's eye view is popular. Referred to by media reporters, management gurus, and politicians alike, a bird's eye view has become a favoured short-hand way of referring to an overview, a new perspective on a scene, or a new way of considering a set of ideas. A bird's eye view implies many things: standing back, looking down from a height, or taking stock of a broad range of ideas. Clearly, as a way of placing things in a broader context, a bird's eye view can be both exciting and liberating. Birds' eye views are, however, not only used as a metaphor. While it must be true that every bird's eye has a real view of the world, people often speak as though we too can see and appreciate such real birds' eye views, and we seem to assume that birds have a special way of seeing the world from a high vantage point.

This assumption appeared to be made concrete by the pioneering systematic use of aerial photographs for military purposes in the Great War of 1914–18 (Barker 2002), which was further developed by archaeologists in the 1920s to reveal traces of human activity and objects hidden below the surface of the earth (Brophy and Cowley 2005). This evidence seemed to show convincingly that a bird's eye view could reveal hidden secrets and to show that in landscapes there are patterns which cannot be detected from the usual ground-based human perspective. Now the runaway success of satellite images, first in posters and books, and from the start of the 21st century everywhere on the internet, means that all of us, at the click of a mouse or the touch of screen can apparently get a real bird's eye view of anywhere on the planet. The worldwide success of Yann Arthus-Bertrand's *Earth from the Air* book and touring photographic exhibitions (Arthus-Bertrand 1999) showed at the end of the 20th century how exhilarating it was for people from many cultures to apparently see the world 'through birds' eyes'.

To free us from a grounded or ground-based existence in order to see more, or gather together ideas in order to understand connections, is clearly a richly rewarding experience. However, the search for such perspectives using the metaphor of a bird's eye view is not new; indeed it has a long history. What is new is that real aerial views are so readily available. Furthermore, because of advances in photography and imaging, they seem more complete, more real, more up-to-date, and so we feel

The Sensory Ecology of Birds. Graham R. Martin: Oxford University Press (2017).
 DOI 10.1093/acprof:oso/9780199694532.001.0001

Figure 1.1 Bird's eye views. To the left an illustration published in 1872 under the title 'Bird's eye view of the city of Raleigh, North Carolina'. An imaginary perspective of how the town would look from the view point of a bird flying over, it is one of a series of similar 'bird's eye views' of towns in the eastern USA published at this time and was presumably inspired by views from a balloon. The collection is held in the Library of Congress, Washington, DC. Nearly 150 years later the same idea is still in use and exemplified by the image on the right, dated October 2014. It is from a press release publicizing a new camera under the title 'The 100% authentic Eagle Eye View'. The camera, mounted on the back of a White-tailed Eagle flying above Paris, looked over the bird's head towards the ground, (Credit SonySWNS.com Creative Commons).

more convinced that they do show us a real bird's eye view. They appear to show us how things actually appear to a bird as it flies over (Figure 1.1).

People have long thought of the view from a high promontory as equivalent to that of a bird, and once artists had mastered the art of perspective their imaginations led viewers to see scenes in new ways. Once flight was mastered by humans, seeing the scene from an ever-changing overhead viewpoint became thought of as even more bird-like, and the moving image captured on film in colour as an even more authentic bird's eye view.

Clearly it is an attractive liberation for humans to think in this way, to imagine that they are birds and can take it all in, to see their world in a broader perspective, and detect patterns that they had not seen before. But while we like to refer to this as a bird's eye view, does it really reflect what birds see? Do birds actually look down and take it all in? Do birds ever see the world as captured by the camera lens looking down from a crane arm, a balloon, an aircraft, or a satellite? Is it that simple?

What are real birds' eye views? How do birds see and experience the world, not only when they are flying over but when they are on the ground, foraging on a tree trunk, flying through woodland, or when they dive below a water surface? Each of these is also a bird's eye view. It is far from trivial to ask what do birds actually see as they fly over? Do they even look down and take it all in, or are they more interested in what is going on above, behind or out to the sides? Do birds attend constantly to what is below, and if so how much detail is evident to them? I will argue that there are in fact very many different birds' eye views. A bird's eye view depends not just upon what a bird is doing but it also depends very much upon the species through whose eyes we hope to view. It will also become evident that the way that we see the world is peculiar and particular to ourselves: the human eye view is as particular as that of the view of any species of bird. It will also become clear that our idea of a bird's eye view is not only quite different to those actually experienced by birds, but also different to those of all other animal groups. The general human-held idea of a bird's eye view is indeed just a metaphor; it would not be understood by any bird.

1.2 Many Birds, Many Views

The overall thesis of this book is not just that there are almost as many birds' eye views as there are species of birds, but also that the vision of each species is finely tuned to the tasks and the perceptual challenges which they face in their daily lives. My argument is that when account is taken of all the other sensory channels through which birds gain information about the world, it soon becomes apparent that each bird has available to them a complex array of information which underpins the performance of the plethora of tasks that constitute the life of birds. Birds do far more than fly and to do it all, they need diverse and complex sources of information.

At first glance this may seem rather bewildering. If there is so much variation between species, can anything be said that has general application? Can we extrapolate knowledge from one species and safely apply it to another? It will be argued here that there are indeed a number of powerful general principles and constraints which have worked to shape how birds gain information about their worlds. At base we must answer the question: 'What have been the drivers of natural selection that have led to the evolution of the many different birds' views of the world?' To understand these drivers we need to determine the functions that the senses of birds primarily serve. Is it flight, as has been commonly supposed, or reproduction, or foraging, or some other key activity in the lives of birds? What aspects of the natural environment really drive differences in vision and other senses between species, and how is vision used alongside, or in conjunction with, other senses in specific situations? In producing this broad perspective of constraints and drivers, we might be led to enquire just what human vision and our other senses are for: what has driven the selection of our own sensory capacities?

In trying to answer these questions, it is necessary to explore not just the diversity of vision and other senses but also the diversity of the environments in which senses have been shaped to operate and how they operate alongside each other. For example, natural light environments in any one place will vary over a very wide range of levels (potentially many millions-fold), the colour (or spectral distribution) of the ambient light regime can also vary dramatically, as also can the clarity of the scene due to particles and water vapour in the air. Furthermore, when a bird moves within a habitat or from one habitat to another (e.g. from air to water), the range and rate of transition between different sensory or perceptual challenges can be very dramatic and rapid. How do eyes, visual systems, and other sensory inputs cope, not only with specific challenges but also challenges that change?

It is also necessary to understand how information from the sense of vision has been compromised or limited by the very nature of light itself and the difficulties of capturing and extracting information from it: in short what are the ultimate limits of vision set by physics? Also how is sensory information from vision traded-off, compromised, or complemented by information gained from another sense? For example, how are vision and hearing, or vision and touch sensitivity in the bill, traded-off against each other to provide complementary information that helps a bird find food while staying safe from a predator?

One of the main problems in choosing to write about the diversity of birds' eye views, and for exploring the reasons for this diversity, is that birds are themselves a very diverse group of animals, a fact that is at the root of their fascination for us all. This general diversity does not, however, occur with respect to basic anatomy and physiology which are remarkably conservative (Baumel 1993; King and McLelland 1985). At a fundamental level, a chicken on a dinner plate is very much like the song bird in the garden, the grebe fishing in the local pond, or the shorebird probing in the mud of a foreshore. They may differ in size but all birds are built around the same basic anatomy, structure, and physiology. This is quite unlike the situation in other major vertebrate groups, such as mammals and fish, in which there are marked differences in all these features (Kardong 2014).

This conservatism among birds has been argued to be the result of fundamental constraints that arise from the requirements that limit all self-powered flying organisms and indeed all flying machines. These requirements are for the ability to achieve high power output combined with low body weight, both of which are necessary for getting off the ground, i.e. overcoming the power of gravity by exploiting the rather meagre properties of air to support weight (King and McLelland 1985). However, the embodiment of those basic conservative features common to all birds has allowed natural selection to shape birds into very diverse forms that are able to reach and exploit a remarkably wide range of habitats and resources across the Earth (Gill 2007).

This exploitation of many diverse resources has been achieved primarily through the evolution of diversity in body sizes and, importantly, through diversity in peripheral body structures: legs, wings, feet, and bills (Perrins 2009). Diversity among these is readily acknowledged and understood. They are explained as being the result of powerful selective pressures for the efficient exploitation of different resources, especially different foods. Indeed all ornithology textbooks contain chapters discussing and explaining the diversity of feet, wings, bills, etc., in terms of the physical and mechanical challenges that they have evolved to meet (Gill 2007; Perrins 2009; Podulka et al. 2004).

Explanations are now available, for example, of how the aerodynamic properties of different wing shapes result in very different flight capabilities and how these evolved to overcome challenges posed by the physics of flight and the mechanical properties of bones and feathers (Norberg 1990). But these different wing shapes are the result of variations within a very conservative forelimb structure, and it is clear that very subtle variations in size, shape, and mechanical properties of bones and feathers can result in quite different flying abilities. Similarly differences between species in their bill structures and shapes are the product of subtle variations in basic anatomy, which give rise to a rich variety of mechanical properties that are analogous to different human-invented materials and tools (King and King 1980; Podulka et al. 2004) (Figure 1.2).

Figure 1.2 A plethora of birds' eye views? This montage shows a wide range of bird species whose vision and other senses have been investigated to various levels of detail. All of these species, and many more, feature in this book. At first sight, this montage captures the striking diversity of bill shapes and sizes and could easily be a prelude to a discussion of the mechanics, functions, and the evolution of bird bills. We would not be surprised by such accounts of the diversity of these bills. However, alongside these more obvious examples of diversity of bird structure is an equal diversity in the structures of the eyes and other sense organs in these same species. The diversity of these sensory structures tell their own story of function and evolution. Especially important is the provision of information for guiding these different bills in their execution of key tasks in the daily lives of these birds.

It is these differences in bill properties that facilitate the exploitation of very diverse food types by different species. There is now good evidence that subtle, but functionally significant, evolutionary changes to bill structure can occur very rapidly within a population in response to the relative availability of different food types. Indeed it has recently been demonstrated that evolution through natural selection can actually occur in 'real time' on these peripheral structures (Grant and Grant 2014; Weiner 1994).

All of these peripheral structures (wings, feet, and bills) can be regarded as functioning principally to ensure the efficient exploitation of environmental resources. This can be through efficient locomotion within and between different environments, and directly in the exploitation of food resources. To the diversity of these peripheral structures, I now add diversity of sensory systems, especially eyes, driven primarily by the informational requirements of efficient foraging.

1.3 The Tuning of Senses

The overall argument of this book is that the sensory systems of birds, and the information that they provide, exhibit differences which are just as subtle and functionally important as the mechanical differences between bills, wings, or feet. Of course, these subtle differences between sensory systems are less obvious and less readily catalogued. Their properties are more difficult to measure than the differences between the structures and mechanical properties of bills or feet. The functional differences between bird species in their sensory systems are less obvious, less readily understood. This book attempts to make these differences in the senses of birds more explicit, more readily understood.

It will be shown that there are many sources of independent variation within sensory systems and these sources of variation can be combined in different ways. Furthermore, this has allowed sensory capacities to become finely tuned to particular sensory challenges so that important variations can occur even between closely related species and, in turn, these can have important influences upon behaviour. For example, two duck species (Figure 1.3) of the same genus differ in a relatively minor way in their visual fields. However, these small differences have clear consequences for the foraging and vigilance behaviours of these birds (Guillemain et al. 2002). This shows that the interactions between tasks and the information used to control them can be significant, but the differences upon which they are based can be very subtle. Differences in the sensory capacities of these ducks, unlike the obvious differences in their bill structures, are not apparent from casual observation. However, differences in the information that these two species have available are as important for understanding their behaviours as the differences in their bill structures.

Figure 1.3 Subtle differences in sensory capacities. Two species of wildfowl which are closely related and placed in the same genus: Northern Shoveler *Anas clypeata* (left), Eurasian Wigeon *Anas penelope* (right). These birds differ in their feeding techniques and can be readily seen feeding in different ways, but often in the same locality. Shovelers and Wigeons differ quite obviously in the structure of their bills. The broader bills of Shovelers function to filter food items from the water's surface. This does not require bill position to be accurately positioned using vision. The narrower bills of Wigeons function primarily for selective grazing in short swards. This does require the bill to be placed accurately to grab small items. These differences in the requirements for bill placement are reflected clearly in differences in their vision and, in turn, this also influences marked differences in other behaviours, especially the time devoted to predator detection and vigilance behaviours (Chapter 8). Subtle tuning of information gathering for the efficient conduct of one task can clearly have important consequences for how information is gathered for other tasks and lead to differences in other aspects of behaviour. (Photo Credits, Pete Blanchard).

Like bills, feet, or wings, there is plenty of scope for variation in sensory systems. Eyes have three main structural and functional components that can be subject to independent evolution: the optical system which produces an image of the world, the retina which starts the analysis of that image, and the position of the eyes in the head. Although they all operate within broad limits, there is much possible variation within each of these components, so much so, that the diversity of eyes and their associated capacities is certainly as large as the possible number of species of birds which exist. Furthermore, as in other peripheral structures, there must be variation between individuals within a species. It is, of course, these within-species variations that provide the material upon which natural selection can act. This is similar to the within-species variations which can be more readily seen in the size, shape, and physical properties of birds' bills, or the sizes and shapes of bones of the foot—features which are routinely measured as an index of variation within a species.

At core, birds may be rather conservative but we readily recognize that diversity in peripheral structures has led to the wonderful diversity that holds our attention and intrigues us all. Diversity in sensory systems, especially eyes, is no less intriguing and important for a full understanding of what birds are. There is, however, also variation in the other main sensory systems, including hearing, olfaction, taste, and touch.

The ears of birds, like those of all vertebrates, are hidden deep within the skull (Podulka et al. 2004). Casual observation might suggest that most birds do not even have ears, in that there are no obvious structures on the head associated with hearing. Even for the tutored eye, birds' ears are indicated only by a small group of feathers that hide the ear openings which are positioned just behind and below the level of the eyes. But ears within the skull may show diversity with respect to the length of the cochlea and the size and configuration of the bones of the middle ear. These can result in subtle differences in hearing capacities. Although most birds lack outer ears, those that do have them show remarkable abilities for the location of sounds (Norberg 1978; Payne 1971). Birds that lack these structures are relatively poor at sound localization. If attention is also paid to other sensory structures associated with the head, such as the clusters of touch-sensitive receptors found round the tip of the bill in some species (Berkhoudt 1980; Cunningham et al. 2010; Cunningham 2010; Piersma et al. 1998), then further diversity in the range of information that can be detected can be added to the information provided by the eyes.

When differences between these three main sensory systems: eyes, ears, tactile receptors, are taken into account, it seems very possible that not only are there different birds' eyes views, but each species actually lives within a unique sensory world, able to detect and respond to different information about the world that surrounds them. Understanding these worlds, why and how they differ, and how they may influence behaviour, is the challenge at the heart of this book.

I intend to take the reader through a series of arguments which I hope will leave them agreeing that we can talk only of birds' eye views, not the bird's eye view. Also I will demonstrate that the idea of a bird's eye view with which the chapter opened is indeed only metaphorical. I will also argue that when it comes to understanding the main factors which have shaped the evolution of vision in birds, the control of flight is perhaps something of an afterthought, something which is achieved within sensory parameters determined primarily by other factors. Flight, which is often portrayed as the very essence of birds, may not be controlled by special adaptations of the sensory systems. In fact, flight is probably controlled within constraints imposed primarily by the requirements for controlling not the position of the whole body, but by the exacting requirements for controlling the position of just a small part of it, the bill. Furthermore, these requirements for controlling bill position are balanced against a usually antagonistic requirement for the detection of predators. The control of general body position, as in flight, must be achieved within these more exacting requirements.

Working through this set of ideas requires a broad comparative context, and there is always need for data from a wider range of species in order to fully explore those general principles which provide the base upon which variation occurs. Once adopted, this comparative approach inevitably casts our own sensory systems, and the information that we use to control our own behaviour, in a new light. The conclusion is that humans too possess a special set of variations in the basic sensory systems of vertebrates, with the inevitable conclusion that the world which we inhabit is just as peculiar or special as that of other species.

In short, all animals share the same planet but they live in different worlds. The human perspective of the world is as specialized or peculiar as is that of other species. One of humankind's most striking primate features, two large eye set in the front of the head, is rather unusual. Furthermore, our retinas have a specialized set of photoreceptor cells arranged in a particular pattern. The result of our unusual eye positions, coupled with the features of our retinas, is that what we consider to be the 'normal' view of the world, a view in which the region of best resolution and best colour vision lies directly ahead, is far from typical. Human senses do not provide a monopoly on the reality of the world, nor do they provide a baseline from which all other species' views can be considered to deviate.

1.4 Epicurus, Sextus, and the Sceptics

That human senses do not provide a fixed reality of the world is not a new notion. The Greek philosopher Epicurus (341–270 BCE) (Figure 1.4) recognized that the senses, especially vision, place very real constraints not only on our overall understanding of the world but also shape how our personal worlds change from moment to moment (Warren 2009). This notion that our senses bring us ever changing information about even the same objects was crucial to Epicurus' system of ideas. It is a notion that has been discussed many times since; it was elegantly rehearsed with reference to the 'reality' of a wooden table by Bertrand Russell in his discussion on 'the uncertain nature of observed reality' (Russell 2001). However, it is a notion that is easily overlooked in the very anthropocentric world of today, a world that for most people is experienced without reference to other animals. It is also a world that for many people is captured by others through a camera lens and viewed on a screen, rather than from direct personal experience. Although Epicurus argued that human reality is always changing and uncertain, he did so from a purely human perspective. The Roman philosopher Sextus Empiricus (Figure 1.4) extended the insights of Epicurus to the world of non-human animals. In so doing, his arguments led to a way of understanding the world and to a philosophical framework whose application underpins the pervasive scientific world view of the present time.

Sextus Empiricus lived between c. 160 and 210 CE (Bailey 2002). He wrote many works, and many of his arguments were built upon ten principal observations (referred to as modes). These modes elaborated upon the ideas of Epicurus and on those of another earlier Greek philosopher Aenesidemus (who lived in

Figure 1.4 The ancient founders of sensory ecology. The Greek philosopher Epicurus recognized that humans and other animals differed in the information that they had available to them about the world in which they lived. He argued that this placed a severe limit on defining 'reality'. Sextus Empiricus expanded these ideas and used them to form the basis of his philosophical ideas of Scepticism.

the 1st century BCE). The writings of Sextus have become recognised as the best argued foundation for Scepticism, a still potent system of philosophy, indeed a whole approach to understanding life and the human condition (Popkin 2003).

Scepticism can be summarized as: 'neither affirm any belief as true nor deny any belief as false', an uncomfortable position for some people, but one which now lies at the heart of using science to understand the world.

The ten modes upon which Scepticism is based all address problems produced by the senses. The first three state:

- The same impressions are not produced by the same object owing to the differences in animals.
- The same impressions are not produced by the same object owing to the differences among human beings.
- The same impressions are not produced by the same objects owing to the differences among the senses.

Clearly the sceptical position is based upon a comparative approach, a cross-species approach, to the senses. It is also based upon recognition that the sensory systems of all animals differ one from another. From this, it is concluded that humans cannot ever be sure about the world. This is because what is known about the world is very different depending upon what information different senses provide, even about the same set of objects. In short, different senses provide different information, and different animals have different senses, so where does reality lie? The sceptics worked from these observations to arrive at a position which argues that it is not possible for humans to either affirm or deny any belief; in modern terms we must 'keep an open mind', because at a very fundamental level we cannot trust our senses to provide a foundation for certain truth about the world.

The sceptical system is underpinned by comparisons with emphasis upon differences among animals, among humans, and among the senses. But while the early sceptics could assert that there were such differences, they based this upon their own observations and upon the application of reason, they could not quantify these differences nor could they say how or why these differences came about. Today we have a range of techniques to enquire about the differences between the same senses in different animals. It is now possible to piece together diverse information to actually quantify differences in sensory information between different animals.

In addition, an evolutionary framework, driven primarily by the idea of natural selection, now gives us a way of understanding why and how these differences have come about. We have also acquired tools to enquire about the types of behaviour that different sensory information controls directly or indirectly. In essence, much of this book is about what has been learnt since Sextus was writing. The sceptical conclusions of Sextus have been reinforced, not diminished, by our modern understanding of sensory capacities and the reasons why they differ between animal species.

1.5 Sensory Ecology

The enterprise of fleshing out these differences, and understanding the reasons for them, falls within the broad subject of sensory ecology, which can be summarized as, 'The investigation of the information that underlies an animal's interactions with its environment' (Dusenbery 1992). David Dusenbery elaborated this approach in his book *Sensory Ecology: How Organisms Acquire and Respond to Information*.

The premises upon which sensory ecology is based clearly have their roots at least two thousand years ago in the writing of Epicurus and Sextus Empiricus. However, really important developments had to wait until the middle of the 19th century when Alfred Russel Wallace and Charles Darwin formulated the idea of evolution through natural selection. This put the enterprise of understanding why and how differences had arisen between animals, including differences in their senses, within a more certain framework.

Early works that looked into describing the differences between the senses of species, and how they were related to the perceptual challenges posed by different environments, first took shape with the publication of Casey Albert Wood's *The Fundus Oculi of Birds* in 1917 (Wood 1917). Wood (Figure 1.7) took a comparative, and what can now be described as a sensory ecology, approach. He was a Canadian-born clinician and Professor of Ophthalmology at the University of Illinois, but

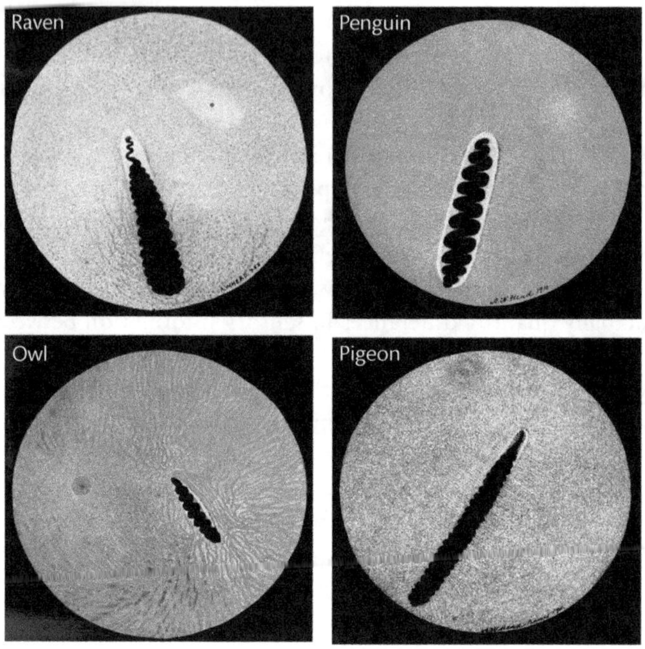

Figure 1.5 Diversity of birds' eyes. Four examples of the retinas of birds as viewed through an ophthalmoscope and portrayed in paintings in the book *The Fundus Oculi of Birds*, which was published 100 years ago (Wood 1917). The large black structure in each illustration is the pecten which extends out of the surface of the retina into the chamber of the eye (see Chapter 2 and Figure 2.1 for an explanation of the function of a pecten). The colours are genuine and modern photographs using a fundus camera show the same patterns. Clearly there is much diversity in all aspects of what is observed in these eyes. However, providing functional interpretations of this diversity is not straightforward. Wood concluded that every eye is different and warrants further investigation of its structure, physiology, and capacities, and in turn how these relate to behaviours and habitat characteristics. Shown here are 4 of the 53 species illustrated by Wood; each one has a different appearance. Raven: Northern Raven, *Corvus corax*; Penguin: African Penguin, *Spheniscus demersus*; Owl: Tawny Owl, *Strix aluco*; Pigeon: Common Wood Pigeon, *Columba palumbus*.

with a passion for birds. He used relatively simple techniques to examine and describe the eyes of a wide range of bird species. He viewed their eyes through a hand lens or ophthalmoscope, and also examined whole retinas of excised eyes using a microscope (Figures 1.5 and 1.6). Wood recorded in drawings and paintings some of the diversity in structures that he found in the retinas, including multiple foveas and elongated areas of high visual resolution. He described the eyes of more than 50 species to provide comparisons across phylogeny and broad characteristics of different environments. Crucially, he proposed ideas as to the functions of the different types of retinal organization that he identified. He suggested

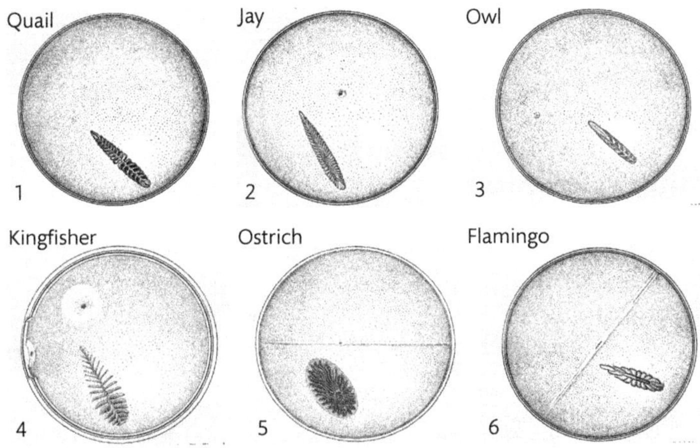

Figure 1.6 Diversity of birds' eyes. Six further examples of the retinas of birds viewed through an ophthalmoscope from Wood (1917). These series of illustrations emphasize structures which provide heightening spatial resolution within the field of view of the eye. Wood examined the eyes of 32 species in this way and reduced the diversity down to six main types in a sequence from least to most complex. Wood's examples of each type are presented here. In Type 1, there does not appear to be any region specialized for higher resolution vision; illustrated by California Quail *Callipepla californica*. Type 2 contains a single fovea in which the density of photoreceptors cells are increased and the retina is thickened (see Figure 2.8); the fovea is more or less centrally placed and so the region of highest visual resolution looks out along the axis of the eye; illustrated by Steller's Jay *Cyanocitta stelleri*. Type 3 is similar to type 2, but the region of higher acuity is displaced away from the centre of the retina and looks slightly more forwards in the field of view; illustrated by Western Barn Owl *Tyto alba*. Type 4 contains two regions of high resolution, one central and the other in the far periphery of the eye's field of view; illustrated by Common Kingfisher *Alcedo atthis*. Type 5 contains a linear area of enhanced resolution which runs in a band across the entire retina with a well-defined fovea placed more or less in the centre; the band is oriented approximately horizontal in some species (as illustrated here) but it can also be oblique; illustrated by Common Ostrich *Struthio camelus*. Type 6 contains a linear area and more than one fovea-like structures situated within the linear area; illustrated by Greater Flamingo *Phoenicopterus roseus*.

retinal organization could be divided into six main types (Figure 1.6). Among his conclusions was: 'The fundus oculi of Birds exhibits a great variety of areas of distinct vision, and these correspond closely to the habits and habitat of these animals—especially their methods of obtaining food, of escape from enemies, of migration, of reproduction, etc.' (p. 7). He also cautioned that 'Domestication or prolonged captivity brings about abnormal changes in the eye-ground of Birds, so that only healthy, wild specimens should be utilized in this or a similar research.' His final conclusion was that 'In future no report upon a particular avian species can be held complete that ignores the visual apparatus.' Thus, Wood was one of the first people to adopt a sensory ecology approach to understanding the function and diversity of birds' eyes, and to recognize that understanding senses is fundamental to understanding the biology of each species. A century after their publication, Wood's conclusions still hold. Furthermore, his illustrations still provide an intriguing catalogue of diversity in the gross structure of birds' eyes, and they continue to be a source of questions concerning how different species extract information from the world.

The next landmark publication was Gordon Lynn Wall's (1942) book *The Vertebrate Eye and its Adaptive Radiation* (Walls 1942) (Figure 1.7). This book was

Figure 1.7 Twentieth century founders of sensory ecology. All four of these scientists played a key role in identifying, describing, and analysing the perceptual challenges posed by life in different environments. They also described and analysed how these challenges had been met by the evolution of differences in sensory systems and sensory capacities across the animal kingdom.

published in the year that Casey Wood died. Although concentrating only upon vertebrate eyes, Wall's book set a standard of breadth and depth in a single volume on senses that has, perhaps, not been surpassed. His volume contained 785 pages with an index taking up 60 of them. Many of the index entries are about individual species and reveal the broad taxonomic base of the comparative approach that Walls adopted. Given the growth in knowledge about how eyes work and what they can do, a modern publication covering the same ground as Walls' book would now have to be multi-volume. Indeed, although taking a more focused approach to particular senses and concentrating much upon humans, a six-volume series, extending to some 4200 pages, *The Senses* was published in 2008 (Basbaum et al. 2008).

Such has been the importance of Wall's book that it is still cited today, not only because it provides clear enunciation of some key hypotheses linking form and function of eyes, but also because it contains the only information available on the eyes of many species. Walls ranged widely and brought together all that was then known about vertebrate eyes, their structure, physiology, and evolution, and very much of what he described still stands. However, Walls was not particularly concerned about what eyes actually did for the animals that possessed them; he was more concerned with relationships between anatomy, physiology, and function. While Walls' main concern was with the features which made eyes specialized for different abilities (e.g. for the detection of colour or spatial detail), he was less concerned with how those abilities facilitated or constrained the behaviour of the particular animals that possessed them.

André Rochon-Duvigneaud published *Les Yeux et la Vision de Vertébrés* in 1943 (Rochon-Duvigneaud 1943), one year after Walls' book (Figure 1.7). At the time of publication Rochon-Duvigneaud was 80 years old. He had retired from clinical practice as an ophthalmologist 17 years earlier and had devoted his retirement to comparative studies of the structure of the eyes of a wide range of vertebrate animals. This book was the result; he published few papers on these topics. Like Walls' study, the book was framed around a broad comparative approach and took what would now be seen as a sensory ecology framework to investigate questions of diversity and function. Also, like Walls', the book was encyclopaedic in approach and ran to over 700 pages. However, Rochon-Duvigneaud's book was never published in English and it did not become as widely known as Walls'. These books complemented each other and a joint publication covering all of the material would have been a marvellous compilation. The two authors were of different generations. Walls graduated with his first degree in mechanical engineering at Tufts University (Boston, Massachusetts), in 1926, the same year that Rochon-Duvigneaud retired from his practice in clinical medicine at the Hotel-Dieu in Paris. It is intriguing that they must have been working on their books in parallel—one at the start of his career as a young man at Wayne State University, Detroit, the other during his second career in retirement in Paris.

A book which helped broaden the framework of sensory ecology, so that it came to embrace a wider range of questions about the information that animals use in the guidance of their behaviour, was Niko Tinbergen's *The Herring Gull's World* (Tinbergen 1953). A pioneering text in ethology, this book has the subtitle '*A Study of the Social Behaviour of Birds*'. It dealt with topics well outside the range of questions posed by Walls, but it brought focus on some of the specific signals that may trigger certain behaviour, and asked whether sensory systems may be tuned to the detection of those signals. The book made it clear that to understand an animal's behaviour, it is necessary to appreciate the world that it inhabits from its own point of view, rather than from descriptions of how humans perceive the world. Tinbergen showed that the world of Herring Gulls is very different from the human world, and also very different from the world as seen by other animals which may be sharing its same environment.

The first work to use the term sensory ecology in an explicit way was John Lythgoe's *The Ecology of Vision* (Lythgoe 1979) (Figure 1.7). This book brought together ideas about the need to describe environments objectively from a sensory perspective. Lythgoe showed that different environments pose different sets of sensory or perceptual challenges, and that these challenges may require different information for their resolution. Importantly Lythgoe also showed that to meet those challenges, natural selections may have favoured the evolution of sensory systems with different properties in different species. Lythgoe's work was focused primarily upon the underwater environment and the visual challenges it posed for fish with respect to such variables as the different spectral distributions of upwelling and downwelling light and turbidity, and how absolute light levels and spectra change with depth and season in different bodies of water. Lythgoe was able to show that there was, perhaps, an optimal or a narrow set of solutions to these various challenges. He also showed that the visual world of fish varied markedly depending upon not just the light characteristics which were typical of the waterbodies that they inhabited, but also what resources the fish were trying to exploit in those habitats, for example whether they were primarily prey or predators.

Sensory ecology finally became established as a field of endeavour when the first international conference on the 'Ecology of Vision' was held at the International Marine Centre, Sardinia, in 1994. The book produced from its proceedings (Archer et al. 1999) brought together contributions from a wide range of authors who, perhaps, had not previously seen themselves as part of a joint sensory ecology enterprise. However, the conference and the book of its proceedings lacked a clear definition of the remit of visual ecology. There were contributions on many aspects of the mechanisms of vision viewed primarily from an ecological or environmental perspective, and these gave an idea of how diversity in sensory capacities and sensory mechanisms could be linked to their function in particular organisms. The contributions ranged across many taxa with speakers dealing with invertebrates, mammals, fish, and birds.

That the term sensory ecology has become adopted is not without significance. Sensory ecology is not just about objective descriptions of the sources of information in the environment, such as light, sound, chemicals, and magnetic and electrical fields, and how these sources vary between environments. Furthermore, sensory ecology is not just about the mechanisms which have evolved to extract the available information. While these topics do contribute important ideas to the study of sensory ecology, they are not, according to Dusenbery's definition, its focus. His definition states clearly that sensory ecology is about the information which guides an organism's interactions with its environment.

Ecology itself has a number of different definitions (Krebs 2009), but all of them focus to some degree upon the interaction of organisms with their environment. Dusenbery's definition picks up on that core definition and focuses upon how organisms acquire and respond to the information potentially available in their environment. While Dusenbery's work focused mainly on sensory ecology questions concerning micro-organisms, he showed the importance of determining what information an animal obtains from its environment (often framed in terms of limitations upon performance), how this information is obtained (the mechanisms of sensory systems), and the uses to which an animal may put that information. Thus, sensory ecology is very much concerned with what animals do with information. It is not just about what information animals acquire but is also concerned with how they acquire it and the uses to which the animals put that information, i.e. why do they acquire it?

Of course, the advances across the three broad areas of scientific investigation, encapsulated in what, how, and why questions, have not been even. Some questions are far better understood than others. There have been many significant advances, particularly with respect to mechanisms viewed from the perspective of environmental challenges. These have been summarized recently by Tom Cronin (Cronin 2008) in his section on *Visual Ecology* in *The Senses: A Comprehensive Reference* (Basbaum et al. 2008). While it may seem disappointing that only 34 pages of this six-volume series are devoted explicitly to issues in sensory ecology, other authors employed a sensory ecology framework within more specific contributions, for example those of Warrant, on the nocturnal environment (Warrant 2008); Martin and Osorio, on birds (Martin and Osorio 2008); Bowmaker and Loew, on fish (Bowmaker and Loew 2008). Further detailed analysis of many aspects of sensory ecology has come in two recent books. *Visual Ecology* (Cronin et al. 2014) expanded comprehensively on the chapters mentioned above by Cronin and Warrant, and *Sensory Ecology* (Stevens 2013) provides a broad analysis across a range of senses and animal groups.

Thus it can be seen that sensory ecology has an impressive pedigree. From foundations laid by Greek and Roman philosophers, sensory ecology is now built upon a mix of ideas from sensory science, behavioural science, and ecology, and it has collected ideas and information from many different types of investigation and questions. At root, it is an endeavour based upon comparisons between species and between environments.

When investigating the sensory ecology of birds, experimental studies have not always been possible, but ideas and information have been patched together to make interesting explanations of what birds do and the information that they use to do it. It is these themes and approaches which are explored in detail in the latter part of this book. They address questions that derive from quite simple observations of bird behaviours that many birdwatchers and naturalists can readily make. It may not always be possible to provide a comprehensive answer which explains the information that guides behaviour in particular species. However, we may, for example, gain insights into the information that guides the plunge dive of a foraging Gannet (Lee and Reddish 1981), or explain why a Griffon Vulture flies into the blades of a large and apparently obvious wind turbine (Martin et al. 2012), or how a Cormorant can find prey in turbid water (White et al. 2007). A sensory ecology framework can provide insights into these and many other behaviours, and these insights may have important applications, for example in the mitigation of birds colliding with human artefacts. Thus understanding the sensory ecology of birds has value as both a pure and as an applied science.

2
Vision

As discussed in Chapter 1, we readily accept the idea that the mechanical properties of the key structures that birds use to interact with their environments have been finely honed by natural selection. Wings, feet, and bills exhibit a fascinating diversity of shapes, sizes, and strengths, which can be readily interpreted as meeting the physical challenges for executing particular tasks in different environments. The overall argument of this book is that the senses of birds have also been honed by natural selection but, in this case, to meet the informational challenges of executing those tasks. What is the basis of this tuning? In what ways can senses vary so that natural selection can shape their attributes in particular species? Which tasks have shaped or driven the senses of birds? How long has this tuning of information extraction been in progress?

The point in the geological record at which a line can be drawn between birds and their reptile (dinosaur) ancestors is much debated, but avian lineages leading to present day bird species probably had their origins about 150 million years ago (Caley 2007). This is about 350 million years after the camera type of eye, which is common to all vertebrates (Kardong 2014), had first evolved (Nilsson and Pelger 1994). In fact, camera eyes and the first vertebrate animals appear in the fossil record more-or-less contemporaneously (Gould 1989). So whatever the factors that may have shaped the vision of modern birds, the vision of their ancestors had been the subject of a 350-million-year period when their sensory ecology could have been progressively shaped by the informational challenges of different environments and tasks. In this chapter, the emergence of vision and the main sources of variation in the components of camera eyes, upon which natural selection can act, are discussed. The tasks which may have driven the natural selection of bird vision are discussed in detail in Chapters 6, 7, and 8.

2.1 The Emergence of Vision

The emergence of vision has been interpreted as a key event in the evolution of animal life. The appearance of animals in the fossil record which have eyes of any kind is seen as an important contributor to the 'Cambrian Explosion' in animal

The Sensory Ecology of Birds. Graham R. Martin: Oxford University Press (2017).
 DOI 10.1093/acprof:oso/9780199694532.001.0001

diversity which occurred around 540 million years BP (Conway Morris 1998; Gould 1989). Indeed, it has been suggested that the appearance of eyes was the primary causal factor in this explosion (Parker 2003), although other explanations for this rapid evolution of animal diversity have been proposed (Conway Morris 2006). Whatever the causes of the explosion, this must have been the time in the evolution of life on Earth that sensory ecology actually began. Because of the rapid evolution of many new kinds of animals, the environment of every animal became more varied, and the task of competing for resources rapidly became more complex. As a consequence, the senses of animals became increasingly demanding for information to execute those tasks.

Prior to the emergence of eyes, the interactions of animals with their environments took place based on information gathered at relatively short range through chemical and tactile senses (Parker 2003). Eyes changed the dynamic. With eyes, animals could interact with their environments, including other animals, over progressively longer distances. For the first time, animals could retrieve information about what was around them but quite remote from their own bodies.

It was during the Cambrian Explosion that nearly all modern-day animal phyla first appeared (Conway Morris 2006). This 'explosion' of animal diversity actually took place over a period of 5 to 10 million years, possibly longer, but this explosion in life-forms was certainly rapid in the context of the 4 billion years that life has been on Earth. What is even more remarkable is that, during the Cambrian Explosion, the evolution of eyes has been estimated to have taken less than 2 million years (Parker 2003). In this time, camera eyes went from their rudimentary beginning, as flat layered structures capable of detecting light, to complex structures that embodied the key features found in the eyes of all modern vertebrates (Nilsson 2009). Furthermore, a conservative estimate has even suggested that a fully functional camera eye could have evolved in as little as 400,000 generations, which could be equal to the same number of years (Nilsson and Pelger 1994), seemingly very rapid evolution for what has become popularly regarded as such a significant and apparently complex structure.

Given the dramatic structural changes that occurred to produce a fully functioning camera eye in such a relatively short period, it may seem reasonable to assume that functionally significant changes in eye structure have continued as the result of further rapid natural selection. The structures could have been continually modified in response to the many and varied new perceptual challenges that must have occurred during the past 500 million years, which encompass the different planet-wide environmental conditions of the Palaeozoic, Mesozoic, and Cenozoic eras, and the two million years of the present Quaternary period. However, continuous gradual change may not have been the case. Nilsson (2009) suggested that evolutionary changes within camera eyes have not been continuous and incremental. He argued that the evolution of eyes has been subject to periods of rapid change, as new visually guided tasks were hit upon through natural selection, followed by relative stasis. He suggests that the evolution of eyes has been the subject of 'task-punctuated evolution'.

Determining what these tasks have been is clearly a key to understanding the sensory ecology of modern animals. However, it seems clear that for most animal lineages selection in favour of vision has been very strong and consistently maintained for the past 500 million years. Therefore, although the main features of camera eyes may have been arrived at through a process of punctuated evolution (long periods of relative stasis followed by shorter periods during which rapid changes evolve), eyes may have become increasingly differentiated and specialized in many and subtle ways, more or less continuously. This adaptive radiation of vertebrate eyes, in both obvious and subtle ways, is seen in extant vertebrates and is what Wood, Walls, and Rochon-Duvigneaud captured so well in their early surveys of vertebrate, especially bird, eyes (Rochon-Duvigneaud 1943; Walls 1942; Wood 1917).

2.2 What Eyes Do

The crucial property of these first camera eyes, and of all eyes since, was that they were capable not only of registering general changes in the light levels of the environment, but they were also able to determine the position of a light source relative to the animal (Land and Nilsson 2012). That is, they had the capacity of 'spatial vision'. At the very least, these first real eyes, as opposed to the simple light sensitive organs which were their precursors, could provide some information on where a light source was positioned relative to the animal.

Since those first eyes evolved, their subsequent evolution has been in essence a series of refinements of spatial vision: first, the determination of object position with increasing accuracy; second, determination of object position within a wider field of view about the animal; and third, determination of object position over an increasingly wider range of light levels. Some eyes evolved to provide spatial information over a very wide range of light levels, but others evolved to function within relatively narrow light level ranges, typically those experienced during daytime and during night-time. Growing out of these refinements of static spatial vision was the ability to determine whether a light source was moving and its direction of travel.

It should be noted that colour vision, which is often thought of as something rather different, something that is additional to 'simple' spatial vision (simple because it can be achieved in what appears to be a simpler world of black and white), is actually an elaboration of spatial vision. Colour vision allows the extraction of finer spatial detail by using differences in the wavelengths of light, not just differences in the intensity of light, which are reflected or emitted from different sources and surfaces in the environment.

It is worth emphasizing that colour itself is not a property of the world but a property of the visual systems which extract information from the world. This observation was first made by Isaac Newton in his *Opticks* (published in 1704) and captured in the phrase 'The Rays, to speak properly, are not coloured' (Newton 2010). The idea has been taken up and discussed from both scientific and philosophical viewpoints many times, for example by Goethe in *Zur Farbenlehre*, published in 1810 (Eastlake 1982), and more recently by W.D. Wright in *The Rays Are Not Coloured* (Wright 1963). Regardless of the aesthetic properties that humans have projected onto 'colour' and the philosophical problems that have been raised about colour (e.g. Stahl 2010; Wittgenstein 1978), colour vision evolved essentially as an elaboration of the mechanism to extract finer spatial detail from the environment surrounding an animal (see 'Colour vision and sensitivity in the spectrum', section 2.6.3).

From their first emergence, the accuracy and precision with which eyes of different types could determine the spatial separation of light sources must have varied greatly. Different eye types still vary greatly one from another in their spatial resolution and sensitivity to light (Land and Nilsson 2012), but there have also been many evolutionary changes within each eye type that have usually been concerned with honing the capacities of spatial vision.

To date, 11 distinct types of eyes have been described across the animal kingdom (Land and Nilsson 2012). There could be more yet to be discovered: a new type of compound eye was discovered only 20 years ago (Nilsson 1988) and new variants in the optical design of eyes in fish are still being discovered (e.g. Partridge et al. 2014). However, there is only one type of eye present in all birds, the simple or camera eye, and all are based upon the same optical design. This is the eye type found in all vertebrates, including ourselves, and in some invertebrates (Land and Nilsson 2012). Although all vertebrate eyes have evolved to perform the same kind of basic functions, there are profound differences in their performances, something to which Sextus Empiricus drew attention two millennia ago.

2.3 Optimal Eyes

It is important to recognize that there is not a single optimal performance for a given eye type. Like wings, bills, and feet, what may be considered optimal or a 'good design' depends on the task that the structure has evolved to achieve. Nuances of the shapes and sizes of wings, or differences in the lengths, shapes, and hardness of bills, are all combinations of functional and structural trade-offs and compromises. A single wing type cannot fly at all velocities, or support a wide range of loads, or carry out all kinds of manoeuvres; a single bill cannot be the all-purpose tool for extracting and handling many types of food. Such observations about the wings, feet, and bills of birds are commonplace and have been investigated in considerable detail (King and King 1980). But functional and structural trade-offs and compromises, although perhaps less obvious, apply equally to the eyes (and other senses) of birds.

Vision is regarded as a multifaceted sense. This is mainly because of the various ways that scientists have devised to probe it. Certainly we, and perhaps all animals, do not experience it as multifaceted. Vision seems to be experienced as an integrated perception, and it is only when we are forced to experience it as fragmented capacities through experimental manipulation that we become aware of vision's different aspects.

There is, of course, nothing unusual about this; splitting things down into various measureable attributes is the way that science investigates the world. Splitting vision into various types of visual performance which can be measured separately (spatial resolution, sensitivity, colour discrimination, etc.) has clarified what vision entails and has given clues to the particular informational capacities that natural selection might have worked upon. However, it is important to stress 'might' in this context. It is dangerous to assume that because humans have been ingenious enough to measure a particular aspect of an animal then, inevitably, there will have been natural selection specifically for that aspect. What humans measure may be a by-product, an epiphenomenon of selection for another capacity, or the capacity that is measured may be a compromise that results from a trade-off between competing sources of selection.

For example, we may be interested in understanding the colour vision of a species and measure how many colours can be resolved in a particular part of the spectrum. This information may be useful when compared with other species for understanding the mechanisms that underlie colour discrimination. However, from a sensory ecology perspective, it is always worth bearing in mind that selection will not have been for colour perception per se. Colour vision will have been for enhancing the spatial information that it can reveal about the presence and properties of certain types of objects that are key in the life of an animal, for example objects used in display behaviours (Endler et al. 2014), particular plumage patterns (Bennett et al. 1997), or particular types of fruits (Burkhardt 1982). Simply seeing more differences between colours is unlikely to have been the driver of natural selection; it is the spatial information about objects that will have driven selection in the direction of detecting finer spectral differences within particular parts of the spectrum.

Splitting vision into various capacities also suggests that visual performance can be considered to be optimal in separate ways. But, while bills or feet can potentially have many different types of optimal performance, for example, strength of grip, flexibility, or speed and dexterity of movement, any one structure cannot be optimal in all of these properties. Similarly, a sensory system may be considered to have different dimensions upon which performance could be optimal, but it cannot be 'good' at all of them simultaneously. Furthermore, the different sources of information available to a particular animal (sights, sounds, odours, tactile sensitivity, taste) may be used together in a complementary way, and there are also likely to be trade-offs and balances within each source of information. This may occur much in the same way that the structure and strength of bills or feet may be considered to be traded-off and balanced against each other in different species.

2.4 A Fundamental Trade-off in Vision

There is an important and fundamental trade-off in vision between the capacities of sensitivity (what is the lowest amount of light that an eye can detect?) and the ability to resolve spatial detail (how close together can light sources be and still be detected as separate?). This was first discussed in detail by Snyder et al. (1977) and the argument refined by Land and Nilsson (2012) who elegantly show that high resolution is not possible at low light levels in any vision system. More pertinent for this discussion is the argument that an eye which has evolved to detect ever lower levels of light, inevitably loses its ability to resolve spatial detail at those low light levels, and vice-versa. That is, an eye which is highly sensitive to low light levels is unable to detect fine detail at high light levels, and an eye that can detect fine detail at high light levels cannot detect such detail at low light levels. This is, in fact, a property of any vision system including man-made ones, such as photographic and video cameras (Land and Nilsson 2012).

These trades-off between sensitivity and resolution arise because of the physical nature of light. For the purposes of understanding vision, light needs to be viewed as occurring in discrete packets of energy: photons. This is because it is individual photons (which each contain a minute amount of energy) that are absorbed by the photosensitive pigment molecules within the receptors of a vision system. Absorption of light photons sets off a cascade of very rapid chemical changes which result in a signal being triggered in the nervous system (Luo et al. 2008). Furthermore, it is well established that in the human eye, and in the eyes of other species, the absorption of a single photon can trigger a signal in the nervous system (Hecht et al. 1942). However, for 'vision' (visual information to be made available in the brain), there is a need to simultaneously receive a sufficient number of these photons to be sure that a genuine signal has been received, as opposed to detecting the random noise that inevitably occurs in any analysing system (Land and Nilsson 2012). The basic questions for a visual system are: Was that a photon or just random noise? Is that a genuine signal about something in the environment? The second question becomes easier to answer as the number of photons received at any one instant increases.

One way to overcome the signal-to-noise problem is to increase the number of photons that are caught at any one instant. This can be done in a number of ways: for example, by making the image as bright as possible, i.e. capturing more light from the scene; stretching what is understood by an 'instant' in time; or increasing the area over which photons are caught before their capture events are combined as a signal that is sent to the brain. All of these mechanisms are found operating within vision systems and they are even used in photographic and video cameras, but there is always a limit to how far they can be implemented and still provide useful information, capable of guiding a particular task. For example, increasing image brightness, i.e. trapping more photons coming from a particular scene into the eye or camera, cannot be achieved without limit. This is because to do so always involves making the imaging system larger in one dimension or another and ultimately physical constraints will be met: the eye becomes too large for its physical integrity to be maintained or its weight and metabolic requirements cannot be sustained by a particular animal. The latter are particularly important in birds where both the power-to-weight ratio and the physical balance of the body are important requirements for powered flight. From this perspective, eyes are heavy liquid-filled spheres placed at one extremity of the body and therefore if too large could interfere with balance when a bird is in flight (King and McLelland 1985).

To increase the area within the image that is sampled at any instant also runs rapidly into very clear trade-offs. This is because ultimately spatial vision depends upon distinguishing between the numbers of photons that are being received simultaneously on adjacent parts of the retina. This solution to increasing sensitivity by pooling receptors to make larger 'effective receptors' inevitably leads to a decrease in the detail that can be detected (Warrant 1999; Warrant 2008).

The importance of all this is that while at low light levels it may be possible for an imaging system to contain information about the presence of an object in a scene, the mechanism to extract that information cannot detect the presence of the object with a high degree of certainty—in fact, the more sensitive the analysing system becomes, the greater the uncertainty. This is something that we can experience directly ourselves. If we allow our eyes to become fully adapted to the dark and we then try to detect the presence of objects in a dimly lit scene, we know full well that we cannot see fine detail. This is a direct result of the inevitable trade-off between sensitivity and resolution; it is not just something peculiar to human eyes. We can also experience how our uncertainty about seeing objects and surfaces fluctuates at low light levels. As we continue to look at any one part of the dimly lit scene, the amount of detail (and the interpretation that our brain can put upon it) changes, sometimes quite dramatically; a crouching cat, a branch, a star, may come and go, all as a result of there being few photons to detect and the uncertainty of their detection and low spatial resolution.

The important thing to note is that this too is not just a quirk of human vision: it is not a flaw in the functioning of our eyes and brain, but a genuine demonstration of the limits of vision. It is a demonstration of how at low light levels it is not possible to achieve high resolution because of the very nature of light itself and the very real problems of extracting information when light photons are scarce (Warrant

1999). Of course, our eyes are not the most sensitive within the animal kingdom and natural selection has driven the eyes of some species to trap more photons and achieve brighter images, but inevitably these eyes will also reach a limit on the details that they can detect and at low light levels they too will be subject to the same uncertainties of image sampling.

The converse of all these arguments is that the highest resolution—the detection of the finest degree of detail in a scene—can be achieved only at high light levels. At these high light levels, the signal is consistently well above the noise level of the vision system. In these conditions, the number of photons that is detected is high enough that signals are sent repeatedly (streamed) to the brain and they can be sampled from a very small area of the image. This requirement for high light levels applies particularly to the use of colour (discrimination between light of difference wavelengths) for the detection of spatial detail. Thus, the fact that we see the finest detail in a scene in bright daylight, not under moonlight or starlight, reflects not only a fundamental property of our eyes, but also a fundamental property of vision. A light detection system that can achieve the highest resolution and can also employ colour discriminations to do so can achieve this only at high light levels; these same systems are not capable of doing so at low light levels. The converse is also true: eyes which are highly sensitive (i.e. are able to detect objects at very low light levels) cannot achieve high resolution. These are trade-offs that arise because of the fundamental properties of light as a source of information in the environment.

It should be emphasized, however, that they are trade-offs, not absolutes. There is scope to balance between these extremes. It is also possible to exploit mechanisms within an individual eye which alter the brightness of images as light levels change, mechanisms which alter the ways that the image is analysed as light levels change, and mechanisms which result in different parts of the image projected onto the retina being analysed to provide different degrees of spatial detail (Warrant 2008). These trades-offs and compromises have been one of the key bases upon which the rich source of variations in eyes have been built. They are the result of refining the properties of eyes and vision through natural selection for the control of different tasks.

The results of these constraints and trade-offs between sensitivity and resolution are manifested in many different ways in the eyes of animals. They will be discussed further in later chapters but it is necessary to note that they are quite fundamental to the problem of using light to extract information from the world around an animal. These different types of trade-offs and compromises have been the raw materials of variation that the process of natural selection has exploited. The results are the many different, and often subtle, solutions and balances of visual capacities in different species. Vision has been tuned to provide information for the control of a wide range of different tasks and in different ecological circumstances. It is this tuning of vision to different tasks that underpins the study of visual ecology.

2.5 The Primacy of Vision in Birds

The evolution of eyes, which began 500 million years ago, changed forever the information that underlies animals' interactions with their environments and with other animals (Parker 2003). It also established the primacy of vision as the source of information used to guide animal behaviour. This primary reliance upon vision as a source of information is particularly the case in birds. While this is easily asserted from the evidence of casual observations, it is also well supported by evidence that relatively large portions of the brains in most species of birds are devoted to the analysis of information from vision, and that the 'intelligent' behaviours of birds are based primarily upon visual information (Emery 2006; Karten and Hodos 1967; Reiner et al. 2005; The Avian Brain Nomenclature Consortium 2005).

Only in a handful of extant bird species is vision not the primary sense. Even in such birds, vision was at one time highly likely to have been the prime source of information. However, vision has become secondary through a process of regressive evolution, while at the same time other senses, particularly olfaction, hearing, and touch sensitivity, have come to take on the primary role of food detection and guidance of locomotion that is underpinned by vision in most birds (Martin et al. 2007). The prime examples of 'non-visual' birds are the five species of flightless kiwi, which probably lost their reliance on vision as they evolved in the absence of mammalian predators on the islands of New Zealand (Wilson 2004). The downgrading, but not complete loss, of vision in these birds is discussed in Chapter 6 where it is argued that it probably attests to the high metabolic cost of vision (Laughlin 2001a; Laughlin 2001b). In other bird species, including some of the petrels (Procellariidae), shorebirds (Scolopacidae), and owls (Strigidae), olfaction, touch, and hearing, respectively, play a key or complementary role to vision, especially with respect to foraging. However, vision is still the primary guide for locomotion in these birds, even though spatial resolution is low (Martin 1986; Martin and Piersma 2009; Nevitt et al. 1995; Nevitt 2008; Piersma et al. 1998).

The evolution of spatial vision allowed an animal to receive information instantaneously about objects and surfaces remote from its body and, because of the broad field of view of the eye, information could be received simultaneously from many directions. Once animals were able to use such spatial visual information to detect the features of food sources, increasing specialization in diets became possible through the detection, procurement, and ingestion of different foods that lay at various distances from the animal's body. In so doing, vision unlocked, or gave a major impetus to, an evolutionary arms race between predators and prey (Parker 2003). Thus natural selection in favour of eyes and refinement of their spatial capacities can be seen to have acted through the consequences of visually guided behaviour towards distant objects. Most probably this involved behaviours associated with food procurement regardless of whether foods were sessile fruits or seeds, or other mobile animals, the latter leading to a role for vision in not just the detection of prey but also in the detection of predators. It will be argued later (Chapter 8) that guidance towards food and the detection of predators were, and remain, the principal drivers of the evolution of vision in birds. Clearly, controlling locomotion is an important role for vision in birds, but it may not have been the prime driver for the diversity in vision that is found in birds today.

2.6 Sources of Variation in Camera Eyes

The camera eyes of vertebrates are not called 'simple' without good reason. Compared to the complexity of the multiple repeated structures which are found in the eyes of most invertebrates, camera eyes are structurally and conceptually simple (Land 1981). However, it took 2500 years of endeavour to arrive at our current understanding of the fundamentals of the 'simple' human eye (Wade 2000); eyes have not yielded their secrets readily. Furthermore, within the simplicity, there is potential for much subtle variation which can profoundly alter vision and the information that it can extract from a scene. To understand this potential, it is appropriate to consider the basic structure of an eye as composed of two functional units, which can vary independently of each other. Added to this is potential for further variation which arises because there are always two eyes in an animal's head. Eyes can be placed in different positions with respect to each other in the skull, thus altering the region about the head from which visual information can be retrieved at any one instant.

2.6.1. The Basic Functional Components

Essentially, camera eyes can be divided into two functional components: an optical system, which produces an image of the world outside the eye, and a system that analyses (extracts information from) that image. These functional components can be matched in a straightforward manner to the main structural/anatomical components (Figure 2.1). These two functional components can vary greatly and one way to appreciate just how profound, and also how subtle, these differences can be is to recognize that the same two functional parts of camera eyes are found in all man-made imaging systems. From large and elaborate astronomical telescopes to the cameras built into mobile phones, these two functional components can be readily identified. As in the eye, all that is necessary is a way of producing an image and a way of analysing it. What differs dramatically between these two extreme examples of man-made imaging systems are the quality and size of the image of the world, and the level of detail that can be extracted from the image.

An essential aspect of this division of functions in both eyes and man-made systems is that each component, optics and analysing system, can vary independently of one another. It would certainly be strange to try to analyse the image of

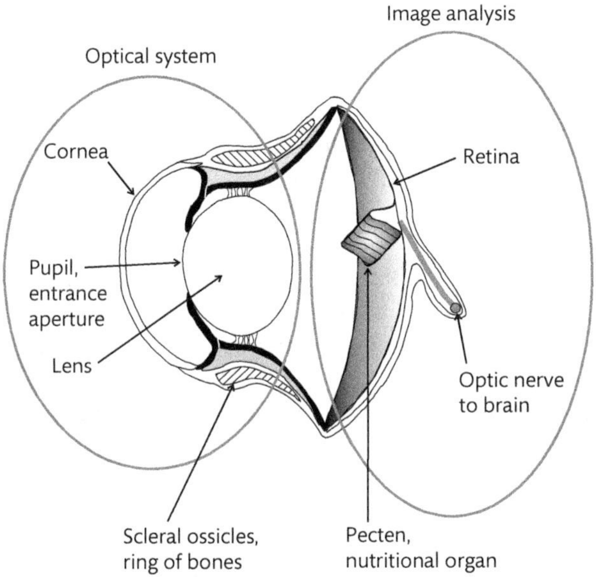

Figure 2.1 The two main functional components of vertebrate eyes, illustrated by a diagrammatic cross section through the eye of an owl. The optical system (lens and cornea) projects an image of the world onto the retina which is the first stage of the image-analysing system. The characteristics of the two systems can evolve independently of each other within broad parameters. However, they are clearly linked together within the overall structure of the eye. Unlike the situations in many vertebrates, in which the eyes are generally shaped like a sphere, bird eyes have a flatter or tubular shape which is maintained and the eye held rigid, by a ring of small overlapping bony plates, the scleral ossicles. In birds, the supply of nutrient to the eye is achieved through the pecten which is a highly vascularized, darkly pigmented, pleated structure with a large surface area. Nutrients supplied by the vascular system to the pecten permeate from the pecten's surface into the fluid-filled chamber and reach the active components of the eye in an apparently passive manner. The pecten sits above the point where the optic nerve exits the retina. The pecten creates a blind area in the visual field of each eye, and the two blind areas in the total visual field typically project upwards and laterally when the head adopts its resting posture (see examples in Figures 6.10 and 8.6). Because of the pecten, there are no blood vessels running across the surface of the retina. This is unlike the situation in mammalian eyes where the retina has a highly vascularized surface and light forming the image has to pass through the blood vessels before reaching the retina.

an astronomical telescope with the small photodiode array of a low-specification mobile phone camera. However, it could be done, although much of the detailed information gathered by the sophisticated lens system could not be extracted. There would be a mismatch between the high level of information that the optics of the telescope can collect and the low level of information extraction that the camera's array can achieve.

If such a system were subject to natural selection, it would be possible to imagine a scenario in which the lens system became progressively less refined, or the image-analysing system became progressively more complex or, more likely, both would change. Certainly natural selection would be unlikely to maintain the mismatch and eventually there would be some degree of matching between the information in the image and the level of detail that the analysing device could extract. But just what the match would be, would depend upon the task that the imaging device is attempting to control.

As in the above example of different types of camera lens systems, there must have been scope within the optical systems of vertebrate eyes as they first evolved to produce images that varied in their properties with respect to a number of parameters, and these same parameters remain central to understanding the role of the image in visual ecology. The two key parameters are the brightness of the image (how much light is captured to make the image) and the quality of the image in terms of the precision with which light from a point in the world is brought to a focus in the image. In effect, how faithfully does the image reproduce the world that it represents? An indication of the importance of image brightness has been discussed already (2.3) and will be returned to later in a discussion of the sensory ecology of nocturnal activity (Chapter 6).

By definition, an image is never perfect: it is a simulacrum which will always lack some information about the world that the image presents for analysis. Furthermore, the quality of the image and hence the information that it contains, may vary within itself. Image quality is usually at its best along, or close to, the optical axis of the imaging system. This axis is the approximate line about which the optical elements of the system are symmetrically arranged. In a simple lens it is the line that passes directly through its centre. In a camera eye it is the direction about which the cornea and lens are symmetrically aligned (Figure 2.1). Moving away from this axis results in an image of progressively poorer quality; it is here where distortions and aberrations of the image occur, something that is readily apparent in cheaply produced camera lenses. To correct for these peripheral distortions requires considerable elaborations and refinements of the optical system, hence the high prices asked for camera lenses which maintain high quality across a broad section of the image. Such considerations about the quality of peripheral optics are not just of interest to optical equipment makers and camera buyers. This is because in many vertebrate animals, including most birds, eyes are placed on the side of the head (Figure 2.2) with the result that the forward direction, the direction of travel, is viewed through the periphery of each eye, while the direction of best optical quality projects laterally from the head (Figure 2.3). The consequences of this for both foraging and locomotion in birds will be discussed in detail later.

Another important property of the image is how much of the world is imaged at one instant (Figure 2.3). Does the imaging device have a wide or narrow field of

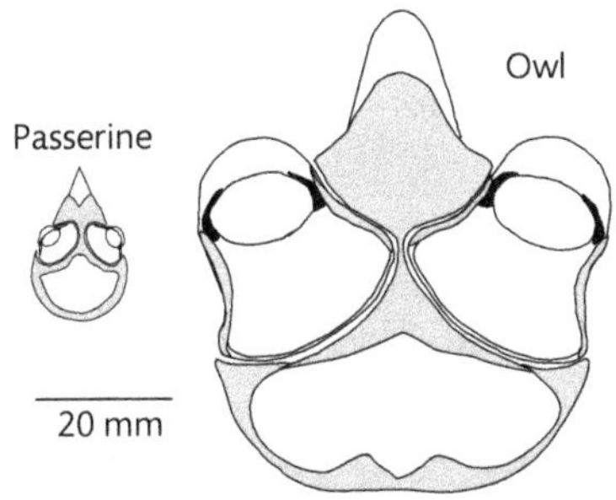

Figure 2.2 The arrangement of eyes in the heads of birds depicted by diagrams of cross sections through the head of an owl and of a small passerine. In no birds do the eyes project directly forwards, all project laterally to some extent. Even among the owls, which are often assumed to have eyes that are frontally placed, the axes of the eyes diverge by 55° (Figure 2.3). In some bird species, the eye axes diverge by more than 120° (see Figure 2.14). These sectional diagrams indicate that in an owl the eyes are very large and do not sit within the protection of the skull; they are held rigidly in place. This is unlike the situation in the majority of birds which is illustrated here by the example of a small passerine species. The difference in eye size found among birds is such that the whole of the head of the passerine could fit, more-or-less, within the eye of the owl. Note than in most birds, the eyes take up so much space in the skull that they almost touch each other. In most birds, they are separated by a thin septum of bone. Diagram based upon original drawings in Wood (1917).

view? This is important since it determines from how much of the world around an animal's head information can be gained at any instant.

2.6.2 Optical Systems of Camera Eyes

The optical system is itself comprised of two main elements (Figures 2.1). The cornea is the relatively simple curved surface at the front of the eye. In eyes that operate primarily in air, the cornea is essentially a boundary between air and the fluid-filled chamber of the eye. The radius of curvature of the cornea is the key to its image-forming properties. A more highly curved surface produces a smaller image than a shallowly curved surface. The lens, which is suspended in the fluids that fill the chambers of the eye, is also relatively simple having two convex surfaces, but these can vary in how curved they are and whether the two surfaces have the same or different curvatures. Furthermore, the interior of the lens is not uniform, as is the case in simple glass lenses, but is made up of a complex structure of transparent layers of different densities. The optical function of the lens is primarily concerned with making relatively fine adjustments to the focus of the image already formed by the cornea, but different lenses can have quite different properties depending upon subtle differences in size, surface curvatures, and internal structure.

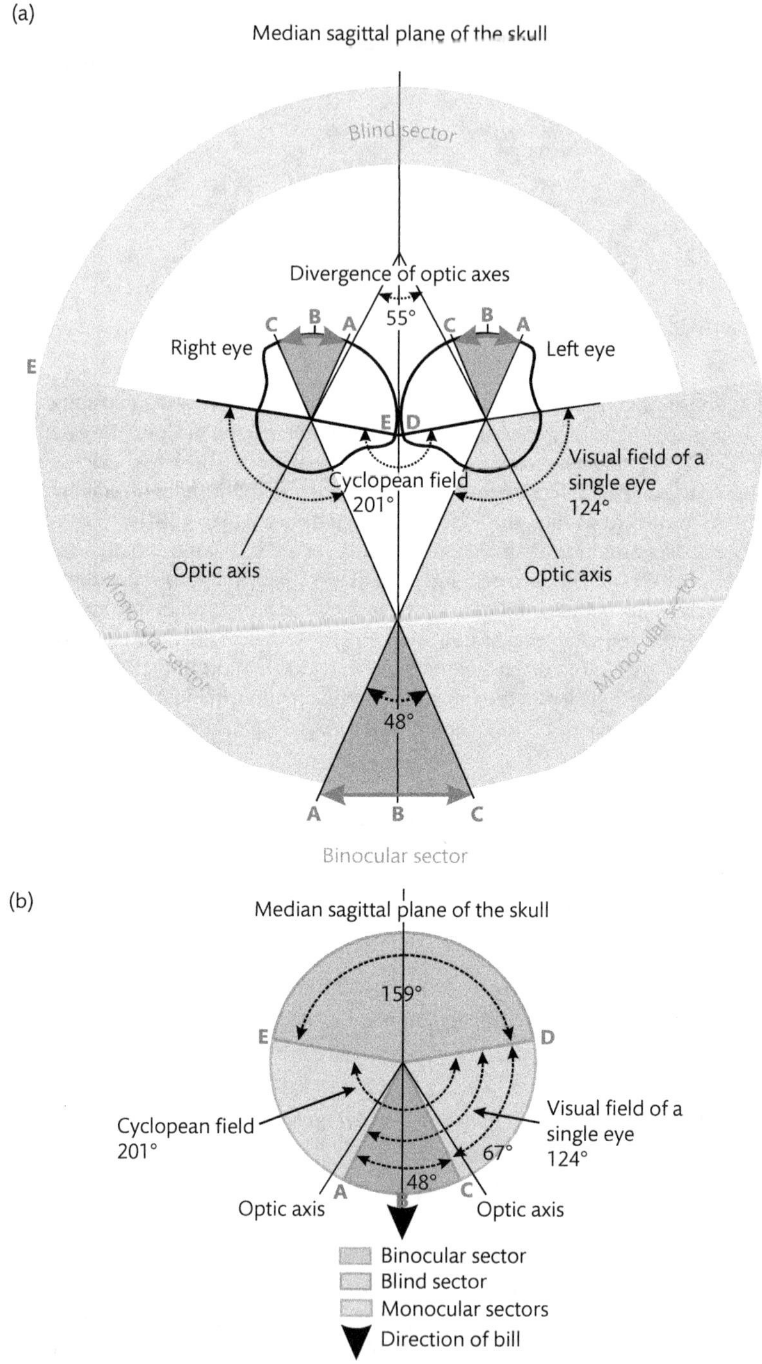

Figure 2.3 How the visual fields of each eye are combined to make the total visual field of a bird. The arrangement is depicted here showing a horizontal section through the eyes and visual fields in a Tawny Owl *Strix aluco*. Diagram A show how the fields of each eye are brought together, while Diagram B shows a simplified way of presenting the resultant visual

field as though it projects from the midpoint between the two eyes. The field of each eye covers a sector of space 124° wide and they overlap to provide the bird with a binocular region that is 48° wide. This gives the bird a wide total field of view (the cyclopean field) of 201°, but there are large regions that project laterally seen by one eye only, while there is a substantial blind area (159° wide) to the rear of the head. The functional interpretation of this kind of visual field is discussed in Chapters 6 and 8. The diagram also makes it clear that forward vision is achieved using the peripheral vision of each eye. Because of this, the image of the same part of the world does not fall on equivalent parts in the retina in the two eyes. Thus, for example, an image of point A falls at the very periphery of the retina of the left eye, but its image falls much more centrally in the retina of the right eye; the reverse is true for point C. The median sagittal plane of the skull is the vertical plane that divides the skull into two mirror-image halves, and is the direction in which the bill projects. From (Martin 2009).

It can be seen immediately that there is much scope for changing the overall image-forming properties of an eye by virtue of small changes in the absolute size, curvatures, and relative positions of these two optical components.

2.6.3 The Image-analysing System

The image-analysing system starts with the retina which extracts and relays basic information about the image for further analysis by the brain. The retina is a curved surface at the back of the eye onto which the optical image is projected (Figure 2.1). Its gross structure is smooth, uniform, and black. Despite this superficially uniform appearance, retinas are, in fact, highly complex structures. Each retina contains millions of individual elements. These are the photoreceptors and the ganglion cells to which they connect and which, in turn, connect to the brain via the optic nerve. Thus retinal ganglion cells are the only link between the eyes and the brain, and it is well established that the spatial density of retinal ganglion cells in an eye places an upper limit on the spatial resolution of an eye (Hughes 1977). Knowledge of maximum ganglion cell density has been widely used, combined with information on the focal length of the optical system (and hence the size of the retinal image), to estimate maximum resolution of different eyes. This technique has been employed to estimate spatial resolution in a wide range of vertebrate species, not just birds, with recent examples among birds including ostriches (Boire et al. 2001), wildfowl (Lisney et al. 2013a), owls (Lisney et al. 2012a), vultures (Lisney et al. 2013b), and parrots (Mitkus et al. 2014) (see Appendix 1).

In all birds, the photoreceptors which feed to the retinal ganglion cells are of only a few types. Photoreceptors can, however, be distributed in highly complex patterns and at different relative densities across each retina. The result is that the retina of every species is probably unique in the information that it extracts from the image projected upon the retinal surface.

It is the photoreceptors which detect the individual photons of the light that make up the image, and it is the ganglion cells, to which they connect, that begin the analysis of the image and thus start the extraction of information about the world outside the eye. The photoreceptors are the well-known rods and cones. They contain photosensitive molecules capable of trapping individual photons of light and signalling their registration.

From one perspective, the retina is relatively simple because it is made up of photoreceptors of just a few types and these types differ surprisingly little between species. From another perspective, however, the retina is immensely complex because of the potential for variation in the relative numbers and distributions of these photoreceptors. It is likely that in each species there is a unique arrangement of photoreceptors and hence a unique pattern in the way that information is extracted from the environment surrounding a bird. The drivers of these different patterns of information probably lie in the control of different keys task in different environments.

Colour vision and sensitivity in the spectrum

We take it for granted that the environment is full of colour. As explained above (2.2) this is not, however, the actual case. Colour is a property of the nervous system, not the environment. This truth was first captured in the phrase 'the rays are not coloured' attributed to Isaac Newton (Newton 2010). We cannot, in fact, know whether birds see the world in 'colour' in the same way that we do. All that can be determined with certainty are the range of wavelengths of light to which a bird is sensitive (the avian visible spectrum) and the ways a bird's eye can subdivide that spectrum into discrete sections.

What is clear is that many birds detect light over a wider range of wavelengths than humans are able to, i.e. they have a broader visible spectrum. Also, it seems that birds may be able to discern more colours within their spectrum, i.e. they can probably make finer colour discriminations, at least in some parts of the spectrum. Some birds can detect light in the ultraviolet (UV) part of the spectrum, light to which human vision is insensitive. Some terrestrial mammals and many invertebrates are also able to detect information using UV light (Land and Nilsson 2012). It is also important to note that not all birds see in the UV; in fact, those species which have true UV vision are found only in the gulls (Laridae, Charadriiformes), ostriches (Struthioniformes), parrots (Psittaciformes), and the oscine passerines (Passeriformes), but excluding the Corvidae (Martin and Osorio 2008; Odeen and Hastad 2013; Odeen et al. 2010; Odeen et al. 2011). Other species may have visual sensitivity that extends into the violet-near UV spectrum but they lack specific UV receptors in their eyes and cannot be considered truly UV sensitive.

The prime function of colour vision is that it allows the extraction of spatial detail by using differences in the wavelengths of the light that are reflected from structures in the environment. While we might find colours aesthetically pleasing this is a cognitive/cultural elaboration; what has driven the evolution of colour vision is the detection of differences in the wavelengths of light reflected from different surfaces. Thus, rather than relying just upon the contrast that results from differences in the intensity of light reflected from surfaces, colour vision can provide additional information about the presence and nature of objects by detecting differences in the wavelengths of light. In essence, the finer the differences in wavelength that can be detected by an animal's vision system, the finer the spatial detail that it can detect in its environment. For example, two patches of plumage could reflect equal numbers of photons but these photons may be from different parts of the spectrum. A bird with colour vision can readily detect this difference, while one without colour vision could not (Hastad et al. 2005). We can readily appreciate this by reference to our own vision; at lower (night-time) light levels colour vision is absent and objects that are readily detectable during the day may disappear. This disappearance is not only because of lower acuity, but also because the differential reflections of light of different wavelengths cannot be detected. These differential reflections are still present in the environment but, in the absence of colour vision, they cannot be detected at low light levels.

It seems unlikely that colour vision abilities have evolved in different species specifically for enhancing the spatial information about the presence and properties of certain types of objects that are key in the life of an animal. It seems more likely that colour vision first evolved to meet a broad range of spatial tasks. It is not, of course, possible to know what kinds of objects and tasks were first detected by the colour vision systems of birds or their ancestors. Today, it is possible to identify examples of objects whose detection is enhanced by colour vision and these include objects used in display behaviours (Endler et al. 2014), plumage patterns (Bennett et al. 1997; Hunt et al. 2001; Hastad et al. 2005), and the detection of particular types of flowers or fruits against foliage backgrounds (Burkhardt 1982; Goldsmith 1980; Odeen and Hastad 2010).

But this does not mean that these specific tasks were the prime drivers for the evolution of colour vision. Indeed, it seems more likely that plumages, flowers, or fruits have evolved to become more (or less) conspicuous in response to the vision of the observers, rather than the vision of the observers evolving to detect particular objects. Furthermore, simply seeing more differences between colours is unlikely to have been the driver of natural selection; it is the spatial information about objects in the environment that will have driven selection in the direction of detecting fine differences within the spectrum.

The ability to detect colour differences within the spectrum has not been determined directly in many bird species. The most detailed knowledge available is from behavioural and electrophysiological studies in Rock Doves *Columba livia* (Emmerton and Delius 1980; Hodos 1993; Wright 1979). However, it is possible to say something about colour vision, and breadth of the spectrum visible in birds in general, from knowledge of the visual pigments found in the cone photoreceptors of bird retinas. This is because it is with the cone receptors that analysis of colour differences in the image begins. Based upon such knowledge, it seems safe to assume that all birds have colour vision. Even the nocturnally active owls seem to have some colour vision, although it is not as sophisticated as that of other bird species (Bowmaker and Martin 1978; Martin 1974). However, genome analysis, but not the determination of visual pigments directly, suggests that colour vision may be absent from kiwi (Le Duc et al. 2015) and they may be the only group of birds which are not capable of some kind of colour discrimination. However, the ability to differentiate colours within the spectrum in the majority of bird species may differ little between species. As explained below, this is because the visual pigments of cone photoreceptors, with the exception of the short wavelength end of the spectrum, show little variation across species, but even at the short wavelength end of the spectrum visual pigment complements fall into just two types (Hunt et al. 2009; Odeen and Hastad 2013).

Photoreceptors and visual pigments

The photoreceptors of all vertebrate retinas are of two functional types—rods and cones—and cones types are classified primarily according to the position in the spectrum of the peak sensitivity of the photopigments that they contain. These pigments trap photons from the image that is projected onto them by the eye's optical system. Once a photon is trapped, a cascade of chemical reactions is initiated which trigger a signal into the nervous system. In human eyes, as well as rod receptors which contain a single type of photopigment, there are cone receptors which contain three types of photopigments (Land and Nilsson 2012). These three types of cone receptors provide the basis for human, trichromatic, colour vision. In birds, there are rod receptors containing a single type of photopigment but the photopigments of the cone receptors are typically of four types. The important functional difference between rods and cones is that rods function primarily at low light levels (twilight and below), while cones function at higher light levels (as experienced in twilight and daytime), that is cones are less sensitive to light than rods (Kelber et al. 2003; Vorobyev and Osorio 1998). There is a range of light levels, for example, the higher levels of natural twilight, at which both types of receptors function, but it is important to recognize that during lower twilight and night-time, spatial resolution is purely a property of the rods, and because they are of one type only, colour vision is not possible. It is this dual function of the rod and cone receptors which underpins the vertebrate eye's ability to function over such a very wide range of naturally occurring light levels.

Spectrophotometric measurements of the sensitivity maxima (λ_{max}) of the photopigments found in bird retinas indicate that these fall into five classes (four in the cone receptors and one in the rod receptors). Four of these photopigment classes (including those found in the rods) show a high degree of similarity across a wide range of species. At first consideration, this may seem rather surprising since it means that there is little evidence of adaptive radiation of visual pigments among birds. That is, photopigments, and hence sensitivity within the spectrum, do not show marked differences between species. This suggests that colour vision arose early in bird ancestry and its properties have been highly conserved. It has been shown, for example, that the visual pigments found in the eyes of a species of pelagic sea bird, Wedge-tailed Shearwaters *Puffinus pacificus* Procellariiformes (Hart 2004), are very similar to those found in a phylogenetically distant species which is terrestrial and lives in open forest habitats, Indian Peafowl *Pavo cristatus* Galliformes (Hart 2002; Hart and Hunt 2007). This suggests that colour vision in birds has general, all purpose, properties, which are not tuned to specific tasks performed by different species. It also seems likely that there may be only two broad types of colour vision system found among birds and these differ primarily by how far into the short wavelength end of the spectrum the sensitivity extends.

The five types of visual pigments of birds are labelled and defined as follows:

RH1, rhodopsin type 1 with a sensitivity maximum at about 500 nm (λ_{max} 500 nm); this type is found in the rod receptors;
RH2, rhodopsin type 2 (λ_{max} 505 nm), found in cone receptors;
SWS1, short-wave type 1 (λ_{max} 365/410 nm), found in cone receptors;
SWS2, short-wave type 2 (λ_{max} 470 nm), found in cone receptors;
LWS, long wave (λ_{max} 565 nm), found in cone receptors.

The SWS1 is differentiated into two types and these types are found in different species. The pigments with λ_{max} at 365 nm are referred to as ultraviolet sensitive (UVS), and those in which λ_{max} is at 410 nm are referred to as violet sensitive (VS) pigments (Wilkie et al. 2000). It is possession of the UVS pigment which gives the oscine passerines, gulls, ostriches, and parrots their visual sensitivity into the ultraviolet part of the spectrum.

To date only Humboldt Penguins *Spheniscus humboldti* have provided a likely exception to the general uniformity of avian photopigments. This species has been reported to have cone pigments with λ_{max} 403, 450, and 543 nm (Bowmaker and Martin 1985), and it is suggested that the penguin's LWS pigment is shifted to a shorter wavelength maximum (565 nm → 543 nm) and the RH2 pigment is absent.

In addition to the five types of photopigments, the retinas of birds have three morphologically distinct types of photoreceptors: rods, single cones, and double cones (Cserhati et al. 1989). Double cones are widespread in vertebrates (Bowmaker and Loew 2008; Walls 1942), but absent from mammals. In birds, the double cones always contain the LWS (λ_{max} 565 nm) pigment.

This apparent complexity of receptor types is, however, complicated further by the presence in each cone receptor of an oil droplet which sits in the proximal part of the outer segment (Bowmaker 1977; Cserhati et al. 1989; Hart 2001). Such oil droplets are absent from most mammals but are common to both reptiles and birds. The importance of the droplets is that they contain pigments based upon carotenoids that are derived from the bird's diet. These carotenoid pigments give the droplet a bright colour (mainly reds and yellows to the human eye) but they are not photosensitive. They serve an important filtering function which has the effect of sharpening the spectral tuning of the cone receptors and shifting the position of peak sensitivity towards longer wavelengths (Hart and Vorobyev 2005).

The sensitivity of the receptors, the photopigments, and the filtering effect of the droplets are shown in Figure 2.4. Such spectral sharpening does not occur

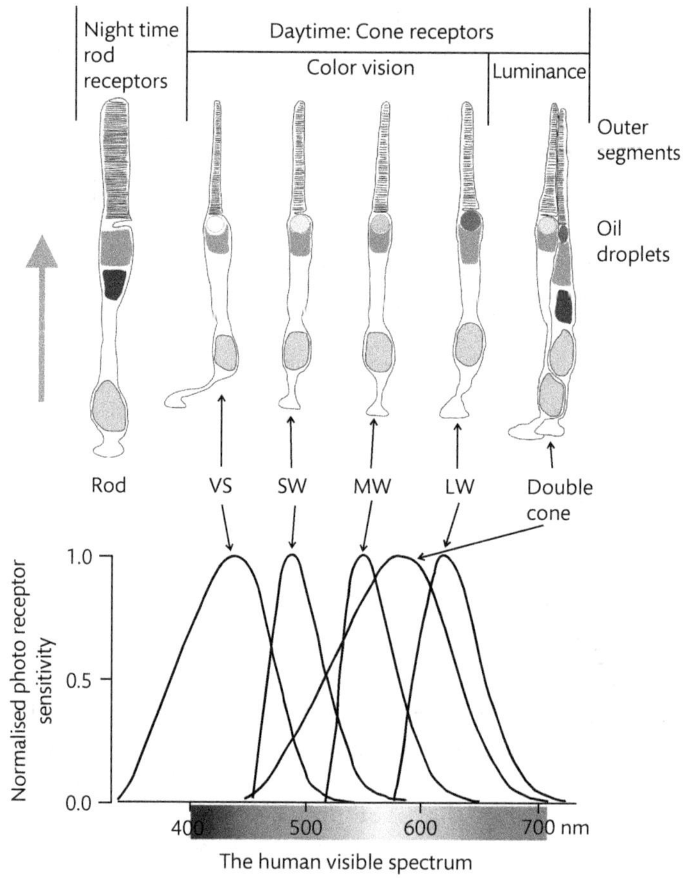

Figure 2.4 The photoreceptors of birds. The top diagram depicts the types of retinal photoreceptors found in bird retinas based upon light microscopy. The cones underlie vision at high daytime light levels while the rods function primarily at low night-time levels. The four types of single cones provide the basic mechanism upon which colour vision is based while the double cones are a channel which is thought to signal luminance. Within each cone type is an oil droplet with different absorption properties and these droplet types are paired with particular photoreceptor pigments housed in the outer segments. The large blue arrow indicates the direction in which light travels; the consequence is that light is filtered by the oil droplets before it reaches the photopigments and the pigments

in the droplet act primarily as cut-off filters (allowing light above certain wavelengths to pass to the outer segment and cutting off light at shorter wavelengths). The combination of photopigments types and oil droplet types results in there being four main types of cone photoreceptors found in birds' retinas. Each type is able to absorb light only within particular parts of the spectrum, although there is overlap between them. The lower diagram shows the resultant photoreceptor sensitivities and the commonly used labels for them are indicated (LW, long wave, MW, middle wave, SW short wave, VS violet sensitive) and explained in the text. Note that the actual sensitivity of the cones may differ only slightly between species and only in the case of the VS type of receptors does this vary by a marked degree. Depicted here is the sensitivity of a photoreceptor having the VS type pigment which has maximum sensitivity at about 410 nm (within the human visible spectrum) and is typical of most non-passerine birds, while the UVS pigment has a maximum sensitivity centred on 360 nm (in a part of the spectrum not visible to humans) and is typically found in oscine passerines and a few non-passerine species (see text for details). Diagrams of receptors are redrawn from illustrations provided by Peter Olsson (Lund Vision Group, Sweden) and the diagrams of normalized sensitivity are redrawn from illustrations provided by Daniel Osorio (University of Sussex, UK).

in the double cones but the oil droplets in the double cones do block UV light from reaching the photopigment. Also the UVS/VS cone oil droplet is transparent which means that UV light can reach the photopigments that they contain and so these receptors are not subject to such sharp spectral tuning as in the other three other single cone receptors. The narrow spectral tuning of the single cones and their comparatively even spacing across the spectrum indicates that they are adapted for coding spectral information, and hence they constitute the foundation of the mechanism of colour vision in birds (Martin and Osorio 2008). Having a broader visible spectrum and receptors that are more sharply tuned in the spectrum means that in some parts of the spectrum birds are probably capable of detecting more subtle colour differences than can be detected by human and other mammalian visual systems. Mammalian systems are generally based upon three cone receptors covering a smaller extent of the spectrum and their cone receptors do not have the benefit of spectral sharpening by the oil droplet filters.

Clearly the retinas of birds are more complex at the level of the receptors than those of mammals, including humans. Thus, whereas humans have three types of cone receptors plus the rod receptors, it seems that most, if not all, birds have four types of cones plus rods and that some, but not all, birds may see into the UV and all will see into the violet part of the spectrum. There is good evidence that all four single cone types are involved in the colour vision of birds and thus they can be considered to have a tetrachromatic colour vision system compared with the trichromatic system of humans (Goldsmith 1990; Goldsmith and Butler 2003; Goldsmith and Butler 2005). This includes direct evidence that the UV/VS cone receptors function as part of the colour vision system rather than as a separate channel (Smith et al. 2002). However, these differences in sensitivity and colour vision apply only at high (daytime) light levels when vision is mediated by the cone receptors. At lower (twilight and night-time) light levels, only the rod receptors are functional and these receptors have very similar characteristics across all birds and mammals. Thus at low light levels similar sensitivity in the spectrum, and the absence of colour vision, will be found across all bird species. Indeed, at low light levels, these aspects of vision will be similar in birds to that of humans and other mammals, and probably to most vertebrates.

Understanding the function of the double cones in bird vision has presented something of a problem. When their sensitivity in the spectrum (Figure 2.4) was first described it was suggested that they were also part of the colour vision system and hence the possibility raised that colour vision in birds was even more complicated and was based upon a pentachromatic system. However, a range of investigative techniques have now indicated that double cones are not part of the colour vision mechanism and may constitute a separate channel which signals luminance at higher (daytime) light levels (Jones and Osorio 2004; Osorio et al. 1999).

Just how the world might look when viewed through the birds' tetrachromatic colour vision system and how it might alter the salience of surfaces and structures compared with the human trichromatic system is of considerable interest (Vorobyev et al. 2001). Certainly knowledge of the tetrachromatic system tells us that the birds' eye views are likely to be different to how humans detect colour information in the world. Of course, it is not possible to see these colours through birds' eyes but statistical methods have been devised to compare entire colour patterns and thus detect differences in the relationships among the colours, and these comparisons have suggested which kinds of differences between natural colour patterns are enhanced by a tetrachromatic system (Endler and Mielke 2005; Endler et al. 2005). Insights into the spectral information available to birds have been gained recently using a modelling technique based upon digital photography (Cynthia Tedore, Vision Group, Lund University, personal communication). Photography through purpose-built filters which mimic the spectral sensitivity of the different cone types in birds, followed by reconstruction of a combined image have provided valuable insight into how different combinations of receptor types might alter the salience of different types of targets (Figure 2.5).

2.7 Variation of Image Analysis

Although there seems to be little variation in the suites of photoreceptors and their associated photopigments found in the retinas of different bird species, there can be very marked differences in the relative abundances of receptor types, and their distributions, within the retinas of different species (Hart 2001; Hart 2004; Hart et al. 1998; Hart et al. 2000b). The result of this is three-fold and of great importance when considering the sensory ecology of birds.

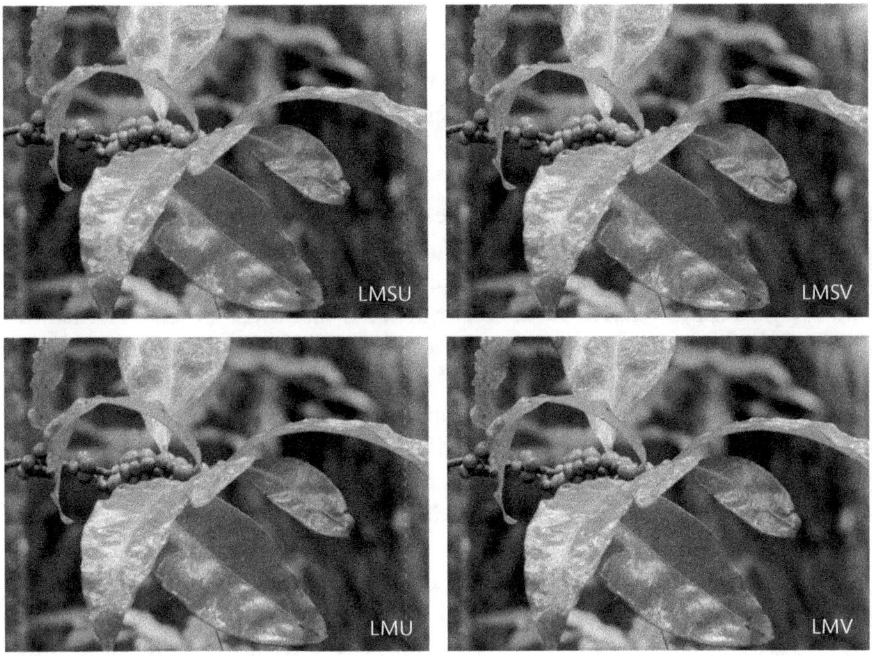

Figure 2.5 The effects on image analysis of using photoreceptors of different types in birds. How different combinations of photoreceptor types can alter the information that is extracted from the same scene is an intriguing question. Basically it is a matter of trying to understand how information about the world may differ when analysed using different types of photoreceptor arrays within the retina. One approach is illustrated here using a digital photographing and modelling technique which simulates viewing the same scene using different combinations of photoreceptor types based upon the normalized photoreceptors sensitivities depicted in Figure 2.4. In the four photographs, the same scene is modelled as seen through eyes having different combinations of photoreceptors (L = long wave, M = medium wave, S = short wave, V = violet, U = ultraviolet, all as referred to in the text and in Figure 2.4). The photograph is of the leaves and fruit of a shrub, *Rapanea subsessilis*, photographed in a Queensland (Australia) tropical rainforest. Differences between the top two modelled views are hard to detect. However, there are obvious differences between the lower two modelled views and also between them and the upper two. It could be suggested that viewing these fruits through an eye that employs photopigments that extend into the ultraviolet part of the spectrum increases the conspicuousness of the fruits. Photographs provided by Cynthia Tedore, (Lund Vision Group, Sweden).

First, the maximum resolution (the finest detail which can be seen), differs markedly between the eyes of different species. Furthermore, in light of the above discussions of the trade-offs and consequences between resolution and sensitivity (2.4), it is not surprising to find that spatial resolution is closely related to light levels, falling markedly as the light level decreases. This is discussed in more detail at the end of this chapter (2.11) and its implications discussed in the later chapters of the book.

The second and third consequences of the marked differences in the relative abundances of receptor types, and in their distributions within the retinas of different species, while no less important are much less easily described and measured. The second consequence is that in any eye the image projected on to the retina is not analysed in a uniform way across the whole of its area. Basically, acuity measures the best performance of an eye (2.11), but this will apply to only a small sector of the whole image or field of view of the eye. Outside of those areas, spatial resolution can be a lot lower. The third consequence is that there are very marked differences in these patterns of analysis between different species. In short, an image of the same scene in the eye of one bird species can be subject to quite different analysis by the eye of another species. Thus, the many different ways that the photoreceptors in bird retinas are configured means that, at any one instant, different parts of the same scene are analysed in greater spatial and spectral detail, and so the information available to birds from two different species apparently looking at the same scene will differ.

Variation in visual ability across the field of view of a single eye has long been recognized in the human eye (Wade 2000) and was discussed by Isaac Newton in his *Opticks* first published in 1704 (Newton 2010). We can readily confirm this variability in image analysis within a single eye in ourselves. For example, it is relatively easy to show that our ability to see detail and colour in a scene decreases quite dramatically in the periphery of our field of view, compared with central viewing. However, to experience this requires some observational discipline. We are all too ready to look at something with our central vision; the use of central vision is a reflex type of behaviour. However, fix your gaze on a distant point and get a friend to present objects in the periphery of your field of view. It is soon apparent that although you can see 'something' it is often very difficult to discern its properties; its shape and colour are often elusive, and your certainty in deciding what it is will vary depending upon the exact position of the object in your field of view.

Although these differences in image analysis within the field of view were easily demonstrated in humans, it was only in the early part of the 20th century that some of the more extensive variation in image analysis in birds was recognized. This was through studies of the anatomy of the retina stimulated mainly by Wood's comparative studies of what he termed the *fundus oculi* (Wood 1917). Wood simply looked into the eyes of live birds with a hand lens or simple ophthalmoscope system and carefully recorded what he saw of the retina (Figure 1.5). He also looked at exposed retinas of excised eyes under a low-power microscope (Figure 1.6). He observed that there were marked differences in the colours of the retinas in different species and was also able to discern small, discrete circular areas in the retinas of some birds. He also showed that such features differed in their position within the retinas of different species. Moreover, in some bird retinas, he could discern linear features, such as a feature running roughly horizontal across the whole retina, and these have become known as horizontal streaks or linear areas. These features were subsequently shown to be indications of areas within retinas where photoreceptors and ganglion cell densities are high compared with the rest of the retina. They are, in fact, slight swellings in thickness of the retinal layers due to the higher density of cells, and they may also contain small pits or foveas where some of the neural layers are displaced exposing the photoreceptors more directly to the image-forming light.

Thus, at the fundamental level of the retina, there are different birds' eye views. It is even possible that the same bird might extract different information from the same scene depending upon whether it is looking at it with its left or right eye. This is because there is evidence of differences in the distribution and abundance of photoreceptors between the eyes of the same individual (Hart et al. 2000a; Mitkus et al. 2014).

2.7.1 Variations in the Distributions of Receptors in a Single Eye

Two major ways have been identified in which variation of image analysis occurs in bird retinas. The first is marked variations in the spatial distributions of cone receptor types. These variations can be striking and it seems likely that they result in differences in colour vision and spectral sensitivity in different parts of the visual field of a single eye. A most striking example has been described in Rock Doves (Galifret 1968) (Figure 2.6). In the retinas of these birds, there is a large area dominated by receptors containing red oil droplets; these are LW sensitive receptors (Figure 2.4). This area looks downwards within the visual field, while receptors containing yellow oil droplets (MW sensitive receptors) predominate in areas which look laterally and upwards. These different areas, known as the red and yellow fields, are obvious even to the naked eye in an excised retina that is illuminated from behind; there is a clear boundary between them (Figure 2.6). However, the visual ecology and function of this striking regional specialization within dove eyes is not understood, although it has been suggested that the yellow field in some way enhances the contrast of objects seen against the blue of the sky (Lythgoe 1979), but this is argued by analogy from a general yellow filter placed across the whole of the image (an effect seen when humans wear yellow filters before their eyes) rather than considering the sensitivity of individual receptors.

Although not so striking as these fields in dove retinas, other work has shown systematic differences in the distributions of cone receptors types (classified by the colour of their oil droplets) in the eyes of other birds. These tend to show a gradient in the relative abundance of different receptor types across the retina. These suggest that marked interspecific differences exist in colour vision and spectral sensitivity within a retina and that these differences are correlated with specific perceptual challenges posed by life in different environments and the conduct of

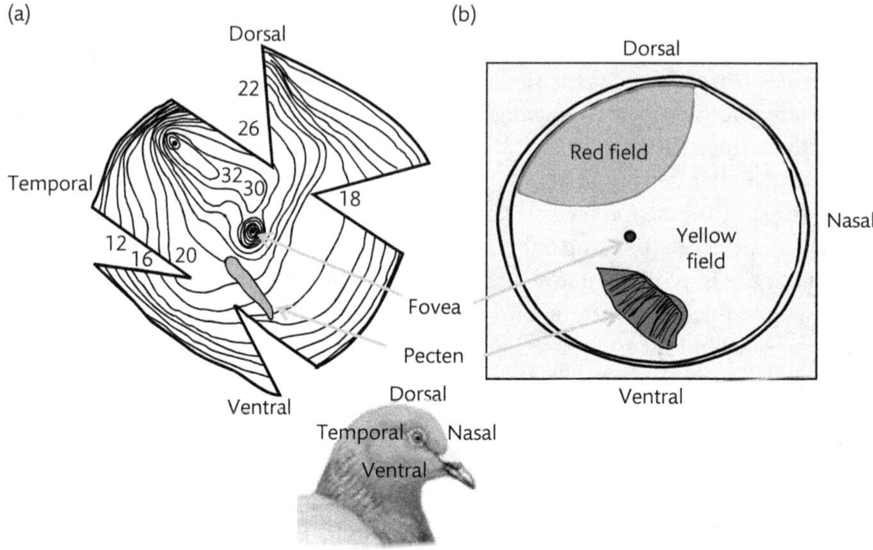

Figure 2.6 Complexity within an individual retina. An indication of the complex way in which image analysis can vary across an individual retina is provided by the retina of Rock Doves *Columba livia*. Diagram (a) shows the topography of ganglion cell density in the right eye with the orientation of the retina as in the sketch of the bird. The lines are isodensity contours that join regions in which ganglions cells have the same density (the numbers by contours indicate ganglion cell density as × 1000 per mm^2). A distinct contour pattern is obvious, with low densities of ganglions cells at the periphery in the ventral and nasal sections of the retina (which looks upwards and backwards), and two distinct areas of very high ganglion cell density, one in the central retina (which projects laterally) and another in the temporal-dorsal retina, which looks downwards, possibly in the direction of the bill. The central region of high cell density is also associated with a fovea (Figure 2.8). Diagram (b) also indicates that the relative densities of the types of photoreceptor cells differ across a retina. The orientation of the retina is the same and so the two diagrams could be overlaid. In the temporal-dorsal retina there is a region, referred to as the 'red field', in which long wave type cones containing red oil droplets (Figure 2.4) dominate. The rest of the retina is dominated by medium wave type cones which contain yellow oil droplets. These fields are very clear when an excised retina is transilluminated. The relative densities of the different photoreceptors are so high that the sections of retina look distinctly red and yellow, similar to the way that they are shown in the diagram. Diagrams redrawn and modified from originals in Bingelli and Paule (1969) and Galifret (1968).

different tasks. It has been suggested, for example, that there is a general association between a high proportion of cone receptors containing red droplets in birds that fly over water and need to see through the surface (Muntz 1972). Cited in support of this are data from such diverse groups as kingfishers (Alcedinidae), gulls and terns (Laridae), and Northern Gannets *Morus bassanus*. However, birds that live on water but apparently do not need to see through the surface (e.g. European Shags *Phalacrocorax aristotelis*, Manx Shearwaters *Puffinus puffinus*) do not have many receptors containing red-coloured droplets. While such interpretations do not seem to square with the presence of a red droplet field in doves, they do suggest that marked interspecific differences exist in colour vision and spectral sensitivity within the eyes of birds and that these differences are probably correlated with specific perceptual challenges posed by life in different environments and the conduct of different tasks.

The second type of variation between species and within individual retinas lies in the striking patterns in the relative densities of receptors and ganglion cells. These differences in the density of cells may be overlaid on patterns of cone receptor types discussed above (Figure 2.6). Very distinct patterns in ganglion cell density are discerned with light microscopy. These have been described in a wide range of species using techniques of increasing sophistication and precise quantification, from the earlier work of Galifret (1968) and Meyer (1977) to the recent work of several authors (Coimbra et al. 2015; Lisney et al. 2013b; Mitkus et al. 2014). All of these studies show that significant patterns of photoreceptor and, in particular, ganglion cell concentrations occur (Figure 2.7). These patterns may be roughly circular with a very high concentration at the centre and a gradual decline of cell density away from the centre. In some species, more than one region of high concentration can occur, while in others, high densities of receptors may occur in linear bands of various lengths relative to the width of the entire retina, and these are typically arranged so that they project approximately horizontally when the bird's head is held in its usual resting or flight posture.

Patterns of these kinds have been described in different species for over a century. Today, more precise analytical techniques have shown considerable variations in patterns in birds from the same species and even between the two eyes of the same individual (Mitkus et al. 2014) (Figure 2.7). The usual interpretation is that the areas of high receptor concentration, of the kinds shown in Figures 2.6 and 2.7, are regions of heightened spatial resolution, which are typically associated with viewing towards the horizon, usually looking out laterally, but sometimes forwards, although not usually forward in the direction of travel. Areas of very high ganglion cell density may also be associated with a fovea, in which the outer layers of the retinal cells are displaced to form a small pit or depression in the surface of the retina (Figure 2.8). Foveas are not unique to birds and have been recorded in all vertebrate taxa, including humans, and are associated with small regions within the visual field where spatial resolution is highest.

New patterns of ganglion cell and receptor densities, and the positions of foveas, are regularly described (Figure 2.9). See, for example, descriptions of these patterns in waterfowl (Fernandez-Juricic et al. 2011b; Lisney et al. 2013a), galliforms (Lisney et al. 2012b), owls (Lisney et al. 2012a), New World vultures (Lisney et al. 2013b), parrots (Coimbra et al. 2014a; Mitkus et al. 2014), passerines

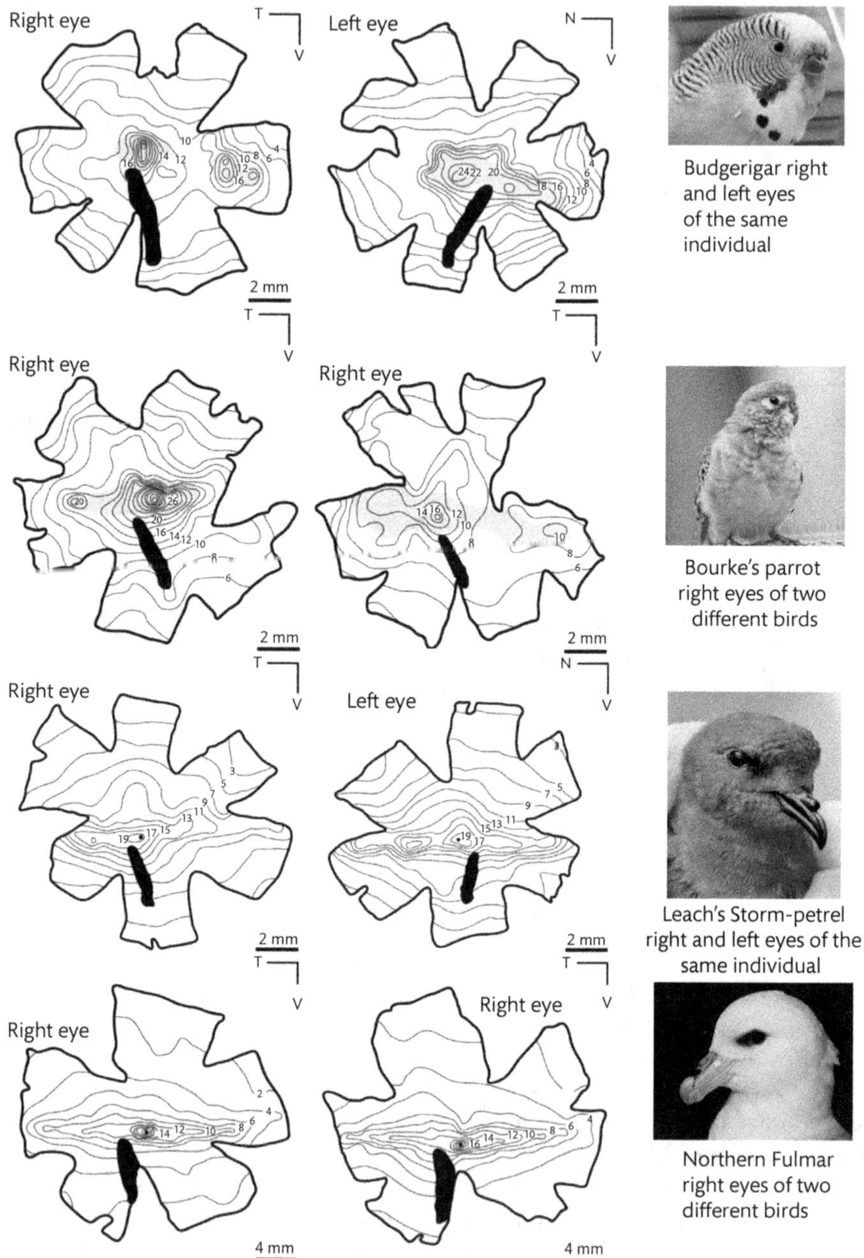

Figure 2.7 Complexity within and between retinas. Retinal ganglion cell density patterns can vary markedly between species and also between the left and right eyes within the same individual bird. This series of pairs of ganglion cell isodensity maps show examples taken from a range of birds which differ in their phylogeny and ecology. Marked differences can

be seen between the two seabirds (Order Procellariiformes) in the lower half of the diagram and the two parrot species (Order Psittaciformes). The seabirds have distinct linear bands of higher density cells running approximately horizontally across the retina and these project out towards the horizon when the birds are in flight. There are also regions of high density in the central part of the retina which project laterally. Significantly higher cells densities are found in the retinas of the two parrots compared with the seabirds. Furthermore, rather than a linear band of higher density, there are two distinct regions of high density in each retina. Note, however, that the two examples of right eye retinas from Bourke's Parrots show markedly different density patterns and the left and right eyes of the same Budgerigar seems to show considerable differences. Both parrots have a region of high density in the centre of their retinas and this projects laterally within the field of view. In the right eye of the Budgerigar, there is a second region of high density in the nasal retina and this must project backwards in the field of view, but this distinct region is absent from the left eye of the same bird. All diagrams from Mindaugus Mitkus, Lund Vision Group, Sweden. Photos credits: Budgerigar (Michael Cole), Bourke's Parrot (Daniela Parra), Leach's Storm Petrel (Fanter Lane), Northern Fulmar (Steve Garvie).

(Coimbra et al. 2014b; Fernandez-Juricic et al. 2011a), and penguins (Coimbra et al. 2012). It is clear that these patterns do vary markedly between species, and there is good evidence of variations between individuals within species, suggesting that these patterns could provide the basis for intense and rapid natural selection tuned to different tasks.

Understanding the function of these ganglion cell distribution patterns in the visual ecology of a particular species is often not straightforward, especially in the context of individual variation. However, the usual functional explanations refer to either a role in foraging (the detection of individual items in the frontal or lateral field of view) (Figure 2.10), or the detection of predators and vigilance behaviour (the detection of predators both laterally or from behind, and usually in the direction of the horizon), rather than a role in the control of locomotion.

The association of linear areas, which are interpreted as a band of heightened acuity distributed approximately horizontally across the field of view, has become known as 'the terrain hypothesis' (Hughes 1977). This refers to the need to detect objects that either lie in the direction of the horizon or predators that approach from that direction. As such, it is expected that a linear area will be a common feature of the retinas of birds which live in open habitats, such as seascapes and open plains, in which the horizon is ever present. This certainly seems to hold for a number of species but not all (Figures 2.7 and 2.9). For example, one particular passerine species of open prairie habitats, Eastern Meadowlarks *Sturnella magna*, which might be expected to have a linear area in view of their use of open habitats and their vulnerability to predators, was found not to have a distinct linear area. Instead it has a centrally placed fovea that projects laterally into its visual field (Tyrrell et al. 2013). Also a linear area has been described in Canada Geese *Branta*

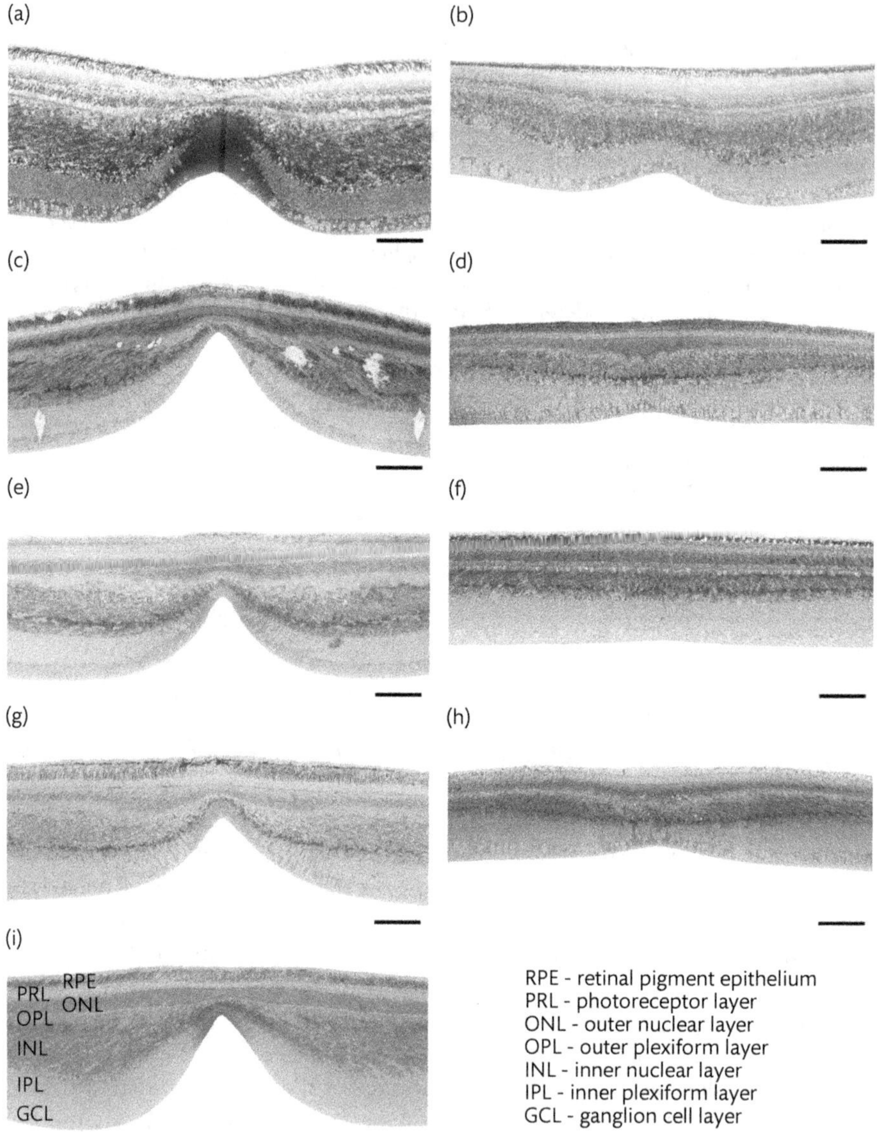

Figure 2.8 Variation within retinas with respect to foveas. These are light microscope images of retinal cross sections through the central fovea (left hand column), which projects laterally in the field of view, and the temporal fovea (right hand column) which projects more forwards in the field of view. In a fovea, retinal outer layers containing the ganglion cells and other intermediate cells are displaced sideways. In these sections, the light forming the image travels upwards through the layers before it reaches the photoreceptors (which are to the top of each section, labelled PRL). This exposes the photoreceptor cells more

directly to the image and consequently these are localized regions of the retina associated with the highest spatial resolution. In all species, the central fovea (left column) takes the form of a deep pit and provides the highest resolution, while the temporal fovea (right column) is shallower and presumably is associated with lower resolution. A and B Common Buzzard *Buteo buteo*; C and D Eurasian Sparrowhawk *Accipiter nisus*; E and F Red Kite *Milvus milvus*; G and H Peregrine Falcon *Falco peregrinus*. I is a guide to labelling the layers of the retina. Scale bar 100 µm. All images courtesy of Mindaugus Mitkus, Lund Vision Group, Sweden.

canadensis) which appears to project diagonally (Moore et al. 2012). Among galliform birds (grouse and their allies), there do not seem to be any specific relationship between retinal ganglion cell distribution patterns and species preferences for more open or more enclosed habitats (Lisney et al. 2012b) (Figure 2.9). Clearly, although ganglion cells patterns have now been described in a good range of species, the taxonomic coverage and habitat preferences of species is not comprehensive, and further comparative studies would be of great interest.

These ways in which receptor cell and ganglion cell distributions can vary within a retina, plus the marked differences between species in these patterns, certainly attest to the fact that birds of two different species placed in exactly the same position at the same time, are very likely to retrieve different information from the world about them. Furthermore, clear patterns in receptor cell density are not unique to birds but seem to occur in all retinas. For example, human eyes have a distinct region of high acuity and enhanced colour vision which constitutes a small portion of the total field of view of the eye. Most importantly, this pattern is quite different to those found among the birds that an ornithologist might be observing. This means that a human observer and the birds that they are watching will certainly all see the world differently and, consequently, they will almost certainly be extracting different information about their surroundings at the same instant.

2.8 Variation of Optical Structure

As explained in 2.6.2, there is much scope for changing the overall image-forming properties of an eye by virtue of relatively small changes in the absolute sizes, curvatures, and relative positions of the lens and cornea, as well as simply changing the overall size of the eye.

A good example of subtle differences in optical structure is exemplified by a comparison of the eyes of Rock Doves and Manx Shearwaters (Figure 2.11). The eyes of these species are a good basis for comparison because they are the same size and casual observation of their eyes would give no clue to their structural differences. Their different optical structures result in important differences in the

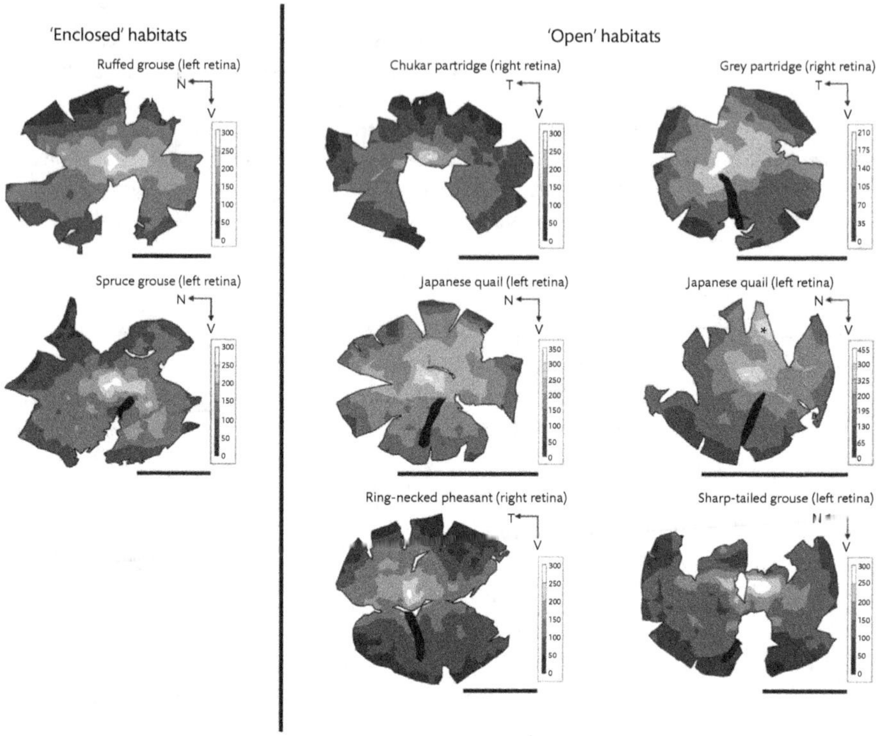

Figure 2.9 Testing the 'terrain hypothesis'. How the patterns of distributions of ganglion cells in different retinas (Figure 2.7) might reflect general perceptual challenges of living in different environments has long been a puzzle. One idea developed by Hughes (1977), from original observations by Wood (1917), is that species living in open habitats are likely to have a horizontal band of higher density retinal ganglion cells running across the retina, as shown in the seabirds in Figure 2.7. Such a band of heightened spatial resolution looking out towards the horizon could allow conspecifics or predators to be detected at greater distance in open habitats. One groups of birds in which some species live in open habitats and others in enclosed woodland habitats are Galliformes (grouse, quail, and pheasants). Analysis of the retinal ganglion cells patterns in a selection of these species, however, has not found a correlation between preferred habitat type and patterns of retinal ganglion cells. This figure shows ganglion cells density patterns in the retinas of eight species of galliforms. The dark shape visible in some of the diagrams indicates the position of the pecten. Drawings courtesy of Tom Lisney and adapted from Lisney et al. (2012b).

images formed in the eyes and these can be related to differences in the ecology of the two species.

Rock Doves and Manx Shearwaters have quite different evolutionary origins being from two well-separated avian orders (Procellariiformes and Columbiformes (Hackett et al. 2008)) with markedly different behaviours and general ecology,

Figure 2.10 An American Robin *Turdus migratorius* foraging on a lawn. These birds frequently, like many other thrush species in this genus, forage on the ground and are often seen to cock their head sideways before rapidly changing position and seizing a small prey item in the bill. Is the bird depicted here using its left eye to seek and detect a food item using a portion of the lateral field of view in which spatial resolution is heightened? Once food is detected, control of the bill towards the item can be switched rapidly to forward, less acute vision. However, it is also possible that this position allows the birds to look upwards with its right eye to scan for predators, or it might be listening for the sounds made by a prey item as it moves at or just below the surface. Photo courtesy of Geoffrey E. Hill.

although a birdwatcher in some locations could see both species from the same view point. Rock Doves are daytime active birds of more open habitats, nesting and roosting (by night) upon cliffs often above the sea. They have, of course, now become naturalized in most major towns and cities throughout the world where they are usually referred to as Feral or Domestic Pigeons. They feed by detecting, and then accurately pecking at, small seeds. Manx Shearwaters, as their name suggests, are birds of the open oceans, coming ashore only to breed and then mainly under the cover of darkness and often feeding at night. They feed primarily by seizing fish and other marine organisms from the surface or during shallow dives.

Simple measurements show that the eyes of these birds are the same size; just under 12 mm long (about half the length of a human eye) and with a maximum diameter of just under 14 mm (Figure 2.11). When looking at the birds in the hand or in photographs, it would be difficult to see any difference in their eyes apart from the fact that the iris in a dove is usually brightly coloured whereas in shearwaters the iris is dark brown. However, careful analysis of the optics of these two eyes reveals quite a different story (Martin and Brooke 1991) (Figure 2.11).

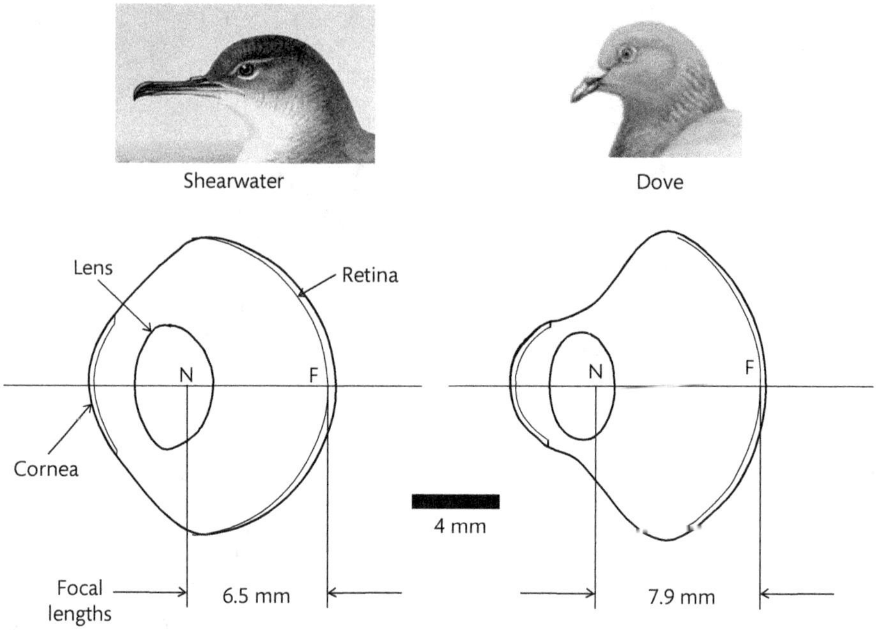

Figure 2.11 Variation in the image-producing structures of birds' eyes. These diagrams show horizontal sections through the optical system of the eyes of two contrasting species (Manx Shearwater *Puffinus puffinus* and Rock Dove *Columba livia*) which have eyes of almost identical size. The diagrams are based upon detailed analysis of the optical systems of the eyes, known as schematic eye models, which describe the essential characteristics and parameters of the eye's optics. The eyes differ in the curvature and power of their corneas and lenses with the result that the focal length of the shearwater's eye is shorter than in the dove's eye. This in turn influences the resolution that is possible in the two eyes such that the dove is likely to have higher acuity than the shearwater. However, the smaller image in the shearwater results in a brighter image (the f-number of the eye is lower in the shearwater) and these birds are therefore more able to operate at lower light levels. There are also marked differences in the retinal structures of these two species (Figure 2.12).

The corneas and lenses, and their relative positions to each other, differ so that although both produce a focused image on the retina, in the shearwaters the image is smaller but brighter compared with the doves'. In fact, shearwater eyes have an image 1.5 times brighter than dove eyes, a difference which is likely to be functionally significant. This difference in image brightness can be related to the nocturnal behaviour of shearwaters or, viewed from the other perspective, to the reluctance of doves to fly at night (Martin and Brooke 1991). These differences are further amplified by the fact that in shearwaters the retinas are rich in rod receptors and have fewer cone receptors than the doves' retinas (Bowmaker, personal

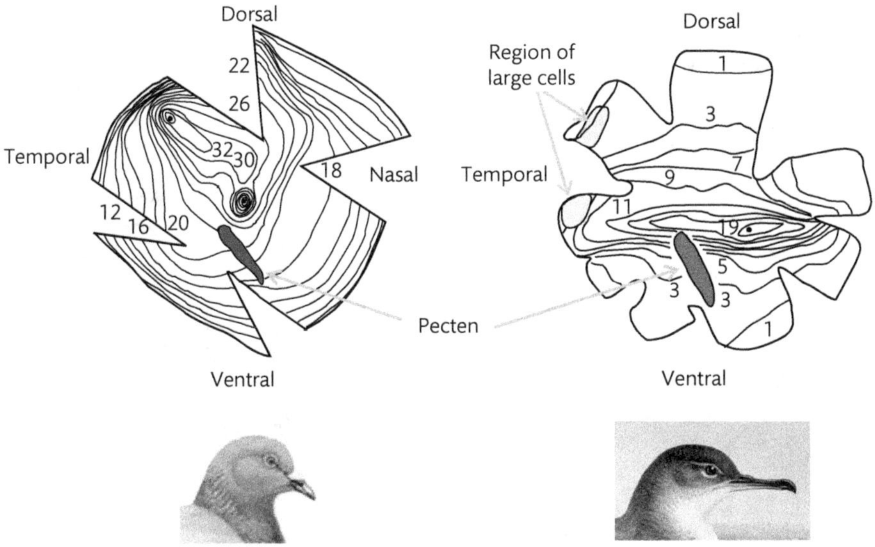

Figure 2.12 Variation in the image analysis in Manx Shearwaters *Puffinus puffinus* and Rock Doves *Columba livia*. Despite the eyes of these two species being of identical size, Figure 2.11 showed that their optics differed in significant ways. The image-analysing mechanisms of these species also differ markedly. This is indicated by these maps of the topography of the retinal ganglion cells in the retinas. The general density of ganglion cells is much lower in the shearwater than in the dove retina, and the shearwater has a linear region of heightened cell density running approximately horizontally across the field of view, while the dove has two localized areas of high cell density. A particular feature of the shearwater's retina is a region in the far periphery in the temporal region (which looks forward in the field of view) containing a small number of widely spaced and large ganglion cells, not found elsewhere in the retina. The function of these large cells is not clear but they may be concerned with the detection of movement in the frontal field of view (see Chapter 8). Diagrams redrawn with modifications from Bingelli and Paule (1969) and Hayes et al. (1991).

communication), and the density and abundance of retinal ganglion cells have very different patterns (Hayes and Brooke 1990; Hayes et al. 1991; Wright 1979) (Figure 2.12).

This example illustrates clearly that, even in eyes of the same size, the optical structures can vary markedly between species. These differences are not trivial and have impact upon the brightness and size of the retinal image presented for analysis by the retinas. Broader comparisons show that the optics of birds' eyes can show great variation. Some of this is captured in Figure 2.13, where the size and optical structure of the eyes of an owl and of a small passerine are compared. Marked

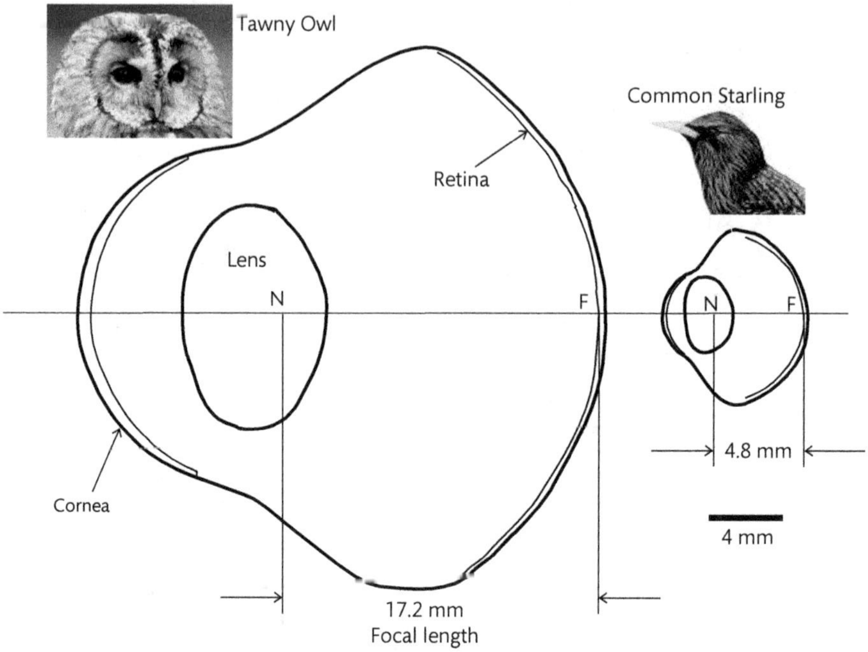

Figure 2.13 Variation in optical structure. Bird eyes can differ markedly in size and optical structure. This is exemplified by this comparison (to scale) of the schematic eyes of a nocturnal predator, Tawny Owl *Strix aluco*, and a ground-feeding passerine, Common Starling *Sturnus vulgaris*. These two eyes show marked differences not only in their absolute size but also in the relative contributions of the cornea and lens in the image-forming mechanism of these eyes. In turn, these differences in optical structure result in marked differences in image size (indicated by the difference in their focal lengths), and also in the maximum brightness of their images. This is reflected in the minimum f-numbers of these eyes which are 1.3 and 1.9 in owl and starling respectively. This indicates that the image of the same scene would be twice as bright in the owl's eye as in the starling's eye. Based upon Martin (1982, 1986b).

differences in optical structure, as exemplified in Figures 2.11 and 2.13, mean that the characteristics of the retinal image presented for analysis by the retina can be significantly different in size and brightness, and probably also in the quality of the image. For example, it has been possible to show that the higher absolute sensitivity of an owl's eye compared to a human eye is attributable to the brighter image in the owl's eye, not to differences in the sensitivity of the retinas. On the other hand, it is also possible to show that the differences in the absolute sensitivity of owls and doves is mainly attributable to differences in retinal sensitivity which is, in turn, attributable to the fact that dove retinas are dominated by cones while rods dominate in owls (Martin 1982).

2.9 Variation of Visual Fields

Differences in optical structure not only affect the size and brightness of the image, but they also determine the extent of the world that is imaged. We are, perhaps, all familiar with how different camera lenses alter how much of a scene is included in the photograph. The visual field of any imaging system is a property as important as the brightness and quality of the image. This is especially so for an eye since it determines at any one instant from how much of the world an animal can gain information and hence how much of the world can guide behaviour. When two eyes are combined the situation becomes more complicated. This is because the two eyes can be combined in many ways to provide markedly different total fields of view about the head (Figure 2.14).

Humans are unusual animals in having two eyes which are placed on the front of the head with the result that much of what one eye sees is seen by the other eye. In humans, each eye has coverage of 160° in the horizontal plane but, because of the frontal position of the eyes, the total coverage is only 200° and about 120° of that is seen simultaneously by both eyes. Compare this with the example of the Kori Bustard *Ardeotis kori* Otididae, in which a single eye also covers a field of view of 162° but the two eyes combined cover 308° and the portion seen simultaneously by both eyes is only 17° wide (Figure 2.14). In White Storks *Ciconia ciconia*, Ciconiidae, the field of a single eye is also about 160° and they achieve total coverage of 288° about the head, and the binocular field has a maximum width of 28°. There are also marked differences in the vertical extent of the region of binocular coverage and where it is centred with respect to the bill (or nose in humans) (Figure 2.14). Storks have vertical coverage of 120°, twice that of Kori Bustards.

It is important to note that in both the bustards and storks as illustrated here, and for that matter all other birds, forward (binocular) vision is, in fact, achieved by the use of the periphery of the optical system of each eye (Figure 2.15). This is in marked contrast to humans in which it is central vision that looks in the forward direction. In birds, the central axis of each eye always looks laterally (Figure 2.3).

Comparison of visual fields and their functional interpretations has provided a rich source of ideas in visual ecology, showing some very marked and also some subtle differences between species. Just what the functions of these differences are, and what might have driven their evolution, will be discussed in more detail in Chapter 8. It is sufficient to note here that the visual field of a single eye, how the pair of eyes are combined to provide the total field, and how much of it is seen by both eyes simultaneously (the binocular field) are susceptible to both marked and subtle variations and are as rich an area for analysis in terms of visual ecology as are optics and retinas. It is also clear from Figure 2.14 just how different from birds is the sector of space about the human head that is available to provide visual information at any instant.

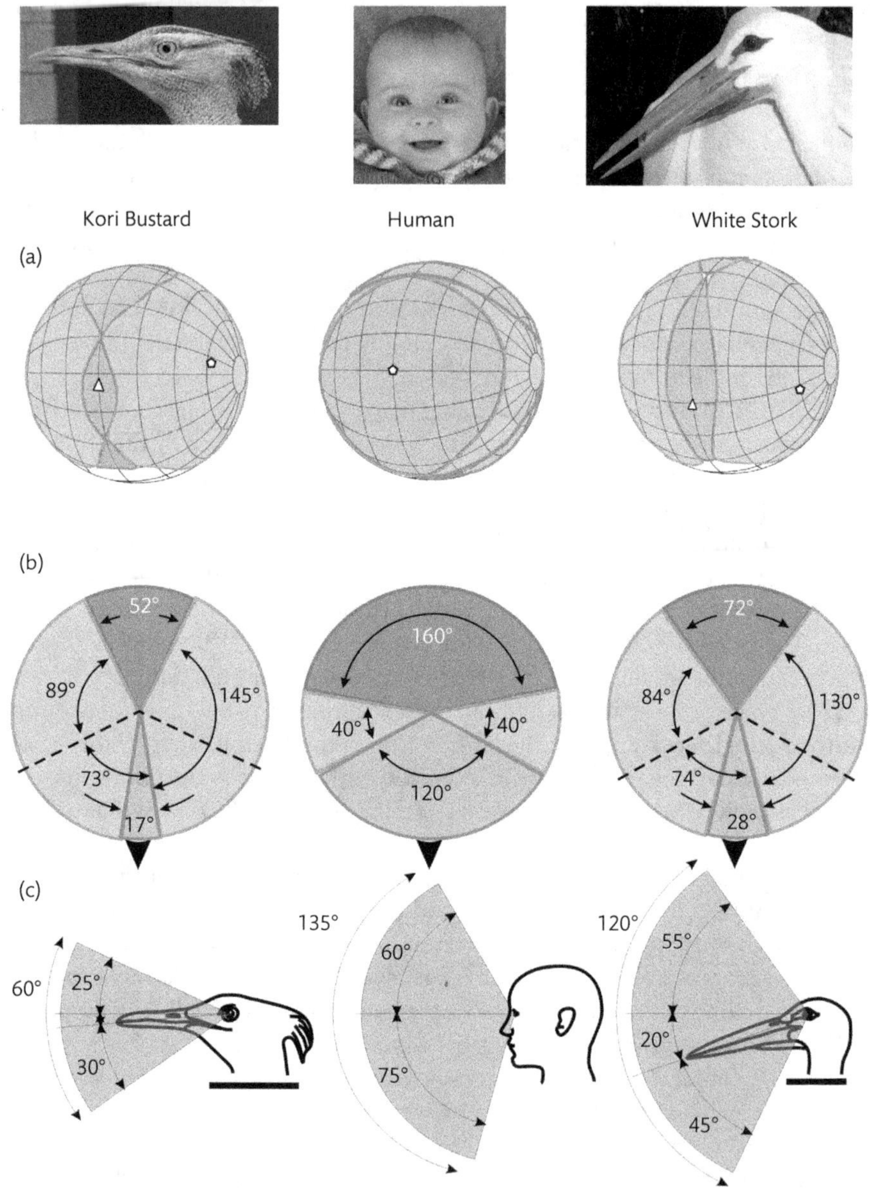

Figure 2.14 Variation in the visual fields of birds. The ways in which the visual fields of each eye are brought together to form the total visual field of an animal can vary markedly, producing significant differences in the size (width and vertical extent) of the binocular field (which is where the fields of the two eyes overlap, indicated in green), the blind area behind and above the head (indicated in blue), and the portion of space that is viewed by each eye alone (indicated in orange). This figure shows the visual field characteristics of two birds, Kori Bustard *Ardeotis kori* and White Stork *Ciconia ciconia*, and the situation in humans. The top row of diagrams (a) shows the visual fields as projected onto the surface of a sphere

surrounding the head. The grid (at 20° intervals) follows conventional latitude and longitude but with the equator aligned vertically in line with the median sagittal plane of the skull. The projections of the bills, and in humans the nose, are indicated by the white triangles. The middle row (b) shows schematic horizontal sections through the visual fields with the black arrow indicating the direction of the bills/nose. The bottom row (c) presents vertical slices through the visual fields in the median sagittal plane of the skull showing the vertical extent of the region of binocular overlap and its position relative to the bill. Very marked differences in these visual fields are apparent with the human differing dramatically from the two birds. However, the birds also show significant differences from each other, this is despite the similar size of the visual fields of the individual eyes in these three species (162°, 158°, and 160° wide in the bustard, stork, and human respectively). Thus small differences in the position of the eyes in the skull have influenced all parameters of the visual fields in these two birds, and they differ dramatically from humans. Functional interpretations of these differences between the two birds refer mainly to differences in the diet, foraging behaviour, and predator detection in the species (Chapter 8). They also have consequences for the collision vulnerability of these birds (Chapter 9). Redrawn and modified from Martin (2011).

2.10 Comparing Doves and Shearwaters: An Example of the Visual Ecology of Optical and Retinal Structures

These comparisons between optical structures (Figure 2.11), visual fields (Figure 2.14), and ganglion cells distribution patterns (Figure 2.12) of doves and shearwaters provide an example of what a sensory ecology approach can reveal about birds. Even in species whose eyes look superficially similar, there are marked differences in all three of the main ways in which eyes can differ. Sensory ecology is built upon knowledge of such differences and their functional interpretation.

This analysis of the eyes of these two birds shows that even if they were sitting side-by-side, they would be experiencing the world quite differently. Their eyes receive and extract different information about the environment. This straightforward comparison between doves and shearwaters provides a clear example of how sensory systems can be interpreted as having become tuned to the informational requirements of particular tasks. But there are always problems in identifying the tasks that drive these differences. These problems arise in part because detailed understanding of the behaviour and ecology of particular species may be lacking, but they also arise because our own imaginations and sensory information may blind us to what is salient to particular species. For this reason, when trying to understand relationships between sensory information and behaviour, comparisons broader than just two species are preferred since they may have more success in identifying common and differentiating factors. Unlike the differences in the wings, bills, or feet, which are readily seen as facilitating their different modes of living, the eyes of birds keep their secrets hidden from the casual observer. However, differences in eyes and their capacities are no less important for a full understanding and appreciation of the biology of birds.

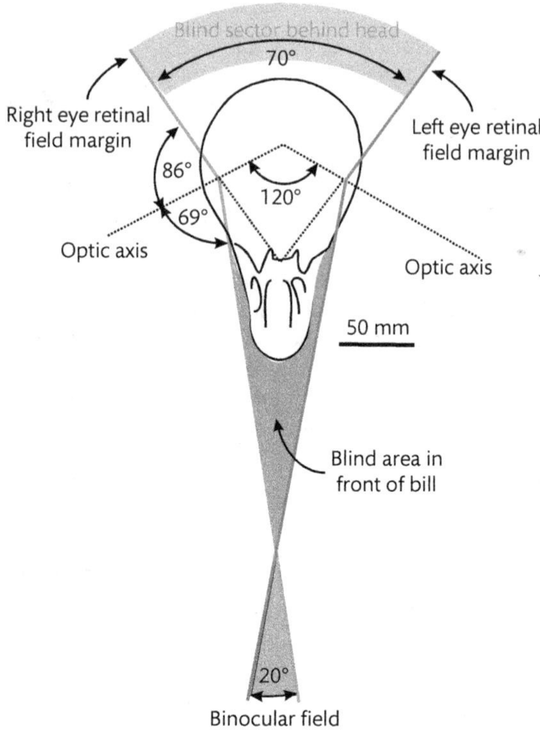

Figure 2.15 How the eyes are brought together. How the two eyes are brought together to form the kinds of total visual fields that are exemplified in Figure 2.14 is illustrated here by the example of the situation in Common Ostriches *Struthio camelus*. The diagram shows a section through the skull in an approximately horizontal plane with the margins of the left and right eye visual fields picked out in red and blue respectively. Each ostrich eye has a field of 155° (similar in width to those shown in the species of Figure 2.14) and they are brought together to achieve a binocular overlap of about 20°. The fields overlap beyond the tip of the bill such that there is a blind area extending for about 100 mm directly in front of the bill. Therefore pecking must be guided by a ballistic final approach to a target before the item is seized in the bill tip. In some birds, however, eyes are positioned so as to allow them to see what they are holding within their bill tip (see Figure 8.6). Modified from Martin (2009).

2.11 Measures of Spatial Resolution

2.11.1 Acuity

The most commonly used measure of spatial resolution is acuity. This is a measure of the ability of the eye to resolve spatial detail when the contrast in the stimulus is high. The best stimuli to be used for such tests of acuity are gratings (patterns of equally spaced black and white stripes of high contrast). Basically, acuity is a measure of how close together lines in a striped pattern can be and still be resolved when presented at a known viewing distance. Although such high contrast stimuli rarely occur in nature, acuity gives a useful indication of the 'best' spatial performance of an eye at a given light level and permits ready comparisons between species in the performance of their eyes when carrying out the same task, and it can be readily extended to investigate how resolution is influenced by light levels. Acuity measures are the basis of the commonly used screening of human vision by optometrists and are a ready way of comparing the performance of a person's vision against normative data.

The best performance always occurs under high, daytime, light levels. Furthermore, in the context of all that has been said above about the possible sources of variation in eyes, it is not surprising to find that acuity shows marked differences across species. Some comparative data are presented in Table 2.1 (also see Appendix 1 for more comprehensive data on the acuity measured in 46 bird species). Table 2.1 shows that from the highest resolution of a Wedge-tailed Eagle *Aquila audax* to the relatively poor resolution of a small passerine, House Sparrows *Passer domesticus*, their highest spatial resolution differs by a factor of 30-fold. It should be noted that the highest acuity so far recorded in any animal is that of a Wedge-tailed Eagle and that this is twice the highest acuity recorded in humans, and about five times that of the standard to which human vision is usually corrected by spectacle lenses (Appendix 1).

It is important to consider the function of high resolution in birds. The highest resolution may not function to aid the detection of very small objects close

Table 2.1. Spatial resolution and acuity in a sample of bird species measured at high (daytime) light levels. They indicate the best performance recorded for each species. The resolution of humans is for young people. Eagle (Reymond 1985); kestrel (Hirsch 1982); dove (Hodos et al. 1976); owl (Fite 1973); sparrow (Dolan and Fernandez-Juricic 2010); human (Land and Nilsson 2012). See Appendix 1 for a table of spatial resolution in 46 species of birds from 12 Orders and 23 Families.

Species	Spatial resolution (cycles/degree)	Acuity (minutes of arc)
Wedge-tailed Eagle (*Aquila audax*)	142	0.21
American Kestrel (*Falco sparverius*)	40	0.75
Rock Dove (*Columba livia*)	18	1.7
Great Horned Owl (*Bubo virginianus*)	7.5	4
House Sparrow (*Passer domesticus*)	4.8	6.3
Human	72	0.4

by, but it is probably particularly important for the detection of larger objects at a great distance. This seems to be the case in the larger raptors. For these birds, food items are relatively rare and widely scattered across the environment, and it can be argued that it is the need to detect objects at great distances that has led to the evolution of vision capable of high spatial resolution. The five-fold higher acuity of a Wedge-tailed Eagle compared with an average human crucially means that these eagles could detect a given object at five times the distance. This performance is about eight times better than a Rock Dove, i.e. a dove would have to be eight times closer to the object for it to be detectable. Clearly doves would gain no advantage in being able to match the performance of an eagle or even of ourselves, since their ecology does not require the detection of smaller objects at large distances. Thus, it would seem that doves, which feed on small items detected on the ground from only a few millimetres away, have evolved much lower spatial resolution. Eagle-eyed acuity would have no selective advantage for doves.

The high resolution of raptors is not without costs. Figure 2.16 shows how measured acuity falls with light levels. As such, these functions depict a key practical consequence of the trade-off and compromise between resolution and sensitivity in vision as discussed in 2.4. As light levels fall, acuity drops rather rapidly and there is evidence that this occurs in American Kestrels *Falco sparverius* (Hirsch 1982) as well as doves (Figure 2.16). In owls, on the other hand, such high acuity is not achieved at high light levels but the decline of resolution with light levels is probably less precipitous than in the diurnal raptors. Some of the implications of these changes in resolution with light level for the sensory ecology of birds are discussed in Chapters 6–9.

It is important to note that there is a link between spatial resolution and the ability to detect UV light (see 'Colour vision and sensitivity in the spectrum', section 2.6.3). The precision with which light can be brought to a focus by an optical system is inversely related to wavelength and this is referred to as the chromatic aberration of a focused point of light. Put simply, it is more difficult to bring light of shorter wavelengths to a sharp focus compared to light in the green or yellow parts of the visible spectrum. Without elaborations of the optical system to overcome such dispersion of light at shorter wavelengths, there is likely to be greater blur in an image formed by UV light. For this reason, there is good reason to filter out UV light and, indeed, this is what occurs in many eyes (such as our own) in which the cornea and/or lens act as UV filters. Thus, it would be predicted that eyes which have the highest resolution should not be using the UV end of the spectrum. This does seem to be the case in that raptors (Accipitridae, Falconidae), which have the highest acuity among birds (Appendix 1), do not have UV sensitivity. In fact, they have optical systems which filter out UV light from the focused image (Lind et al. 2013; Lind et al. 2014). Birds which have UV sensitivity (e.g. Passeriformes, ostriches) on the whole have relatively lower acuity (Appendix 1). It should be noted that the claim that falcons have UV

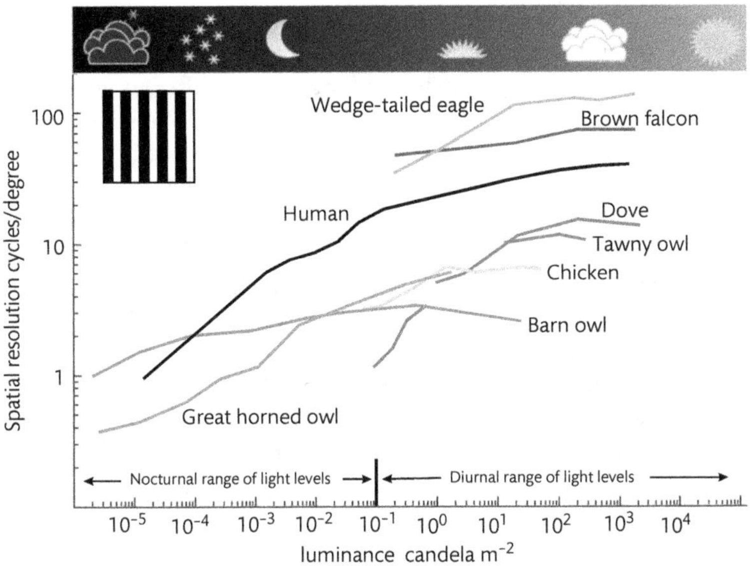

Figure 2.16 The effect of light levels on spatial resolution. Only in a few bird species has visual spatial resolution (acuity) been measured across the full range of naturally occurring light levels. It has been determined across a narrower range of light levels in the daytime-twilight range in a larger group of bird species. In all species, except Western Barn Owls *Tyto alba*, acuity has been found to decrease significantly as light levels decrease. Spatial resolution is expressed here by reference to the highest frequency of a high contrast grating that can be resolved at different light levels, usually employing a behavioural training technique and gratings of the kind shown in the inset (see Appendix 1). In such gratings, frequency describes the number of pairs of black and white stripes per degree of visual angle. In the insert, there are five cycles and ten stripes. The narrowest stripe that can be detected defines the acuity threshold at each luminance level. Wedge-tailed Eagle *Aquila audax* (Reymond 1985), Brown Falcon *Falco berigora* (Reymond 1987), human (a combination of data from Shlaer 1937 and Pirenne et al. 1957), Great Horned Owl *Bubo virginianus* (Fite 1973), Tawny Owl *Strix aluco* (Martin and Gordon 1974), Rock Dove *Columba livia* (Blough 1971), Chicken *Gallus gallus* (Gover et al. 2009), Barn Owl (Orlowski et al. 2012). Figure redrawn and modified from Martin (1986a) and with permission from Mindaugus Mitkus, Lund Vision Group, Sweden (Mitkus 2015). See also more extensive data on spatial resolution in birds in Appendix 1.

vision which they use to guide their foraging (Viitala et al. 1995) is no longer supported (see Chapter 8, 8.12).

2.11.2 Contrast Sensitivity

Natural stimuli are rarely black and white, and many real tasks, in effect, require the detection of grey targets against grey backgrounds, and a way of describing the ability to detect stimuli which vary in this way is to measure 'contrast sensitivity functions' (Ghim and Hodos 2006). In effect, these determine the minimum amount of contrast that can be detected for stripes of different widths presented as gratings of different spatial frequencies. From this, it is possible to determine how wide stripes have to be for the smallest contrast to be visible. Studies have suggested that birds have surprisingly low contrast sensitivity compared to mammals, including humans. That is, at a given stripe width contrast has to be higher in birds than in mammals before they can be detected (Ghim and Hodos 2006; Harmening et al. 2009; Lind and Kelber 2011; Lind et al. 2012), and this is made clear in Figure 2.17, which shows that contrast sensitivity of birds sits 'inside and below' that of humans.

Figure 2.17 shows that contrast sensitivity declines as stripe widths both increase and decrease away from the widths where contrast sensitivity is highest, that is there is a middle range of stripe widths where small amounts of contrast differences are relatively easily detected. Most importantly these studies demonstrate that when stimuli are of low contrast, the optimal size at which they are most likely to be detected is relatively large; both very broad and very narrow striped patterns require a high contrast for them to be detected.

As will be discussed in Chapters 6 to 9, these general findings on how spatial resolution declines with ambient light levels and with contrast have clear consequences for the sensory ecology of birds. This is especially so when birds function in situations in which ambient light levels are low and/or variable, or when tasks involve the detection of low contrast targets. Such conditions are, of course, more typical of the real world than the high contrasts and high light levels that are employed in experimental studies.

2.12 Conclusion: Vision in Birds

This chapter has been long and complex. This reflects both the complexity of vision and the long pedigrees of detailed analyses of vision in birds. I have not tried to provide a comprehensive review of bird vision; that would need a volume of its own. I have endeavoured to structure the chapter to bring out both the basic simplicity of bird eyes and the very wide range of diversity that has evolved in its two basic mechanisms, the imaging systems and the analysing systems.

Built around common design and basic units, these systems show great diversity between and within species. In the retina, photoreceptors of only a few types are found, but these occur in a very wide range of patterns of abundance between the retinas of different species. The relatively simple combination of a lens and cornea, by virtue of small changes in their physical sizes, positions, and curvatures, provide different eyes with images with markedly different properties that are presented for analysis by the retina.

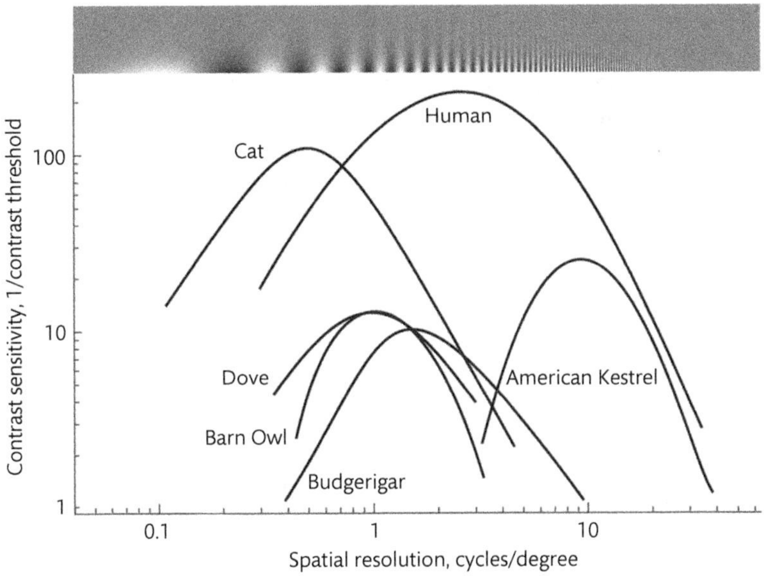

Figure 2.17 The effect of contrast on spatial resolution. In only a few bird species has the relationship between contrast and resolution been studied in detail. In these types of investigation, rather than use strongly contrasting black and white grating stimuli (see inset in Figure 2.16), the stimulus varies between different shades of grey in a sinusoidal fashion, as indicated in the band across the top of the diagram. For each spatial frequency the minimum contrast at which the bird can detect the stimulus is determined. In all species, this produces a curvilinear function indicating that at both very high and at very low spatial frequencies contrast has to be high for the pattern to be detected, while for a mid-range of spatial frequencies contrast can be relatively low. One general finding is that birds are poorer at detecting such stimulus patterns than mammals. That is, birds require higher contrasts in patterns for them to be detectable. Budgerigar *Melopsittacus undulatus* (Lind et al. 2012), Barn Owl *Tyto alba* (Harmening et al. 2009), Rock Dove *Columba livia* (Hodos et al. 2002), American Kestrel *Falco sparverius* (Hirsch 1982), cat (Bisti and Maffei 1974; Berkley 1976), human (Berkley 1976). Redrawn and modified with permission from Mindaugus Mitkus, Lund Vision Group, Sweden (Mitkus 2015).

The descriptions presented here of what is known about the eyes and the vision of birds are far from comprehensive. However, the examples used show that eyes can vary significantly one from another and that it is possible to interpret much of the diversity by reference to the provision of information for the conduct of specific tasks in different environments.

How this information can be used in understanding the behaviour of birds in different environments and what are the main tasks which have driven the diversity of eyes are considered in later chapters. Before that is possible, however, it is also necessary to gain understanding of the other senses which provide birds with information about their environments. Vision is important and complex, and shows considerable diversity in its ability to extract information from the environment. However, vision alone is not sufficient to guide the behaviour of any bird. It will be seen that vision always works alongside other senses to provide the wide array of information that underpins the behaviour of all birds.

3
Hearing and Olfaction

In birds, vision is often assumed to play such a dominant role in gaining information about the world that it is easy to overlook the fundamental roles played by other senses. The sought-after 'bird's eye view' betrays our bias in assuming that the all-important sensory aspect of a bird is what it sees, not what it hears, tastes, smells, or feels. Non-visual senses are often considered to be secondary sources of information, serving primarily to supplement information provided by vision.

This is, however, far from the case. Birds continuously receive diverse information from their other senses. These sources of information are exploited not only when vision is no longer capable of providing information but are often key to guiding behaviour. In fact, in the later chapters, which discuss how birds cope with specific environmental challenges and behavioural tasks, it will be shown that non-visual information can be dominant. Even when vision is important, non-visual information often provides vital information for the conduct of key behaviours, especially foraging. As with vision, there may be both marked and subtle variations between species in these non-visual senses and their functions are best explained through the framework of sensory ecology.

In this chapter, the basic mechanisms and capacities of hearing and olfaction are described. Like vision, these two senses can retrieve information about objects that are remote from the receiver—they are telereceptive or 'far receiving' senses. They are senses which provide information about objects in the broader environment in which a bird lives. The spatial scale, accuracy, range, and temporal scale, over which the information is retrieved, can differ markedly between species and between situations. Hearing and the sense of smell are discussed both to demonstrate the fundamental information that they can provide and to highlight their crucial role in some species.

Chapter 4 considers touch, taste, and magnetoreception. The first two senses provide crucial information about a bird's intimate environment, about objects and substances that are in contact with or very close to the bird's body, and even inside a bird's body. They are 'near receiving' senses which provide information that may be used to guide foraging and also information that is crucial to the well-being of a bird, especially with respect to the identification and the qualities of its food.

The Sensory Ecology of Birds. Graham R. Martin: Oxford University Press (2017).
 DOI 10.1093/acprof:oso/9780199694532.001.0001

Magnetoreception, unlike vision, hearing, olfaction, touch, and taste, is neither telereceptive nor is it concerned with objects in contact with, or inside, the body. Magnetoreception stretches our imaginations in that it provides information about ambient conditions, about properties of the environment that are intangible to us. It provides information about a property of the environment in which all organisms are immersed, a property that changes constantly with position in the environment. Magnetoreception does not provide information about objects or remote events. Rather, magnetoreception enables birds (and other organisms that possess it) to determine their own location, not with respect to objects, and perhaps not with high precision. Magnetoreception seems to provide birds with something akin to a low resolution GPS (global positioning system), something which we have all come to see the benefits of in the last two decades, and which many people could not do without. However, for humans to use GPS information requires a mapping system derived from other information. Magnetoreception may provide birds both with the map and the location within it.

3.1 Hearing

Hearing provides a unique suite of information that is not available through other senses. Of particular importance is auditory information that allows birds to respond to sound signals produced by other birds and to the sounds produced by the movements of other animals. Crucially hearing can provide this information when the sources are hidden from view. Hearing, or something akin to hearing, may even provide information about weather events that are occurring hundreds of kilometres away, beyond the curvature of the Earth.

There has been a degree of coevolution between sound reception and sound production and many different types of sound are produced deliberately by a bird to send specific information to others. These sounds include the ubiquitous songs and calls produced by the specialized vocal apparatus of a birds' syrinx. They may also involve sounds produced by direct mechanical means, for example by woodpeckers (Picidae) drumming on wood with their bill, Flappet Larks *Mirafra rufocinnamomea* using specialized wing feathers and snipes *Gallinago* spp. using specialized tail feathers to produce sounds, pigeons *Columba* spp. clapping their wings together in flight, and storks *Ciconia* spp. making sounds by exaggerated bill clapping. In some species, produced sounds may be amplified by elaboration of anatomical structures, such as the long and coiled tracheas found in some cranes (Gruidae) and swans (Anatidae) (Catchpole and Slater 2008).

To be effective as a means of communication, all of these sounds must match to a broad degree the hearing abilities of the intended targets of the communication. These deliberately produced sounds have become elaborated so that the receiver may extract a complex of information about the signaller, and this can include its species, sex, readiness to reproduce, and position within a social hierarchy

(Catchpole and Slater 2008), information that human listeners (ornithology fieldworkers and birdwatchers alike) have also learnt to extract from these sounds.

Sounds are well suited for these communication functions. Sounds are transmitted through the medium of air more rapidly than odours but more slowly than light. While odours can communicate a rich diversity of information between birds (Chapter 4), sounds have the advantage that unlike odours they do not linger, meaning that information is received and communication occurs in real time. Although sound travels more slowly than light through air, as a mean of communication between individuals at close range it is, for all intents and purposes, instantaneous.

Sounds, however, do not propagate very far since the power transmitted away from a sound source is absorbed by the medium that it moves through and its power decreases with the square of the distance. This means that at twice the distance the sound level drops by a quarter and at three times the distance by one ninth, etc. The power of a sound can also be attenuated and its frequency spectrum distorted by atmospheric conditions, particularly by the presence of water vapour and wind, or the sound may be distorted as it is reflected from surfaces (Catchpole and Slater 2008). However, unlike light, sounds have the advantage that they can transmit around an opaque object; thus information can be gained about the movement of an object, or the presence of another animal, when it cannot be seen. This is one of the key advantages of sound for the extraction of information from an animal's environment. Sound also has the advantage that practically every movement that an organism makes can potentially make a sound. This makes sound an effective means for detecting the presence of many types of prey. However, because these sounds are typically of low power it is effective for prey detection only at relatively close range.

Hearing is widely distributed among vertebrates, and there are some impressive examples of particular specializations of the hearing apparatus, especially among the mammals (Gridi-Papp and Narins 2008). Particularly important are specializations that allow the detection of sounds of high or low frequencies. These sounds are outside the hearing range of human ears and we cannot appreciate them without the aid of technologies. For example, in aquatic environments cetaceans of the Odontoceti suborder (toothed whales, which include dolphins and porpoises) have hearing that extends to very high (so-called ultra-sound) frequencies, and these are used for accurate object location and identification. Conversely the hearing of the Mysticeti suborder species (baleen whales) extends to very low (so-called infra-sound) frequencies, and these are used for long-distance communication. In terrestrial environments, the microbats (Microchiroptera/Yangochiroptera) have hearing that extends to very high frequencies which are used for object location, while elephants (Proboscidea, Elephantidae) detect sounds of very low frequency which are employed for communication using sounds propagated through the ground. However, mammal species which live permanently underground, such as Blind Mole Rats *Spalax ehrenbergi*, have hearing which is described as degenerate and probably has little, if any function (Fritzsch 1992; Heffner and Heffner 1982; Heffner and Heffner 1992; Lewis et al. 2006).

These specializations in the range of frequencies that mammalian hearing systems can detect illustrate how hearing has evolved so that across the animal kingdom sounds of a wide range of frequencies are used to communicate and to determine the positions of objects, in the air, underwater, and in the substrates on which an animal might stand (Gridi-Papp and Narins 2008). Among birds, however, hearing is sensitive only to sounds transmitted through air and is restricted to a relatively narrow range of frequencies. Although the hearing of birds may be considered impoverished compared with that found in mammals, hearing is used by some species as a very important source of information about objects within a bird's environment and for most species as an important means of communication between individuals.

3.1.1 The Sound Stimulus

All physical movements of objects cause the medium that surrounds them (air or water) to oscillate, i.e. the movement produces momentary compression of air or liquid molecules. These oscillations are propagated away from the object through the surrounding medium in the form of compression waves. It is these oscillations that hearing organs detect, and which we call sound. Properties of the oscillations are correlated with properties of the object that caused the disturbance, for example whether the object has hard or soft surfaces, how fast and how far it moves, all influence the nature of the oscillations that are emitted as a result of its movement. Furthermore, these oscillations can be used to determine the direction and distance of the sound source from the observer. Thus, sounds present a rich source of information about moving objects and events in the environment. These events may vary from some relatively slow but very large amplitude oscillations of air produced by weather systems, to the very small scale oscillations produced by the vocal apparatus of an organism, or the rustle of vegetation caused by an animal moving across a forest floor.

The amplitudes of the oscillations of air molecules, detected as 'sound', are very small and of low energy. Sounds are produced by oscillations in the air that are smaller than the wavelength of light, and the ear's detection system clearly has to be extremely sensitive to register them. Close to the threshold of hearing in air, which in humans is approximated by the faint sound of leaf litter rustle heard from a couple of metres (described as having a sound pressure level of between 0 and 10 dB), the displacements of air molecules that are detected at the ear are about the diameter of a hydrogen atom ($\approx 10^{-10}$ m; visible light has wavelengths between 400 and 700 nm, i.e. 4×10^{-7} m). Even sounds which we perceive as very loud, for example a car horn sounding a couple of metres away (which will have a sound pressure level of about 100 dB), are produced by air molecules whose movements are approximately only 1 μm (10^{-6} m). Thus birds' ears are extracting information through their ability to detect extremely small movements of air. Hearing spans air oscillations which have an amplitude range of approximately 1 million-fold, but at the highest sound levels (140 dB) damage to the ear is highly likely to occur.

Ears not only detect these vibrations over a wide range of amplitudes, but also detect vibrations over a very wide range of frequencies of oscillation. The frequency range of hearing among vertebrates can extend from approximately 1 Hz (1 cycle per second) to 100 kHz (100,000 cycles per second). However, no one ear can detect this full range of frequencies that can occur in the environment. The ears of different species of animals are capable of detecting only certain ranges of those frequencies. The range detected by birds will be discussed below. The important point to note is that all ears are selective of the sound frequencies that they can detect.

3.1.2 The Hearing System of Birds

Detecting the oscillations of air molecules (acoustic vibrations) is achieved in birds, reptiles, and mammals by an ear of three main functional parts (Gridi-Papp and Narins 2008). The minute movements of air molecules are detected at an interface between the outer ear and the middle ear structures. In birds, the outer ear is usually a simple tube whose entrance is positioned just below and behind the eye. It leads from the surface of the skull into the head where the middle ear is positioned, protected by the skull. The opening of the outer ear in most birds is hidden beneath a specific group of small feathers which have a rather open structure, the ear coverts. In some birds, for example Vultures (Accipitridae) and Guinea-fowls (Numidae), the entrance is unfeathered and the entrance to the outer ear can be seen clearly (Figure 3.1). The structure which detects air molecule movements is

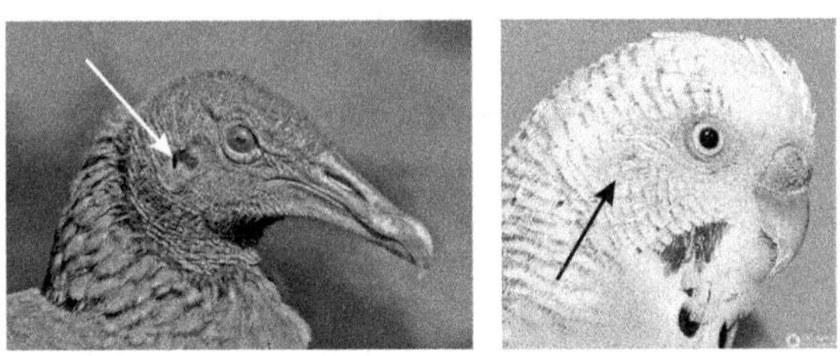

Figure 3.1 The ear openings of birds. In the majority of birds the entrances to the ears are just simple holes leading into the skull; there are no external structures akin to the pinnae which extend from the surface of the head in mammals. The ear entrances in birds are positioned behind and below the eyes and are typically hidden by a small group of feathers, the ear coverts. These are indicated by the black arrow in this photograph of a Budgerigar *Melopsittacus undulatus.* In some birds with bald heads, the ear openings are more conspicuous, as in the photograph of a Black Vulture *Coragyps atratus.* Photo credits: Black Vulture Bryan William Jones; Budgerigar, Ian Junor.

the ear drum (tympanic membrane). It is a very thin and taut membrane which separates the outer ear tube from the middle ear, which is also air filled. The thinness of the membrane allows it to follow airborne pressure fluctuations (average molecule movements) from instant to instant.

Vertebrate ears are unable to respond directly to these airborne vibrations since, as a legacy of the evolution of vertebrates from aquatic ancestors, vibrations are detected in the inner ear by receptors that are surrounded by fluid. There is a mismatch between the impedance of the thin membrane of the ear drum and the fluid surrounding the receptors, and this mismatch is overcome by the mechanism of the middle ear.

The function of the middle ear is to increase the amplitude of the vibrations of the ear drum so that they are sufficient to make the fluid filling the inner ear vibrate with the same frequency as the air molecules striking the ear drum. In mammals, the middle ear mechanism is composed of three small bones which articulate and serve as levers to amplify the movements of the ear drum. In birds, reptiles, and amphibians, this amplification is achieved by a single bone, the columella. It is believed that the three-bone system of mammals is a key reason why their hearing sensitivity can extend to sounds that are of much higher frequency than can be detected by animals which have a columella mechanism (Gridi-Papp and Narins 2008).

The individual receptors which detect the vibrations that have been transmitted to the fluid of the inner ear are called hair cells and are bathed in the fluid of the inner ear. They occur in bundles and are fixed rigidly at one end to a stiff membrane while their other ends are embedded in a flexible membrane. The flexible membrane distorts as vibrations pass through the fluid and these distortions cause the hair cells to bend. It is their bending which triggers nerve impulses that are sent to the brain. Hair cells vary in many ways both within and between species, and it is these differences, as well as the transmission properties of the middle ear, that are primarily responsible for differences in the information that different animal species are able to detect from a given sound stimulus.

The tubes of the inner ear are longer in birds and mammals than in reptiles and amphibians (Manley and Clack 2004). In birds, the tube is usually in the form of a single curve, while in mammals the tube may be coiled. The curved or coil structure allows for the packing of a longer tube within a small space within the skull. The length of the tube and the packing of hair cells in the membranes that run along its length are highly correlated both with the frequency sensitivity of the ear and its ability to distinguish between different frequencies of vibration (Fay 1992).

3.1.3 Hearing Sensitivity

The principal tools for describing and comparing the hearing of different animals are audiograms (Figure 3.2). They capture in one graphic presentation the range

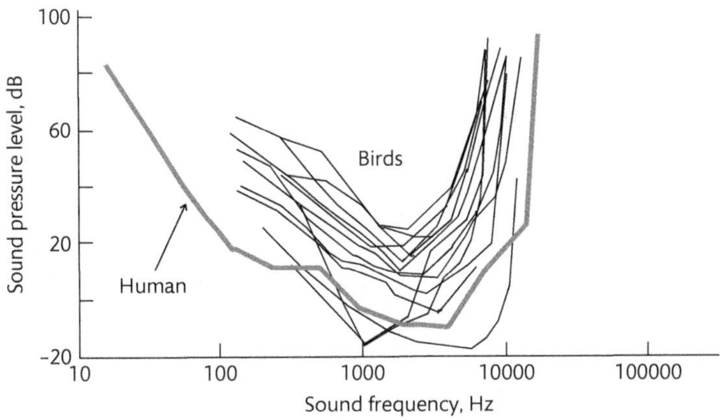

Figure 3.2 The hearing of birds depicted by audiograms. These show the threshold sound pressure level as a function of sound frequency. The lower the sound pressure level that can be detected, the higher the sensitivity. Audiograms for a number of different bird species are shown together with the standard audiogram for young humans. The typical 'U-shape' of vertebrate audiograms is clearly evident; there is rapid loss in hearing sensitivity at higher frequencies and a less steep loss at lower frequencies. It is clear that the hearing of birds sits within the overall envelope of hearing of young humans indicating that whatever sound a bird can hear, a human will also be able to detect it, and that humans can hear a wider range of frequencies than birds. Highest sensitivity in both humans and birds is in the region of sounds with a frequency of about 4000 Hz. (Redrawn from a diagram of the audiograms of birds from Fay (1992) and the typical audiogram of humans is redrawn from Heffner and Heffner (1998)).

of frequencies that an animal can hear and the minimum sound levels necessary for detecting sounds at selected frequencies. Audiograms can be constructed using information gained by a number of different techniques but the most valuable audiograms are those derived from behavioural studies. By using training techniques, a bird can be made to indicate that it can detect sounds of known frequencies and amplitudes. However, such studies are difficult and laborious and so it is not surprising that there are few detailed behaviourally determined audiograms available for birds. Furthermore, audiograms show considerable individual variation within a species but comparison of sample means allows differences between species to be described. Audiograms can also be constructed using physiological techniques, for example direct recording from the auditory nerve or recording the auditory brainstem response (Walsh et al. 2015). On the whole, audiograms determined in this way support data from the behavioural audiograms.

A broad comparison of audiograms across the five major classes of vertebrates shows significant variations. These attest to both phylogenetic constraints (similarities with and between major vertebrate classes) and ecological constraints upon hearing, especially those associated with hearing systems which function in either air or water (Fay 1992). Among birds, however, comparison of average audiograms shows considerable similarity across species (Figure 3.2). All audiograms are U-shaped: there are rapid declines in hearing sensitivity for both high and low frequency sounds, and a broad range of frequencies where sensitivity is similar. The general pattern among the audiograms of birds indicates that highest sensitivity to sounds is in the frequency range 1–4 kHz. Towards the high frequencies, there is a rapid decline in sensitivity with a cut-off in hearing at about 8 kHz; while at low frequencies the cut-off is at about 300 Hz. The absolute threshold (the lowest intensity sound that can be heard) is at a sound pressure level of about 20 dB in the 1–4 kHz frequency range.

There is, however, behavioural evidence that some birds can detect infra-sounds. These are sounds with frequencies below 20 Hz, which are inaudible to humans (infra, simply indicating sound frequencies below those which humans can detect). Sensitivity at such low frequencies was first reported in Rock Doves in which it was shown that they could detect sounds with a frequency as low as 1 Hz (Kreithen and Quine 1979). This has been backed up by circumstantial evidence for infra-sound detection in the same species when performing homing flights (Hagstrum 2000).

The function of this low frequency sound detection has been linked to navigational mechanisms in Rock Doves, but it may also serve to warn of advancing intense weather systems. This was suggested to have occurred in some Golden-winged Warblers *Vermivora chrysoptera* in which satellite tracking showed that some birds left their breeding grounds ahead of a tornado event and returned when the event had passed (Streby et al. 2015). The explanation is that low pressure weather systems are known to emit low frequency sounds at high intensity. These transmit over distances of hundreds of kilometres, and have been used in human weather forecasting. Infra-sound detection has also been reported in Chickens *Gallus gallus* and in Helmeted Guinea-fowl *Numida meleagris* (Hill et al. 2014; Theurich et al. 1984), but the function is unknown since these birds do not travel large distances or even fly readily. However, it is not clear whether the detection of these infra-sounds by any birds is, in fact, an instance of hearing. It could be that such low frequency sounds, at the high intensities at which they have an effect, are detected by somatic receptors, perhaps by the groups of somatic receptors which provide information about internal conditions within the limbs, or the cardiovascular and the respiratory systems (Chapter 4). Thus while these birds may sense these low frequency sounds, they may actually 'feel' them, rather than hear them.

The hearing of birds is significantly different from that of mammals. Hearing in many species of mammals, including rodents and bats, extends to high frequencies, typically up to 100 kHz, while in other mammal species hearing extends routinely into the infra-sound range, for example elephants show sensitivity to such sounds (Heffner and Heffner 1982). The audiogram of young humans is the typical U-shape with frequency sensitivity cutting off at 20 Hz and 20 kHz); this compares with the 300 Hz to 8 kHz range of most birds (Gridi-Papp and Narins 2008) (Figure 3.2). Thus young humans can certainly hear all the sounds that a bird can hear plus sounds with frequencies both above and below the birds' range. In terms of absolute sensitivity, birds are generally about 10 dB less sensitive than young humans. However, there is variation between individuals and some species or individuals appear to have sensitivity equal to that of some humans.

It should be noted, however, that human hearing declines with age and there is typically a loss of both sensitivity and a narrowing of the frequency range. This starts from about the age of 18 years, but most humans will have a hearing range even in advanced years which embraces the frequency range of most birds. Thus, with the possible exception of some birds being able to detect infra-sounds, it is safe to conclude that the hearing range and sensitivity of birds sit within the majority of humans who have not suffered specific hearing loss, associated with prolonged exposure to loud sounds, or to advanced age.

3.1.4 Locating Sounds

Locating the source of a sound is as important as detecting it. This is true irrespective of whether the sound is used as a cue for the detection of prey or as a means of communication. However, sound location has two components, direction and distance, and these are determined by different mechanisms and involve different degrees of accuracy. Determining direction is a purely sensory ability, but determining distance (often referred to as sound ranging) is primarily a cognitive ability in that it appears to depend upon experience and familiarity with the characteristics of known sounds.

Determination of sound direction

The primary problem of determining the direction of a sound source is that it requires two ears. This is unlike an eye in which the direction of a light source is determined primarily by where its image falls on the retina and hence the direction of a light source can be determined accurately by a single eye. A single ear, however, can signal only the presence of a sound. Accurate location of a sound source requires near simultaneous detection by both ears. It is small differences between the signals received at each ear when detecting the same sound source that are used by the brain to compute the direction from which that sound originated.

Sound localization mechanisms exploit two major sources of information: differences in the intensity and differences in the time of arrival of the same sound at each ear (Klump 2000). These differences arise because the ears' entrances are separated and so the sound is slightly attenuated when it arrives at the ear furthest from the source, and it will also take longer to arrive at that ear. However, in birds the head is small and not very dense (bird skulls do not have the thick bone of mammals skulls) so sound differences between the two ears are also small, both in intensity and time. Furthermore, intensity differences are dependent upon the frequency of the sound such that between-ear intensity differences in general increase with frequency (Klump and Larsen 1992). In Common Starlings *Sturnus vulgaris*, for a given sound source between-ear intensity differences vary from 2 to 8 dB over a sound frequency range of 1–8 kHz. Differences in the time of arrival of sounds at the two ears are about 100 microseconds. The two ears of birds may not be totally independent (as they are in mammals) because the middle ears are joined by an air-filled tube within the skull. It has been suggested that this structure could enable the two ears to also function as a 'pressure difference receiver'. However, this is disputed and the ears may in fact be functionally independent (Klump 2000). Whatever the mechanism, these time and intensity differences between the two ears when detecting the same sound source are very small. In fact, they are much smaller than in most mammals, including ourselves, where the ears are more widely separated and the head itself is dense. This provides greater between-ear sound attenuation and time differences in mammals than are found in birds.

These small between-ear differences in birds have the effect of greatly reducing the accuracy with which the directions of sounds can be determined compared with mammals. This is borne out by a number of different studies and in different species. For example, in Great Tits *Parus major*, depending upon frequency of the sound source, direction accuracy in the horizontal plane varies between 20° and 26° (Klump et al. 1986), in Zebra Finches *Taeniopygia guttata* between 71° and 180°, Budgerigars *Melopsittacus undulatus*, between 25° and 69° (Park and Dooling 1991), Atlantic Canaries *Serinus canaria* between 49° and 71° (Park and Dooling 1991), and Common Starlings between 19° and 27° (Feinkohl and Klump 2013). In all these species, sound direction accuracy was studied in a similar way, involving the use of a single sound source and sound position was altered only in the horizontal plane. However, using another approach which aimed to measure the accuracy with which Eastern Towhee *Pipilo erythrophthalmus* could tell a difference in position of two similar sounds presented simultaneously, the accuracy was 7° (Nelson and Suthers 2004).

Analysis of the possible between-ear cues used by Common Starlings has suggested that birds use both time and intensity cues depending upon the frequency of the sound. Thus time cues may be used for sounds of up to a frequency of about 1 kHz but as frequency increases it is between-ear intensity cues which are used (Feinkohl and Klump 2013).

None of these sound location performances are impressive. If they are translated to physical distances their inaccuracy is shown to be rather telling. For example, a 20° localization accuracy (the best performances recorded in Common Starlings and Great Tits) means that a sound 20 m away from a bird could be anywhere in a horizontal arc 7 m wide, while a sound at 50 m distant could be anywhere within an arc 17 m wide. If localization accuracy is 50° (the best recorded performance of a Canary) then the comparable distances are 17 and 43 m.

In summary, these results suggest that among songbirds the highest accuracy of sound localization is about 20°, and that many birds have a much worse performance than this. Clearly, however, these performances must have great utility for these birds since they use songs and calls as a key means of communication. It should be remembered that these are highly mobile bird species. They can respond to important sounds, especially the songs and calls of other birds which are typically repeated many times in succession, by using short flights to home in on a sound until visual contact with the singer is possible. Thus, while these birds may not know initially the direction from which a sound is coming to them with high accuracy, they can soon find it, especially if that sound is repeated many times or is given continuously, conditions which often apply to the songs and calls of many bird species.

There is, however, a group of birds, owls (Strigidae) and barn owls (Tytonidae), which can achieve much higher sound localization accuracy than the songbirds. These birds detect the locations of sounds that are produced for relatively long periods (at least a few seconds) but are not stereotyped and repeated like songs and calls. They are the sounds that prey animals cause as they move through leaf and grass litter.

Species of owls and barn owls have been shown capable of taking prey guided only by sound cues in total darkness (Payne 1971). This requires accurate sound localization. Using bursts of broadband noise (such sounds are unlike songs and calls in that they contain simultaneously a wide range of frequencies changing randomly and sound to humans like a hiss) that are more than 1 second long, Barn Owls were shown to able to locate the sound source to an accuracy of 3–4° (Bala et al. 2003; Knudsen and Konishi 1979). This degree of accuracy is, in fact, comparable to that achieved by humans using similar sounds. There has been much interest in the mechanism that owls use, especially as owls have been shown to be able to locate sounds with a similar degree of accuracy both vertically and horizontally.

The key to the owls' accuracy lies not in the ears themselves but in the fact that owls have elaborate outer ear structures (Figure 3.3). They are the only group of birds to have ear structures at the entrances of the outer ear canals (Norberg 1968; Norberg 1978). These structures enlarge the distance between the ear openings and take the form of flaps of skin placed both before and behind the ear openings. To them are attached specialized small hard feathers which are capable of reflecting sounds, unlike the feathers which cover the ear openings in most birds which are more or less transparent to sound (Figure 3.1). The positions of these flaps of skin, and hence the feathers, are under the bird's control and changes in their position can result in dramatic alterations to the appearance of an owl's head, from narrow and vertically elongated to round and broad. The presence of these outer ear structures, in fact, accounts for the generally broad head shape of owls. Despite their skulls having basically the same general shape as that of other birds, owl heads

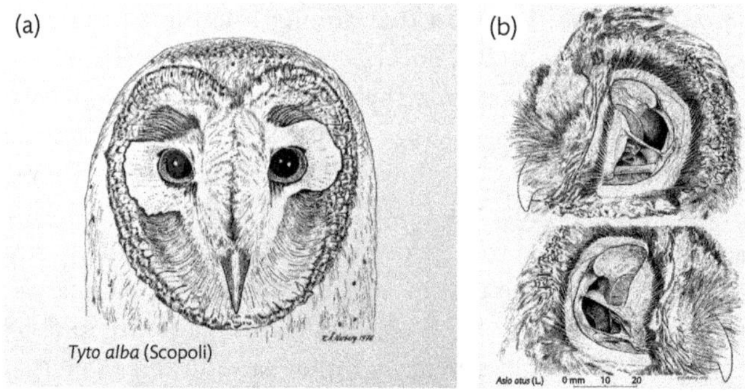

Figure 3.3 The outer ears of owls. Unlike most bird species which do not have outer ear structures (Figure 3.1), owls do have elaborate structures surrounding the entrance to the ear canals. (a) shows a Western Barn Owl *Tyto alba* in which the feathers of the facial disk have been removed to show the flaps of skin that are positioned in front of the ear openings which are positioned just behind the eyes. These structures are of different size and are in different positions on opposite sides of the head. (b) looking into the ear openings of a Long-eared Owl *Asio otus*, the feathers at the edge of the facial disk have been parted to reveal the entrances to the ears. The edges of the ear openings are large flaps of skin which are under muscular control and the flap to the front is the equivalent structure to the skin flaps that are shown in the Barn Owl. Note that the ear openings in the Long-eared Owl are also of different size and shape on the two sides of the head and that the side of the eye can be seen when looking into the ears. Original drawings by courtesy of Rolf Åke Norberg (Norberg 1977).

appear much broader because the outer ears are positioned just behind the eyes and are covered in feathers.

The outer ear structures in a number of owl species are asymmetric, being both higher and larger on one side than the other (Norberg 1978) (Figure 3.3). Analysis of how these outer ear structures distort sounds before they arrive at the eardrums have led to a detailed understanding of how they enhance the accuracy of an owl's ability to determine the position of a sound, in both the vertical and horizontal planes. The accuracy is such that an owl can learn to capture prey in total darkness by dropping from a perch onto small mammal prey moving beneath, guided by the sound of the leaf litter rustles emitted as the prey moves (Knudsen and Konishi 1979). Tawny Owls *Strix aluco* have even been shown capable of detecting and catching earthworms apparently using the sounds that are produced as worms move through leaf litter or when they drag leaves towards a burrow (Macdonald 1976).

This particular development of auditory localization in owls is intimately linked to the nocturnal hunting habit of these birds in which hearing is used to complement vision in the location of prey and is discussed in detail in Chapter 6.

Determination of sound distance (sound ranging)

Determining the distance to a sound source requires a different mechanism to that of determining its direction. Information on the distance to a sound source is very important in determining whether a sound source should be responded to and what the response should be. As with direction, the ability of birds to determine the distance to a sound source has been measured in a range of passerine birds and again the performance is not impressive. In these studies, experiments have used situations that evoke territorial behaviour in response to the playback of a song of the same species. A male territory holder will approach the sound source in an attempt to localize its presumed rival, and the distance covered and its direction of travel are used as a measure of the bird's sound ranging ability (Holland et al. 2001; Morton et al. 2006; Naguib et al. 2000; Nelson and Stoddard 1998).

Manipulation of the playback sound allows the teasing out of cues that the bird might use to estimate the distance to the sound source. However, it is important to note that the sounds used are not novel sounds but are species-specific calls and short phrases of song. These studies show that simple reduction in the overall amplitude (volume) of a familiar sound is used by these birds as a cue for distance (Naguib 1997; Nelson 2000); that is, the quieter a familiar sound is, the further way a bird estimates it to be. However, sound amplitude is not a very reliable cue to distance because naturally produced sounds can vary in amplitude, and the direction in which the calling or singing bird is facing, or head movements during singing, can all alter the amplitude of the sound (Catchpole and Slater 2008; Nelson 2000). Thus, it seems that this kind of sound ranging is a cognitive ability as it depends not upon a specific hearing ability of the listening bird, but upon the bird's knowledge of the signal's original frequency spectrum and intensity (Naguib and Wiley 2001). Therefore, such sound ranging depends crucially upon the use of signals that are familiar to the listener. This is, of course, met through species-specific songs and calls that are used in the interactions between territorial neighbours, which are the situations when accurate sound ranging is most likely to be important, especially to the passerine species in which these studies have been conducted.

In support of this reliance upon familiar sounds for accurate ranging are a number of field studies which have demonstrated that the degree of familiarity with a specific song type will strongly influence a male's ability to discriminate between degraded and undegraded playback songs, and the ability to assess the distance of a sound source (McGregor et al. 1983; Morton et al. 2006; Naguib 1998). There are, however, some field studies that have suggested that the distance to unfamiliar sounds can be effectively determined in certain situations (Naguib 1997).

In summary, evidence from a range of studies indicates that sound ranging is most accurate for familiar sounds, especially for species-specific songs and calls. However, one problem with studies using unfamiliar sounds is that they involve a number of trials with the same sounds. Thus, what may have been unfamiliar at the start of the investigative trials could rapidly become familiar. Much depends upon how rapidly birds can learn something of the characteristics of the stimulus sounds and what reinforcement they receive for making 'correct' distance estimations. Therefore, how rapidly an 'unfamiliar' sound can become 'familiar', and how rapidly its patterns of degradation with distance and environmental types can be learnt, may be crucial in sound ranging.

A complication is that territory holding birds may not in any case make fine sound ranging discriminations. Their responses to territorial songs may in fact be categorical, basically 'near' or 'far'. Common Chaffinches *Fringilla coelebs,* for example, showed a categorical response to the playback of degraded songs within a range of sound transmission distances between 0 and 120 m. Thus, the birds seemed to distinguished 'near' sounds (0, 20, and 40 m away) from 'far' sounds (80 and 120 m away) (Naguib et al. 2000). In the context of territorial defence, this might be all that is required of the bird's ability to determine the distance to another bird singing in the vicinity. A bird may simply categorize the song of nearby males as potentially threatening when 'near' ('inside' its territory), or of no threat when 'far' ('outside' its territory). However, Great Tits may be able to categorize sounds as coming from several different distances (Pohl et al. 2015), but nevertheless their response may still be categorical and thus express a rather crude determination of distance to a sound source.

The idea that a bird needs to be familiar with the characteristics of a sound source to be able to determine its distance is supported also by studies of the ability of owls to capture prey in total darkness using sound cues emitted as the prey moves. Observations clearly show that the birds which pounce on prey in total darkness may, indeed, have reasonably accurate information on the distance to the sound. However, to achieve this, an owl has to have a high degree of familiarity with the experimental setup. Thus the birds need to have had a lot of exposure to the sounds of leaf litter rustles caused by prey or dummy prey in the experimental setup. They also need to learn their way around the experimental situation and get used to capturing prey from fixed perches at low light levels for a couple of weeks before they will readily do this in darkness (Payne 1971). Thus, in owls there would seem to be a significant cognitive, as well as sensory component, to using sound sources to catch prey.

3.1.5 Echolocation (active SONAR)

A specialized use of hearing is echolocation, also known as 'active SONAR' (SOund NAvigation Ranging). As in the location and ranging described above, active SONAR depends to some extent upon knowledge of the source sound's characteristics and knowledge of how its reflections are modified by surfaces and objects in the environment. Thus, active SONAR has a cognitive as well as a straightforward sensory means of extracting information from the environment. Humans are certainly capable of using active SONAR for the location of objects and for humans the cognitive or learning component is of key importance (Kellogg 1962). Echolocation involves the detection of echoes from sounds emitted by the birds themselves (their own vocalizations) hence the label 'active', as opposed to the 'passive SONAR' which is what was described in the previous section; that is, the bird determines the range of a sound that is emitted by the object. The key component of echolocation is the time delay between the sending out of the signal and receiving its echo. There is no information on the extent to which the interpretation of such delays has to be learnt by birds that are able to echolocate. This is because not only would such experiments be intrinsically difficult to devise but also because of the difficulty of working with these species in experimental situations.

The accuracy and functions of echolocation are well understood in mammals. Species which use active SONAR include bats, toothed whales (Cetaceans, Odontoceti), tenrecs (Tenrecidae), and shrews (Soricidae) (Brinklov et al. 2013; Johnson 1986). In the bats and cetaceans, active SONAR functions to provide fine spatially detailed information about the nature, position, and movements of objects. The extraction of fine spatial detail by these species is achieved through the production and detection of high frequency sounds (up to 100 kHz with most energy in the range 30–70 kHz). This is because it is only high frequency sounds that can provide coherent reflections from objects of small dimensions; low frequency sounds either are not reflected from small objects or are scattered in an incoherent manner (Pye 1979). However, not all sounds used by mammals to echolocate are ultrasounds well above the auditory range of humans (>20 kHz). Some species of Old World fruit bats, tenrecs and shrews, for example, use echolocatory sounds within the human audible range and as a result these species cannot discern such fine spatial details as are achieved by bats.

Because the hearing of birds is within an even narrower and lower range of frequencies than mammals (Figure 3.2), the size and the position of objects which can be located by active SONAR in birds are rather coarse. They are, however, sufficient to provide key information about objects and surfaces in a small number of species.

Oilbirds *Steatornis caripensis* (Caprimulgiformes, Steatornithidae) (Konishi and Knudsen 1979; Roca 1994; Thomas 1999) and cave swiftlets (Apodiformes, Apodidae) (Chantler et al. 1999; Medway and Pye 1977) are the only birds in which active SONAR has been demonstrated (Figure 3.4). These two groups of birds are, in fact, the only non-mammalian animals known to employ active SONAR. These echolocatory bird species have in common the use for nesting and roosting in deep but typically large caves where there is complete darkness, or where very few light photons penetrate. The interiors of these caves are, however, relatively open locations with few obstacles. Echolocation does not seem to be used by these birds outside of these caves. Oilbirds are a species of large nocturnal nightjars found in northern South America and in Trinidad in the Caribbean

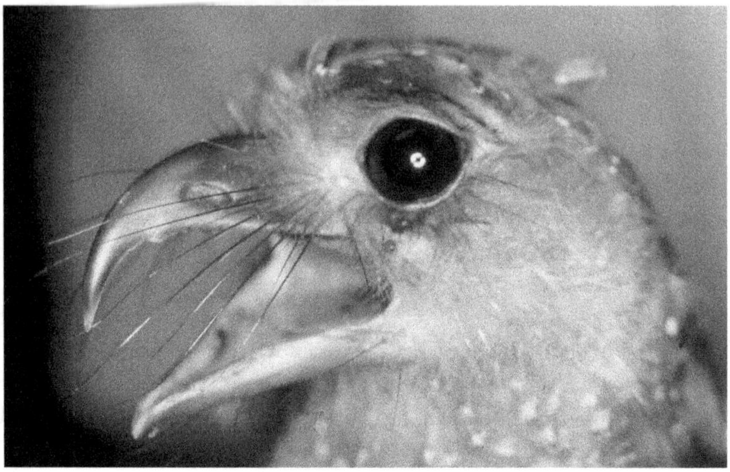

Figure 3.4 The head of an Oilbird *Steatornis caripensis*. Oilbirds and cave swiftlets are the only birds known to employ echolocation (active SONAR) to locate surfaces and objects. They do so inside the totally dark interiors of deep caves in which they roost and nest throughout the day, emerging at night to forage on fruits in the tropical rainforest canopy. Echolocation is achieved using relatively low frequency vocalizations (clearly audible to humans) which are emitted as rapid clicks. There is no indication that the ears of Oilbirds show any specializations associated with echolocation and their hearing seems to lie within the typical range of the audiograms of birds (Figure 3.2).

(Thomas 1999). The echolocating swiftlets are a group of about 16 species, which inhabit tropical regions from the Indian Ocean, through the Far East, northern Australian, and western Pacific Ocean; they are not nocturnal but may fly regularly at dusk and dawn (Chantler et al. 1999).

Not all swiftlet species have been shown capable of echolocation, the exceptions include Giant Swiflets *Hydrochous gigas* and Glossy Swiftlets *Collocalia esculenta* (Chantler et al. 1999; Medway and Pye 1977). The echolocatory abilities of both swiftlets and Oilbirds were first investigated systematically more than 50 years ago (Griffin 1958), but information is not available on the abilities of all species due to the difficulty of working with them. They are difficult, if not impossible, to keep in captivity for long periods and furthermore they are unlikely to show natural behaviours requiring the use of their active SONAR in captivity (Brinklov et al. 2013).

The vocal sounds produced by these birds when echolocating are quite distinctive. As in many of the echolocating mammals, the echolocatory sounds of birds are sharp single or double clicks, or trains of clicks, i.e. each sound is short with a clear start and finish, but unlike the echolocating mammals they are produced at frequencies within the human hearing range (Figure 3.2), with most of the energy between 1.5 and 2.5 kHz in Oilbirds (Konishi and Knudsen 1979). In swiftlets, clicks were reported to be repeated with increasing rapidity when approaching obstacles or when entering caves (Griffin and Suthers 1970), although Fullard et al. (1993) did not find such an effect.

Experiments designed to determine the smallest objects which swiftlets can detect in free flight have involved the birds flying through a space in which wooden or metal rods, or wires were arrayed. These experiments produced a range of results. In one study involving a number of cave swiftlet species, the smallest targets that were avoided had a diameter of between 4 and 10 mm (Griffin and Suthers 1970), while the smallest objects that could be detected by White-rumped Swiftlets *Aerodramus spodiopygius* were reported to be between 10 and 20 mm diameter (Smythe and Roberts 1983). In *A. hirundinacea* a threshold of below 10 mm was reported (Fenton 1975). These performances are not impressive compared with what bats can achieve in comparable tests in which wires as small as 0.1 mm diameter could be detected. It should also be noted that these tests simply involved investigating the size of objects which the birds could avoid as they flew towards them. Nothing is known in detail about the distance, shape, and volume of the space around the birds in which targets can be detected.

The minimum sizes of objects that Oilbirds can detect using active SONAR are even larger than those detected by swiftlets. When plastic discs of various diameters were hung in the passageway leading out of a nesting cave, 'All birds hit 5 and 10 cm diameter discs as if nothing had existed in their paths. The first signs of avoidance appeared when 20 cm discs were presented and all birds avoided 40 cm discs' (Konishi and Knudsen 1979, p. 426). However, Suthers and Hector (1983) reported that Oilbirds could detect obstacles as small as 3.2 cm in diameter. They also reported that the sound levels of the echolocating pulses are as high as 100 dB recorded at a distance of 1 m from the birds, with vocal clicks produced up to 12 times per second, but usually at a lower rate. Certainly the interior of Oilbird caves is very noisy with large numbers of birds simultaneously and continuously producing streams of loud clicks when birds are preparing to leave the cave for their nocturnal foraging forays.

The performance of all these birds in terms of the threshold size of object which can be detected would seem surprisingly poor, but they are well within what is predicted by the relatively low frequencies of the sounds that are emitted (Pye 1979; Pye 1985). However, the results of all the testing, plus field observations (Snow 1961), clearly indicate that active SONAR can be used to guide these birds with respect to large objects. These include other birds and cave walls. Clearly such low spatial resolution is sufficient to guide the flight and landing of Oilbirds within complete darkness.

The interiors of Oilbird roosting caves are typically very noisy. It seems likely that specific nest sites, which are on ledges on the cave walls, may also be located using passive SONAR with flying birds using the calls of mates or young on nest ledges to guide them (Snow 1961; Thomas 1999). Oilbirds also have large, high-lift wings, which give them the ability to fly slowly and to hover for short periods, so any collisions which may occur within caves are not at high speed. Oilbirds seem to live highly predictable lives within the darkness of their roosting and nest caves, apparently being attached to the same nest ledge throughout their relatively long lives (Snow 1961). Thus there is plenty of time for learning about how sounds are degraded and be used as a source of spatial information regardless of whether this is using active or passive SONAR.

3.1.6 Conclusion: Hearing in Birds

Acoustic signals in the form of songs, calls, and instrumental sounds, such as bill drumming, provide sources of information that mediate the social behaviour of birds. There is an ever-growing and impressive body of research findings on the intricacies and functions of acoustic communication in birds (Catchpole and Slater 2008; Kroodsma and Miller 1996). The frequency and timing of these sound signals have been the subject of intense natural selection, and there is compelling evidence that, depending upon species, signals have been shaped to communicate specific details about the species, sex, and physiological and emotional state of individuals. There is also evidence that the acoustic properties of signals have been shaped to optimize high levels of sound transmission in different environments. These including adaptations to enhance sound transmission in different natural environments (Marten and Marler 1977; Morton 1975) and also recent evidence for the rapid selection for improved sound transmission in environments made noisy by human activities, suggesting a high degree of flexibility in the structure of songs and calls (Kight and Swaddle 2015; Luther and Gentry 2013; Marten and Marler 1977; Morton 1975).

It is important to recognize, however, that all acoustic signals used by birds employ a relatively narrow range of sound frequencies (most songs and calls are within the frequency range 1–5 kHz (Catchpole and Slater 2008)) which are all audible to humans, at least to younger humans who have not suffered any hearing loss. The hearing of a bird thus 'sits inside' the hearing of humans. Hence, there are unlikely to be 'secret' sound signals used by birds that humans cannot detect with their own hearing. This is quite different to the situation in mammals and insects. If humans wish to study their acoustic signals then technology must be employed to detect sounds within the full frequency ranges that are used. It is also important to note that the sensitivity to sounds in most birds is lower than that of humans, with the consequence that humans should be able to hear the sounds produced by a bird at a greater distance than another bird can.

It may also seem rather surprising that most bird species have a relatively poor ability to locate the direction and distance of a sound source. Even among songbirds, the ability to determine the source of a sound is not very impressive and is quite inferior to that of humans. Indeed, as described above, it has been suggested that the ability of songbirds to determine the distance to a sound source is so poor that location maybe coded only categorically (Naguib et al. 2000). The birds respond as though sounds are perceived as originating as simply 'near' or 'far' from them. The actual position of a signalling bird (if it is important to the listener) is determined by approximations and approaches in a series of successive homing type movements. However, it seems possible that for many communication instances between birds, it is not necessary for the birds to determine with any precision where the signalling bird is, and so crude sound localization is sufficient.

The direction and distance of a sound source can, however, be determined with greater accuracy by owl species. But in this instance, the range is limited to just a few metres and a sound that can be located needs to have rather specific features. Thus, for the most accurate localization, the sound needs to contain a mixture of a wide range of frequencies (approximating the hiss of white noise) and be of a relatively long duration, at least 1 second. This is the sound of leaf litter disturbance produced by the movements of an animal through or over it, or the sound of a leaf being dragged across a surface. Thus the location of sounds in this instance seems to have been selected to provide information for the quite specific task of prey detection and capture. There is no evidence on the accuracy of owls in locating distant sound sources. Furthermore, it seems highly likely that prey capture based upon this cue alone can be achieved only in specific situations with which the individual is familiar (Martin 1986a) (Chapter 6).

In a handful of bird species, active SONAR, that employs bursts of relatively low frequency and stereotyped vocalizations, underpins their use of caves as safe roosting and nesting sites. This may also require high familiarity with the specific situation in which the active SONAR is used and outside of these quite specific situations sound localization cannot be used to gain spatial information about the environment.

Finally, there is the intriguing possibility that some birds may be able to employ very low frequency sounds as a means of predicting weather patterns, but it is not clear that this is achieved through hearing rather than the detection of low frequency pressure changes possibly through somatic (touch) receptors.

3.2 Olfaction

It is only in the past 20 years that the importance of the sense of smell as a source of a broad range of information has become well established in birds. For many years, avian olfaction was considered something rather special, a sensory capacity found within relatively few species in which it served specific functions in foraging. Birds have often been considered to lack a functional sense of smell or it has been considered rudimentary compared to the sense of smell in mammals. Recent research, however, now makes it possible for authors to write about the excellent sense of smell of birds, its wide distribution across taxa, and its importance as a source of many different types of information (Caro et al. 2015; Corfield et al. 2015).

In some birds, this olfactory information has become seen as equally crucial for individual survival and reproductive success, as the information derived from vision or hearing. It seems possible that information gained through olfaction may eventually be understood to play crucial roles in a wide range of species. Olfaction in birds is not fully understood. There is still much to learn about the subtle tuning of olfactory abilities in different species, the behaviours in which olfactory information plays a key role, and the specific functions of identified chemical compounds in the social life of birds, as well as in foraging. As with the other senses, the understanding of avian olfaction has been pieced together from various types of evidence: behavioural, morphological, neuroanatomical, and physiological.

3.2.1 Organization of the Olfactory Systems of Birds

The anatomical organization and function of the olfactory system in birds is similar to that of all vertebrates. Air, which potentially contains a huge range of chemical compounds, is taken in through the nostrils and the olfactory system signals to the brain the presence of certain compounds that are detected in the sampled air. Olfaction does not involve a complete chemical analysis of the sampled air; only certain compounds are detected by specific receptors. In effect, the olfactory system searches for specific compounds and signals their presence to the brain.

In the majority of birds, the paired nostrils are positioned towards the base and above the bill and the nostril entrances are always open. In kiwi species (Apterygidae), the nostrils are positioned just behind the tip of the upper bill with their openings projecting laterally. These birds probe in leaf litter and soft substrates, and they can be heard sniffing (or expelling air to remove material blocking the nostrils) as they forage on a forest floor. In some plunge diving birds (Suliformes; gannets and cormorants), the nostrils open inside the mouth. Closing the mouth seals the nostrils and this prevents the rush of water into the nasal chambers when birds enter water at high speed. In surface divers (e.g. Auks, Alcidae), the nostrils are narrow slits low down just above the cutting edge of the bill close to the mouth hinge and can, apparently, be closed shut by a flap of skin. In penguins (Spheniscidae), the nostrils are in the maxilla towards its base.

In all species, inhaled air passes from the nostrils and then successively through three chambers (conchae) which serve to filter out small debris, warm, moisten, and finally chemically sample the air. The third concha is lined with mucosa which is innervated by dense neural fibres. These are the olfactory receptors which detect the presence of specific chemical compounds as the air swirls around the chamber before passing into the respiratory system. From these receptors, neurons reach the brain at the olfactory bulbs and the presence of specific compounds is signalled depending upon the suite of olfactory receptors that are present.

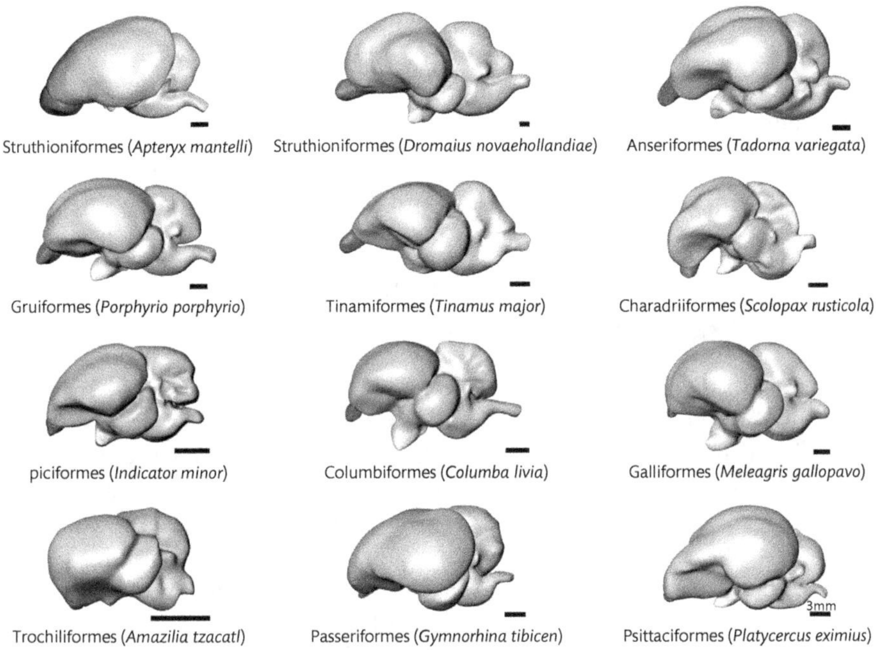

Figure 3.5 Size and shape of the olfactory bulbs in birds. Lateral views of three-dimensional models of the brains of 12 bird species with the olfactory bulbs highlighted in blue. The sequence runs from largest to smallest olfactory bulbs that have been recorded to date. Scale bar = 3 mm. Drawings courtesy of Jeremy Corfield (Corfield et al. 2015a).

Olfactory bulbs vary markedly across bird species in both their shape and size. They are usually paired structures situated at the front of the brain; in some species the bulbs are fused. A lot of attention has been paid to the olfactory bulbs in birds. This is mainly because they are a readily identified brain structure whose dimensions can be measured and compared (Corfield et al. 2014; Corfield et al. 2015) (Figure 3.5). Comparing olfactory bulbs is, however, not the same as comparing the function and structure of the olfactory organs. Comparing olfactory bulbs is equivalent to comparing the areas of the brain devoted to the analysis of visual information rather than comparing and understanding the structures of eyes and how they gather and constrain information.

The internal organization of olfactory bulbs in birds shows the same laminar structure across all species; this includes both palaeognathous and the neognathous species which are thought to have diverged at least 70 million years ago (Benton 2005). This suggests, therefore, that this olfactory bulb structure was a feature of the earliest birds and that this structure has been highly conserved. The size of olfactory bulbs in birds varies markedly across species. The smallest reported bulb has a volume of just 0.06 mm^3 in Spotted Pardalotes *Pardalotus punctatus*

(Passeriformes), while the largest has a volume of 217.63 mm^3 and is found in Emus *Dromaius novaehollandiae* (Casuariiformes) (Figure 3.5). This is a greater than 3500-fold difference in size of the portion of the brain that receives information about the chemical components of air (Corfield et al. 2015).

3.2.2 The Importance of Olfaction and Olfactory Bulb Size

Focusing on those species which have relatively large olfactory bulbs, behavioural studies first showed that olfaction is used for food location in kiwi species (Wenzel 1968) and New World vultures (Cathartidae) (Graves 1992; Houston 1984; Houston 1986). The specific role of olfaction in food location in petrels (Procellariidae) has been demonstrated more recently (Cunningham et al. 2003; Nevitt 2000; Nevitt 2008). Studies have now also shown that olfaction can play a key role in behaviours other than foraging. These include the detection of predators by some passerines (Amo et al. 2008; Amo et al. 2011; Leclaire et al. 2009), recognition of various odours in domestic Chicken chicks (Bertin et al. 2010; Jones and Roper 1997), and navigation in a range of species from different avian orders (Gagliardo 2013). These latter species include Procellariiformes (Cory's Shearwaters *Calonectris borealis*), Columbiformes (Rock Doves), Apodiformes (Common Swifts *Apus apus*), and Passeriformes (Common Starlings, Grey Catbirds *Dumetella carolinensis*).

This range of bird species in which behavioural evidence has demonstrated a specific function of the sense of smell includes those with both relatively large and relatively small olfactory bulbs. This demonstrates that although a greater proportion of the brain is devoted to the analysis of smell in some birds, the presence of a small olfactory bulb does not mean that smell is non-functional or of little importance. Recent analysis of olfactory bulb size in a sample of 135 bird species (Corfield et al. 2015) concluded that the relationship between absolute size and proportional size of olfactory bulbs is complex. Cross-species comparisons showed that bulb size scaled allometrically with brain size, but that there are also close links between olfactory bulb size, and both the ecology and phylogeny of species. This comparative anatomical study also supported the conclusion that olfaction is an important sense in all avian species. Thus, how much of the brain is devoted to analysis of olfactory information does not seem to be a good guide to the importance of olfaction in the behaviour of birds.

3.2.3 Olfactory Information and Foraging for Specific Items

The best examples of the use of specific olfactory cues in foraging for individual prey items come from various species of kiwi (Cunningham et al. 2009) and New World vultures (Cathartidae) (Graves 1992; Houston 1986) (Figure 3.6). The phylogeny, behaviour, and ecology of these birds are quite different yet it has been possible to

Figure 3.6 Olfactory guided foraging. Turkey Vultures *Cathartes aura* are one of the species of New World vultures (Cathartidae) which have been shown to use olfaction to locate decomposing carcasses. The large and prominent nostrils are clearly evident. Photo courtesy of Maggie Smith, Arroyo Grande, CA.

show that olfaction plays a key role in the location of specific food items which are important in their diets. In kiwi, olfaction is used in the detection of soil invertebrates, and in New World vultures, it is used to detect decomposing carcasses. In both species olfactory cues are important in finding food sources that cannot be readily detected by information from other senses. In kiwi, this is because their key food items, mainly earthworms, are buried in the soil (Cunningham et al. 2009), and in the vultures, the use of olfaction is particularly valuable when carcasses are hidden from sight under vegetation (Houston 1986). Evidence that other birds use olfaction to find individual food items is lacking but has rarely been sought. There is, however, good evidence that birds use olfaction to find good foraging locations in which there is likely to be an abundance of preferred food items.

3.2.4 Detection of Foraging Locations using Olfaction

Olfactory cues may be used by many birds to find profitable foraging locations, and once an area has been located other senses may be used to find individual items. The presence of one particular chemical compound, dimethyl sulphide (DMS), has been shown to be used to indicate profitable foraging locations for pelagic seabirds. This compound does not indicate directly the presence of the birds' preferred prey items, rather it is a surrogate. DMS is produced by marine phytoplankton when it is grazed by zooplankton. Therefore when it occurs at high concentrations, DMS is associated with areas of the ocean with high primary productivity. In turn, these are areas where the larger animals, which exploit the food chain that is based upon this high productivity, concentrate. It is these animals which form the preferred prey of particular seabirds. Pioneering work has shown that the presence at

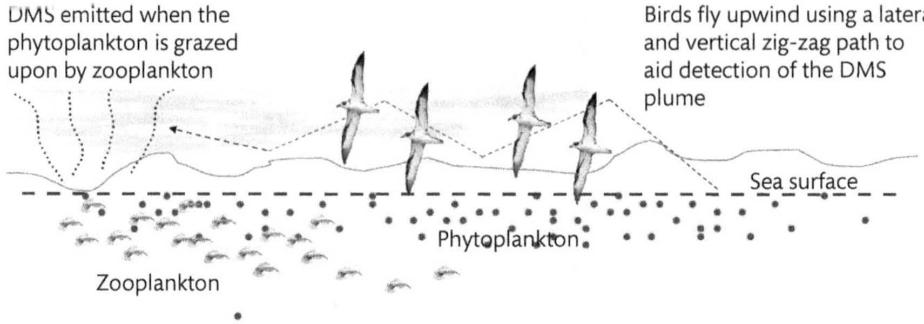

Figure 3.7 Seabirds exploiting an odour landscape to detect prey concentrations in the open sea. Concentrations of dimethyl sulphide (DMS) indicate efficient foraging opportunities. Antarctic Prions *Pachyptila desolata* are an example of a species which fly in zigzag paths across the open ocean to detect an odour plume. Once the plume has been detected, the birds fly upwind still maintaining zigzag trajectories until the source of the odour concentration is located. Illustration courtesy of Gabrielle Nevitt.

high concentrations of DMS are, indeed, detected and exploited by various petrel species (Procellariiformes) and large concentrations of these birds may be drawn upwind to DMS concentrations. Furthermore, these birds exhibit particular flight patterns, zigzagging across the ocean, aimed at detecting plumes of DMS and once a plume has been detected the birds fly upwind towards the source (Nevitt 2000; Nevitt 2008) (Figure 3.7). It has also been suggested that these intense sources of DMS are exploited as a means of olfactory navigation since they may act as relatively fixed beacons in an otherwise uniform seascape (Nevitt and Bonadonna 2005).

A remarkably similar use of odour concentrations to locate a profitable foraging area has also been demonstrated in Great Tits, a small passerine of mainly woodland habitats (Amo et al. 2013). These birds may be attracted to concentrations of odours that are produced by tree foliage following its attack by herbivores, particularly the caterpillars of moths, which are a preferred food of these birds, especially for the provisioning of young. The chemicals that are released by the attacked plants provide a signal that there is a concentration of these birds' preferred foods. Thus specific odours signal a profitable feeding opportunity for Great Tits much in the same way that DMS concentrations signal good foraging areas in some seabird species. This provides an explanation of how Great Tits are able to detect concentrations of caterpillars that infest trees particularly during the spring when tits are breeding and in need of abundant food for provisioning nestlings. It was shown that when Great Tits were offered chemical cues or visual cues (tree branches) alone, or in combination, the birds were attracted to the caterpillar-infested trees and this attraction was controlled exclusively by the volatile chemicals emitted by the damaged trees, not the signs of defoliation (Amo et al. 2013).

3.2.5 Body Odours and Semiochemicals

As well as these general sources of odours which can indicate foraging opportunities, the chemical environment also contains a rich variety of odours emitted by animals themselves. Many of these have been found to be 'semiochemicals'. These are chemical substances, produced by an organism, which are used for communication between individuals of the same species. The existence of semiochemicals has long been established in mammals and insects and includes the well-studied group of compounds known as pheromones which are chemical substances capable of directly eliciting a social response in members of the same species (Wyatt 2003). In particular, they may play a crucial role in initiating and sustaining behaviours involved in reproduction. But not all semiochemicals are pheromones in that they do not necessarily elicit a specific behavioural response. The fact that odours produced by birds could function as semiochemicals has become clear only recently. Furthermore, the extent to which they are actually exploited is less clear, but there is evidence that at least some birds employ semiochemicals and that they serve important functions in reproduction.

Odour-based recognition of species

In birds, odours that are produced by the body mostly originate from secretions of the uropygial gland which is situated low on the back, above the tail. This gland produces waxy fluids (preen oils) that a bird picks up in its bill and spreads on its feathers during preening. The primary function of preen oils is to maintain the flexibility and waterproof properties of contour and flight feathers. Thus any odour compounds (volatile chemicals) produced by the uropygial gland will be spread over much of the feather surfaces during routine preening and potentially released into the air over a sustained period.

There is good chemical evidence that the secretions of uropygial glands vary markedly between species, and there is evidence that odours associated with these differences in chemical composition are detected by birds across a range of avian orders. For example, there is circumstantial evidence that odour information is used for species recognition in certain petrel species (Procellariiformes) which nest underground in mixed species colonies and emerge only at night when visual recognition may be difficult (Bonadonna and Mardon 2010).

Behavioural evidence of species recognition has also been found in some passerines (Dark-eyed Juncos *Junco hyemalis* (Whittaker et al. 2013), Waxwings *Bombycilla* spp. (Zhang et al. 2013), and in a parrot, Psittaciformes (Budgerigars (Zhang et al. 2010)). Furthermore, species and sex recognition have been found in Spotless Starlings *Sturnus unicolor* (Amo et al. 2012a). Odour signals that differentiate even between subspecies may also exist as, for example, between the Atlantic and the Mediterranean Cory's Shearwaters *Calonectris borealis* (Caro et al. 2015). (Note that these subspecies have recently been raised to species status

(*C. borealis, C. edwardsii, C. diomedea*) (Gill and Donsker 2016)). However, it is not yet known whether these shearwaters exploit these sources of information. Similarly body odours may differ strongly between closely related species, for example two species of Estrildidae, Zebra Finches and Diamond Firetails *Stagonopleura guttata* (Caro et al. 2015), which have overlapping distributions in Australia, produce odour signals that chemical analysis suggests could be used to differentiate between these populations. However, to date behavioural evidence that these birds actually exploit this source of information is lacking. This evidence does suggest, however, that there is a rich diversity of odour signals potentially available to birds that could function, and in some species do function, in species recognition.

Odour-based recognition of individuals

Evidence that birds can recognize individuals based upon odour cues alone has come from procellariiform seabirds. The chicks of European Storm Petrels *Hydrobates pelagicus* can differentiate between individuals on odour alone (De Leon et al. 2003), and the same has been shown among adult Antarctic Prions *Pachyptila desolata* (Bonadonna and Nevitt 2004), Wilson's Storm Petrels *Oceanites oceanicus* (Jouventin et al. 2007), and Blue Petrels *Halobaena caerulea* (Mardon and Bonadonna 2009).

An interesting source of evidence that birds could potentially discriminate between individuals based upon odours has come from experiments using mice as odour detectors. These experiments showed that mice can differentiate between the odours of individual Chickens showing that chickens do have individual odour signatures (Karlsson et al. 2010). To date, however, there is no evidence that Chickens themselves are able to detect this information, but the appropriate experiments do not appear to have been conducted.

Odour-based recognition of individual quality and mate choice

Olfactory signals of individual quality have been clearly demonstrated to be detected in seabirds of the Order Charadriiformes, in particular among species of the auk family (Alcidae). Crested Auklets *Aethia cristatella* produce a seasonally released scent, described by humans as 'citrus-like' (Hagelin 2007) (Figure 3.8). Production of this odour is associated with a display behaviour called a 'ruff-sniff' which involves a bird rubbing its face in the scented nape of a displaying partner. This odour is not, however, associated with secretions from the uropygial gland but comes from feathers in the interscapular region high up on the back. This secretion is thought to have a function as a parasite chemical repellent (Douglas 2006; Douglas 2013), and it has been suggested (Caro et al. 2015) that exchange of these repellents through ruff-sniffing could also function in the exchange of information on the health status of the individuals and therefore these odours could be contributing to mate choice.

Figure 3.8 Semiochemicals in birds. Semiochemicals are substances produced by an organism that carries a message for the purpose of communication between individuals in a species. Personal odours may function in the mate choice behaviour of Crested Auklets *Aethia cristatella*, and in male-male competition in House Finches *Carpodacus mexicanus*. Photograph of Crested Auklet, Public Domain, www.fhwa.dot.gov/byways/photos/58710, photograph of House Finch, Geoffrey E. Hill.

In House Finches *Carpodacus mexicanus*, individual or personal odours may be important in male–male competition (Figure 3.8). Evidence for this comes from experiments which show that males avoided the scent of other males which were in better body condition than themselves (Amo et al. 2012b).

Uropygial gland secretions may play an important role in the control of copulation in Chickens (Hirao et al. 2009). Males preferred to mate with females whose uropygial glands were intact as opposed to females who had had the uropygial gland surgically removed. But this difference disappeared in males who had had their olfactory bulbs removed. This strongly suggests that the behaviour of the males was based upon the detection of odour cues. However, copulation is multifaceted behaviour and just what is being detected from the odour is not clear; it could be recognition of the species, the sex, or the reproductive status of the female, or indeed all of these.

3.2.6 Odours and Nests

The role of olfactory cues during nest building has been demonstrated in a number of bird species, especially the role of odour cues in the selection of specific nest materials. Some species of passerines, tits (Paridae) and starlings (Sturnidae), are well known for incorporating specific aromatic plant materials in their nests. These plant items may be added during the life-time of the nest, not just in the initial building. It has been suggested that this plant material functions to repel ectoparasites and ultimately functions to enhance the growth rates of nestlings (Deeming and Reynolds 2015). The fact that starlings use odour cues to select

Figure 3.9 Olfactory nest finders. Leach's Storm Petrels *Oceanodroma leucorhoa* have been shown to use olfactory cues to locate their own nests which are placed deep in burrows or under rocks, in crowded colonies. The birds enter and leave colonies under the cover of darkness. These birds also use olfaction for the location of suitable foraging areas in the open seas (Figure 3.7). The prominent nostrils, which are characteristic of procellariiform birds, are clearly evident. Photographs: close up of head, courtesy of Fanter Lane; bird in flight, courtesy of Peter R. Flood.

suitable material is well established (Gwinner 2013) and in tits the parents appear to use odour cues to determine when to replenish the nest with fresh aromatic herbs (Petit et al. 2002).

Pioneering work on recognition of specific nest sites was conducted in Leach's Storm Petrels *Oceanodroma leucorhoa* (Figure 3.9). Like other petrels, these birds breed in dense colonies which they enter at night. To humans, these nest burrows have a strong musky odour. It has been shown using choice experiments that individual petrels are able to identify their own burrow by smell alone and probably use this to help in the location of their nest burrow under the cover of darkness (Benvenuti et al. 1993; Bonadonna and Bretagnolle 2002).

3.2.7 Conclusion: Smell in Birds

Diverse data, collected using a wide range of techniques, show clearly the importance of olfactory information for birds. The sense of smell should no longer be viewed as something that has specialized uses in a small number of special species, notably procellariform seabirds and kiwi. The use of odour information has now been found in species from a wide range of orders and families, including species from either side of the major evolutionary division between palaeognathous and the neognathous species. Information obtained from odours has been shown to be important in species from among the passerines, auks, chickens, New World vultures, and pigeons and doves. The olfactory apparatus and the olfactory lobes of the brain show great diversity of size but the functional significance of these differences is not clear. However, the internal organization of the olfactory lobes indicates the ancient origin and highly conserved nature of the olfactory apparatus in birds.

Clearly, odours do play a role in foraging for specific items, and in the location of profitable foraging areas in procellariiform seabirds, kiwi, New World vultures, and passerines. But olfaction can also play a role in social behaviour and reproduction across a range of species including Chickens, auks, and finches. Evidence for the importance of odours in birds does, however, come from a small sample of the total diversity of birds and this seems to be primarily because the use of odour information has not been investigated, not because it does not occur. Because of this we may be only at the beginning of a full understanding of the roles that odours play in the lives of birds. Available evidence demonstrates that olfaction should be considered a key telereceptive sense in birds, and that it has a complex sensory ecology providing information that is used to control a wide range of tasks.

4

Touch, Taste, and Magnetoreception

Touch and taste are the 'near receiving' senses. They are exclusively concerned with gaining information about a bird's intimate environment; information about objects and substances that are in contact with or, at the most, within a few millimetres of a bird's body. They also provide information from within the body. Touch and taste provide information crucial to the well-being of a bird, especially with respect to the identification and qualities of foods. They may also detect invasion of the body and the presence of potentially damaging situations, such as high temperatures. In some species, however, touch and taste sensitivity have evolved to mediate the actual detection of food items. In these birds, touch and taste function in the active exploration of the environment rather than just passively registering the near environment.

Magnetoreception is a telereceptive sense but it has quite a different role to that of any other sense. It does not provide information about objects that are either remote from or in contact with a bird. Rather, it enables a bird to determine its location within a co-ordinate system, usually at a large spatial scale. There is evidence, however, that magnetoreception can be used by some birds to aid their orientation within a local area, and this ability might be common to all birds. Magnetoreception may be the key source of information that enables birds to make predictable movements towards target locations at large spatial scales, sometimes between locations that are many thousands of kilometres apart. However, on both local and large spatial scales, and in the homing movements of displaced birds, magnetoreception may function alongside visually based information to mediate movements towards particular locations. The information extracted from vision may include the recognition of landmarks, a daytime compass based upon the position of the sun, and a night-time compass based upon the position of stars (Beason and Wiltschko 2015; Gauthreaux 1980; Guilford and Biro 2014; Guilford and Taylor 2014).

4.1 Touch or Somatic Sensitivity

Mention was made in Chapter 2 that in certain bird species there is a trade-off between information derived from vision and information derived from 'touch'

The Sensory Ecology of Birds. Graham R. Martin: Oxford University Press (2017).
 DOI 10.1093/acprof:oso/9780199694532.001.0001

sensitivity in the bill. Like vision and hearing, 'touch' is a multifaceted sense. What humans refer to generically as 'touch' is the amalgam of a set of information with each component having separate receptors, sensitivities, and thresholds. Like vision, we cannot access each component separately other than through carefully controlled experimental conditions. It is very difficult to make ourselves aware in everyday experiences of the different dimensions of touch, very much in the same way that we cannot separate acuity and colour vision in our everyday visual experiences. Like vision, the components of touch can evolve independently of each other, and this means that the components can vary in their importance in different species. Furthermore, the distributions of the receptors for each component of touch may be distributed across the surface of the body at different densities so that some parts of the body more readily detect touch information.

Touch receptors are not usually concentrated or grouped in special ways within a specific sense organ, rather they are widely distributed as individual receptors, or clusters of receptors, about the body, but their relative numbers vary with location on the body surface. Specialized concentrations of touch receptors do, however, occur in some species. These concentrations are found in their bills and because of their distinctive nature these have become known as 'bill tip organs'. They are seen to provide information used in the control of specific tasks. However, these are not organs in the same sense as applied to eyes or ears in that they do not prepare or present the primary stimulus in a particular way before it is detected by the receptors.

It is possible to understand interspecies' differences in touch sensitivities in an ecological context and understand the part they play in the execution of particular tasks. Unfortunately, such studies are few but they have shown that there are specializations of touch, and that touch can be seen as being finely tuned to specific tasks, most notably foraging tasks. Touch also has important roles in the guidance of more general tasks but these have not been investigated in detail.

Humans usually think of 'touch' in a general way, as a unified sensation. Touch, however, is the product of several quite different modalities and often information in each modality can be received simultaneously from the same object. Because of this, rather than referring to 'touch', the term 'somatic sensitivity' is often used. Somatic sensitivities include pressure, skin stretch, vibration, noxious stimuli, and temperature. Each involves different receptor types which can show marked differences in their densities and distributions within and between species. The stimuli of somatic sensitivity are detected mainly through or at the surface of the skin. All of the main somatic receptor types found in mammals have been reported in birds and it seems possible that all birds have all of these different receptors, although detailed studies are restricted to just a few species.

The general function of somatic sensitivity is to provide information about the environment in contact with or close to the body surface, including the inside of the mouth. Somatic sensitivities also provide information about the physical conditions of the internal environment of an animal (Gottschaldt 1985). The next section emphasises the information the somatic senses can provide about the external environment.

4.1.1 Somatic Sensitivities

Somatic receptors in the skin detect mechanical, thermal, and noxious stimuli coming from outside the body. Other somatic receptors provide information about internal conditions within the limbs, gut, cardiovascular, and respiratory systems. In birds, there is a multitude of somatic receptors, and several classifications have been proposed which are based upon the sensory modality that they respond to. The relationships between structure (morphology) of the receptors' endings and their function are complex, and much of what is assumed to be the case in birds is derived from knowledge of mammals (Gottschaldt 1985).

It is the receptors in the bills of birds which have received most recent and detailed study concerning their form and function (Cunningham et al. 2007; Cunningham et al. 2010; Cunningham 2010; Piersma et al. 1998). These are particularly interesting since they have been shown to be sufficient for prey detection in certain bird species.

Mechanoreception

Mechanoreceptors respond to the movements of an object or fluids with which they are in direct contact. These movements can have different properties and the different types of mechanoreceptors, listed below, are capable of responding to these properties separately. Acceleration (vibration), velocity, and amplitude of movement of stimuli are the key properties that are detected and each is signalled to the brain separately.

Herbst corpuscles

Herbst corpuscles detect the acceleration component in vibrations and are named after Ernst Herbst who first described them in bird legs in 1848. These receptors can respond directly to a vibrating stimulus in a one-to-one manner for vibrations at frequencies between 100 and 1000 Hz, but they do not detect either the amplitude or velocity of the vibrating stimulus (Gottschaldt 1985). The actual structure of these receptors varies markedly depending upon their location on the body and also with the bird species so that there is not one type of receptor, rather they are a class of receptor. Clear criteria have been identified which allow the function of these receptors to be understood across a range of species and bodily locations (Gottschaldt 1985). Most is known about Herbst corpuscles that are located in the bill, particularly in ducks.

Herbst corpuscles are the most widely distributed mechanoreceptors in a bird's body occurring everywhere in the skin, along the large bones of the legs and wings, in tendons, muscles, and joint capsules, and near large blood vessels. In feathered skin, Herbst corpuscles are located primarily at the base of the feather follicles. These have been interpreted as vibration detectors that enable birds to detect the vibrations of their feathers and hence they may provide birds with information on the stresses on feathers, particularly as feathers are moved by air currents during flight.

The distribution and number of Herbst corpuscles in the bills of birds vary greatly. High numbers and concentrations in particular locations on the bill are related to the manner in which the bill is used by certain species as a tactile exploratory device, especially during foraging (Cunningham et al. 2007). For example, in those species of shorebirds (Charadriiformes) which probe with their bills in mud or sandy substrates when foraging, the tip of the bone of the upper jaw (the premaxilla) may bear countless small cavities (pits) which are packed with Herbst corpuscles. These concentrations of somatic receptors are known as 'bill tip organs' (4.2). Similar groups of these pits are found in the distal portions of the bills of ducks and geese, and in kiwi, ibises, and parrots. In grain feeding songbirds, especially the Finches (Fringillidae), Herbst corpuscles and other types of mechanical receptors are located at places in the bill which are mechanically involved in seed-opening. In the long tongues of woodpeckers (Picidae), which are used for extracting invertebrates from narrow holes, Herbst corpuscles are also numerous.

Grandry corpuscles

These mechanical receptors detect the velocity component of a mechanical stimulus. Grandry corpuscles are found particularly in feathered skin. There is a degree of uncertainty about the types of Grandry receptors responsive to velocity in the skin of birds, but it seems that there are at least two types. Grandry corpuscles have also been identified in the skin of the soft part of the bill covering (rhamphotheca) of geese. Since they cannot follow vibrations, they are thought to be primarily velocity sensitive, and tangential movements which cause crumpling of the skin seem to be the main type of stimulus that causes these receptor units to respond (Gottschaldt 1985).

Thermo-sensitive receptors

Although there seems to be good general evidence that the skin of birds is sensitive to temperature, the existence of specific thermo-receptive structures has not been demonstrated unequivocally (Gottschaldt 1985). There appear to be what are termed 'thermo-sensitive units' which respond to both cold and warmth, with input that responds to both the amplitude and rapidity of a temperature change. Although their presence has been demonstrated in very few species, they may be widely distributed among all bird species and they may play a role in the control of behaviours involved in regulating nest temperature during incubation and chick rearing.

Cutaneous nociceptors

These somatosensory units are probably very important for the survival of an animal, but little is known about them in birds (Gottschaldt 1985). Nociceptors respond to invasion of the body (e.g. cutting or pinching) and their responses probably elicit changes in the condition of the body, such as an increase in blood pressure, heart rate, or respiratory frequency. In birds the receptors involved probably have similar properties to those found in the skin of mammals.

4.1.2 Bill Tip Organs

The most interesting aspect of somatic sensitivity in birds is the concentration of groups of receptors in the bills which have become known as 'bill tip organs'. Bill tip organs occur only in certain bird orders and families in which they are readily interpreted as having a specific role in the foraging or bill exploratory behaviour of particular species (Cunningham et al. 2010; Demery et al. 2011). Bill tip organs turn the bill from being just a mechanical structure primarily concerned with seizing and manipulating objects into a 'tactile exploratory organ'. They have been described by Gottschaldt (1985) as an avian equivalent of the mammalian sinus hair system in moles (Eimer's organ, Eimer 1873) whose properties have been most recently studied in star-nosed moles *Condylura cristata* (Catania and Remple 2004) and found to be crucial for guiding the underground behaviour of these animals. Like Eimer's organs, bill tip organs allow the bird to detect, and possibly identify, objects which are buried or at least are out of sight. Bill tip organs seem to allow these probing birds to distinguish between stones, molluscs, and crustaceans, without the need to bring them to the surface.

Bill tip organs do not seem to have any special receptor types but have high concentrations of known somatosensory receptors, mainly Herbst corpuscles positioned at or close to the bill tip. The link between bill tip organs and foraging seems to be clear. If the bill is used to search for, catch, and select food items, as in waterfowl and shorebirds, a well-developed bill tip organ is present. The structure of bill tip organs of geese and ducks are known in most detail (Gottschaldt 1985), although recent research has revealed details of the structure and mode of operation of bill tip organs in shorebirds (Piersma et al. 1998), kiwi (Cunningham et al. 2007), ibises (Cunningham et al. 2010), and parrots (Demery et al. 2011).

Bill tip organs in waterfowl

In geese and ducks, the outer surface at the tip of the upper bill and the outer and inner surface of the lower bill are covered by a horny plate. This is relatively flat and is shaped somewhat like a human fingernail. Hidden underneath the hard outer horn of the surface of the nail are many 'touch papillae' which protrude from the deeper part of the skin around the flat rim of the bill. The tips of the receptors reach the surface. Even the naked eye can see the small pits or surface irregularities which

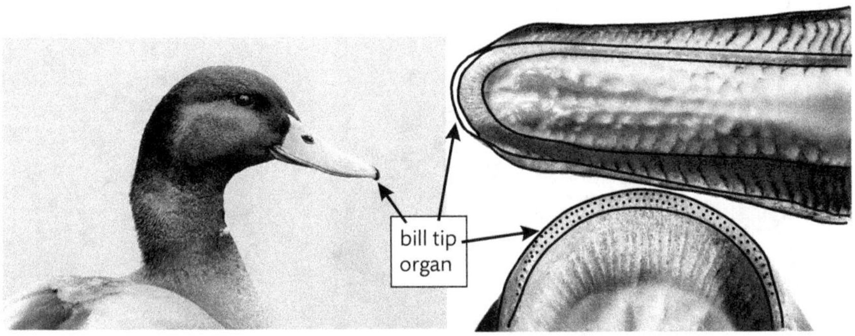

Figure 4.1 Bill tip organs in waterfowl. Touch-sensitive receptors are clustered in pits arranged in rows with the tactile receptors protruding through the surface of the hard keratin inside the upper bill, around the rim of its tip. The number of touch receptors and their actual arrangements vary between species and possibly between individuals. In this diagram, the location of the bill tip organ in a Mallard *Anas platyrhynchos* is highlighted in white on a photograph of the inside of the upper mandible. The enlarged photograph of the organ shows the pits arranged schematically in rows. Photograph of Mallard by Bryan William Jones; the base photographs of the bill are from Gottschaldt (1985).

occur in continuous rows around the rim of the inside of the bill. It is these pits which contain dense clusters of touch receptors (Figure 4.1).

The number, size, and shape of individual touch papillae vary between wildfowl species. These interspecific differences suggest a fine tuning of the structure of the bill tip organ to meet the sensory demands of detecting various types of food items using touch cues alone. Thus, there is highly likely to be a sensory ecology of bill tip organs in wildfowl that is yet to be understood. Neighbouring touch papillae are mechanically isolated one from another and it is argued that this should allow fine spatial tactile discrimination by the bill tip organ (Gottschaldt 1985). Wildfowl with such organs may well be able to discriminate between different objects and/or the surface structures of objects using tactile information alone. Indeed, there is experimental evidence that Mallards *Anas platyrhynchos* can distinguish between real and model peas buried in soft substrates using only cues derived from the bill tip organ as sight of the buried peas was not possible (Zweers and Wouterlood 1973). However, taste may also be involved in this behaviour (Berkhoudt 1985); see 4.2.1.

Bill tip organs in parrots

The bill tip organs of parrots differ from those of wildfowl. Rather than continuous rows of touch receptors arranged around the bill tips, the touch receptors in parrots occur in groups of discrete bundles well separated from each other in a symmetrical pattern along the edges of the highly curved maxilla and mandible (Figure 4.2). In Senegal Parrots *Poicephalus senegalus*, for example, the bill tip organ in

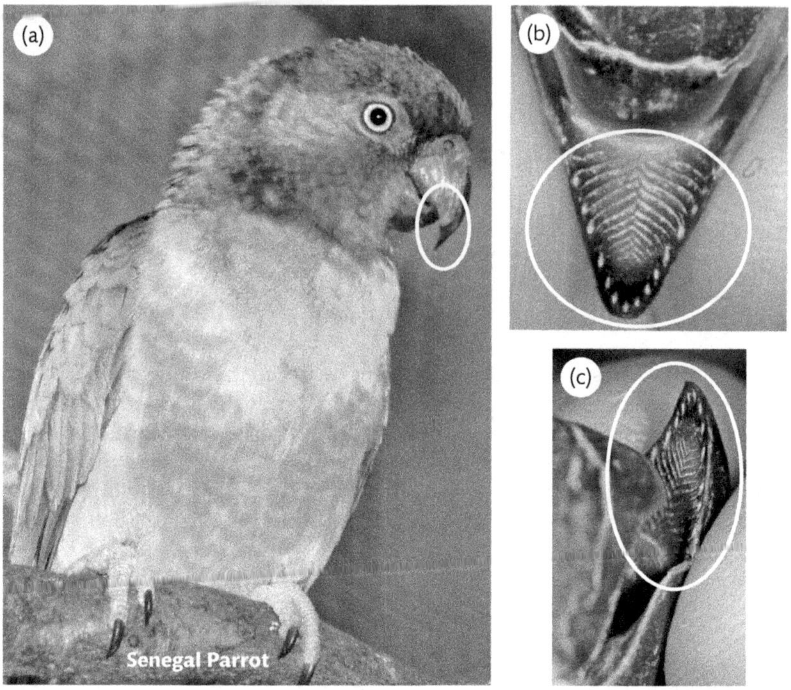

Figure 4.2 Bill tip organs in parrots. The bills of parrots have clusters of touch receptors arranged in discrete groups inside the upper bill in the curved tip part, highlighted in (a). It is this tip that is used prominently in the manipulation and examination of objects and is also used as a third limb when climbing. Eye position and visual fields do not allow parrots to see what they are grasping in their bill tip. Photographs (b and c) show the inside of the bill tip. The clusters of touch receptors can be seen to be symmetrically arranged just inside the edge of the bill. There is no bone in this part of the bill and the touch receptors are embedded in the horny keratin. The pattern of lines inside the bill are structures that help grip objects and are not part of the bill tip sensory organ.

the maxilla consists of seven pairs of pit clusters spaced out along the two edges of the inside of the bill and with a single pit cluster at the bill tip (Demery et al. 2011). The capabilities of such arrangements of touch receptors are not known although it seems that, because of their spaced distribution, they are unlikely to provide the kinds of fine-grained spatial information that is thought to be available through the bill tip organs of wildfowl. In parrots, bill tip organs may function in the manipulation of objects held in the bill tip and also in the positioning of the bill when it is used as a 'third limb' in climbing. The visual fields and bill shape of parrots preclude them from seeing their own bill and so somatosensory control of bill position with respect to twigs and branches when climbing may be important when the bill is used in this specialized way (Demery et al. 2011).

Bill tip organs in shorebirds, kiwi, and ibises

A distinctly different type of bill tip organ occurs in shorebirds (Scolopacidae), kiwi (Apterygidae), and ibises (Threskiornithidae) (Figure 4.3). All of these birds probe with their long, thin bills into soft substrata, which may be covered in water, in search of invertebrate prey. This type of bill tip organ is capable of detecting vibrations, within the substratum, produced by moving prey. It can also detect pressure patterns as the bill is thrust into the substrate and water is displaced away from it. If this displaced water encounters an object it creates a back pressure and this may be detected by the receptors of the bill tip organ. This capability is referred to as 'remote touch' because birds are capable of detecting objects within a cylindrical volume around the bill tip. The bill tip does not have to make direct contact with an object in order to extract information about its presence in mud or fluid sand substrates, although further probing may be necessary to determine the exact position and identity of the object (Piersma et al. 1998).

This type of bill tip organ consists of clusters of pits within the bone around the tips of the maxilla and mandible (Figure 4.3). The pits contain bundles of Herbst and Grandry corpuscles; however, unlike the other bill tip organs of wildfowl and parrots, the receptors are not exposed directly to an external stimulus but lie beneath the soft pliable skin that covers the bill. Thus the movements of mechanical stimuli are sensed through the skin, much in the same way that most somatosensory receptors can pick up information through skin rather than being stimulated directly by objects.

A notable feature of this type of bill tip organ is that it occurs in quite distinct avian lineages; indeed, it is found in species in the two living superorders of birds: the Palaeognathae (Kiwi) and the Neognathae (Scolopacidae), which are thought to have diverged at least 70 million years ago (Benton 2005). Because of the deep taxonomic divide between these bird clades, the occurrence of a similar bill tip organ in them has been interpreted as an example of convergent evolution rather than retention from a common ancestor. That is, while Herbst and Grandry corpuscles have deep evolutionary origins, their concentrations into similar bill tip organs probably have quite separate evolutionary origins (Cunningham et al. 2007).

In ibises, there is indication of a subtle tuning of bill tip organ structure in relation to differences in foraging tasks (Cunningham et al. 2010). The length along the bill that sensory pits extend from the tip and the total number of sensory pits in the bill, two parameters of bill tip organ structure, show a clear relationship with the type of habitat in which the birds probe their bills. The more aquatic the habitat, the greater the number of sensory pits and the further up the bill that they extend. Thus, more extensive bill tip organs are found in Glossy Ibises *Plegadis falcinellus* while less extensive bill tip organs are found in Buff-necked Ibises *Theristicus caudatus*. Members of the former species forage almost exclusively in standing water, while the latter are dry land foragers. A functional interpretation of

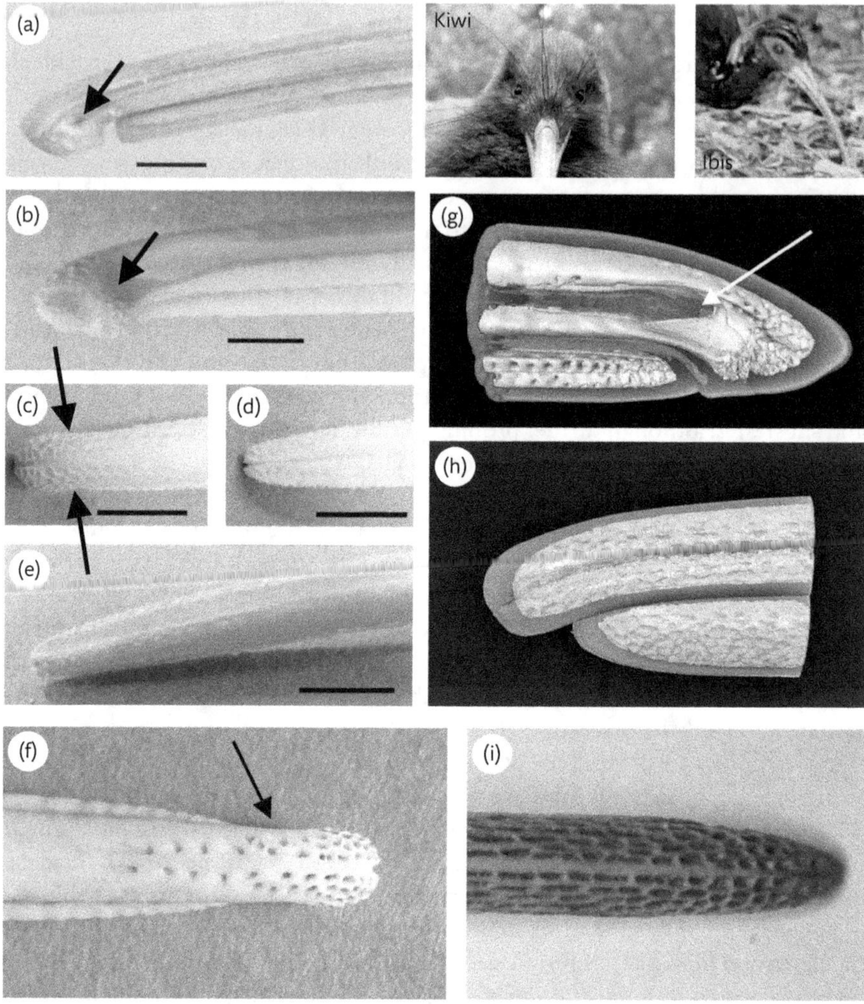

Figure 4.3 Bill tip organs in kiwi and ibises. Photos (a–g) show different views of the bill tips in Southern Brown Kiwi *Apteryx australis*. (a) the intact bill with the keratin sheath in place; (b–f) the structure of the bone beneath revealing the numerous pits where clusters of touch receptors are housed. (b, c, and f) show the upper bill, (d and e) show the lower bill, (g) is a three-dimensional construction from a μCT scan. In all of these photos the arrows point to the opening of the nostrils which, uniquely among birds, open close to the bill tip; they open laterally just behind the tip. (h and i) bill tip in an ibis (Madagascar Crested Ibis *Lophotibis cristata*), showing the same kind of honeycomb of pits in the bone in which clusters of tactile receptors are housed. The distant phylogenetic relationship between kiwi and ibis species suggests that these similar touch-sensitive systems are the result of evolutionary convergence, not common ancestry. Photos a–e, the author (Martin et al. 2007c); (f–i) are courtesy of Susan Cunningham and Jeremy Corfield (Cunningham 2010; Cunningham et al. 2010; Cunningham et al. 2013).

these differences in the number and distribution of receptors in the bill tip organs is that for the dry land foragers prey items occur only near the bill tip, while an ibis foraging in water and softer substrates may encounter prey at a wide range of depths. Therefore, touch sensitivity further along the length of the bill will enable the location of prey at a range of depths.

4.1.3 Conclusion: Somatic Sensitivity in Birds

The information that somatosensory receptors provide is about objects at the body surface including inside the mouth or in the case of woodpeckers from the tongue extended outside the mouth. Somatosensory receptors almost certainly provide information for monitoring internal body conditions, and although this information is vital it has not been investigated in detail in birds. The presence of bill tip organs in certain species shows that the somatosensory system can take on highly specialized functions that have become refined to provide information specifically for the detection of prey. These bill tip organs make the bill into a device for actively exploring the environment, not just a device for manipulating objects. In parrots, it seems that the bill tip organs can also provide a vital source of specialized information that guides the use of the bill as a 'third limb', thus facilitating these birds' ability to climb both upwards and downwards in structurally complex situations.

4.2 Taste

When experienced by humans, taste is generally perceived as a unified sensation that is triggered by the chemical properties of an object that has been taken into the mouth (Berkhoudt 1985; Roura et al. 2013). As with touch and olfaction, taste arises from the responses of populations of different receptors which detect separate properties of a stimulus, and people can learn to discriminate the presence of certain key stimulus properties in an item of food or drink. Similar to olfaction, taste is a sampling or screening process that detects the presence of specific groups of compounds; it is not a full chemical analysis of what is present in an item taken into the mouth. Taste receptors are of relatively few types but, as in other sensory systems, they can be numerous. It is the combined responses of these different receptors which, in humans, give rise to the overall perception of taste. Also, as with olfaction and touch, there are marked differences between species in the absolute and relative numbers of different receptors and in their distributions within the mouths of difference species. This strongly suggests that natural selection has shaped both the detection of different tastes and the ways that the chemical compositions of specific objects are sampled in different species. Thus, there is a sensory ecology of taste.

It is clear that taste is as vitally important to birds as it is to mammals. Taste in birds has evolved primarily to identify nutritious foods through the detection of nutritionally relevant compounds: carbohydrates, amino acids, lipids, salts, calcium, and also toxic compounds. The distributions of taste receptors in the mouths of birds seem to follow ingestion pathways which allow items taken into the mouth to be screened for a range of chemical properties which are particularly relevant in the diet of individual species, as they are transported through the mouth cavity. Much of what is known about taste in birds is from domesticated breeds of chickens and ducks. This is because understanding their taste preferences is valuable for understanding how to maximize the efficiency of food intake in the commercially exploited breeds. Another applied aspect of taste research in birds has been the determination of tastes which birds find aversive with a view to developing bird repellents (Roura et al. 2013).

4.2.1 Taste Buds

In all vertebrates taste signals are generated in taste buds. These are barrel- or flask-shaped groups of cells embedded within the skin of the tongue or other structures within the mouth. Taste buds communicate with the oral cavity through a pore and sample the chemical composition of objects or liquids resting against the skin surface at the pore entrance. The number of types of taste buds in birds is not clear but at least three are found, classified primarily by the shape of the pore and the types of receptor cells which extend through the pore entrance.

The distribution of taste buds in birds was first described in Mallards in 1906 (Bath 1906) and has subsequently been refined to show that taste buds in this species not only occur at the base of the tongue but also at the bill tip, just behind the bill tip organ, and in the floor of the mouth, just behind the bill tip. This positioning of taste buds shows that in Mallards taste discrimination can take place when an object is held at the bill tip. This means that an object does not have to be brought further into the mouth for its chemical composition to be sampled. This finding correlates well with studies in Mallards of the way in which food items are selected and held prior to ingestion (Zweers and Wouterlood 1973). However, in all birds so far investigated, taste receptors occur on the tongue, particularly towards the back, but they do not occur widely over the dorsal tongue surface as is typical in mammals. Clearly, much sampling of chemical composition occurs when objects and fluids are well inside the mouth, not just at the bill tip, but bill tip taste buds, where they do occur, may be particularly important for the initial screening of foods.

4.2.2 Taste Genes and Taste Receptors

Much has been learnt in recent years about the presence of taste receptors in birds from genetic analysis, with the identification of 'taste genes' in a wide range of vertebrates and the detection of the presence of these genes in bird species. These genes seem to be highly conserved across vertebrate classes attesting to the critical role that taste has played in the survival and adaptation of species throughout their evolutionary history (Shi and Zhang 2006). Genomic evidence also indicates that taste systems are related to aspects of the dietary regimes of vertebrates in general (Jiang et al. 2012). This work has also led to the identification of taste receptors, as opposed to the taste buds, in which the receptors are housed (Hoon et al. 1999; Matsunami et al. 2000). This work has also shown that in mammals taste receptors occur not only in the mouth but also at other locations that are involved in nutrient absorption and metabolism, and that they are found in tissues such as the gastrointestinal tract, liver, adipose, and hypothalamus. There is some evidence that taste receptors occur in similar locations in birds. However, because of the high concentration of taste receptors in the mouth, the oral cavity should perhaps be considered a sensory organ in the same way the clusters of somatic receptors can be considered to form a bill tip organ.

4.2.3 Relative Numbers of Taste Receptors

In the past, it was argued that birds have an inferior taste acuity compared with mammals. Mainly because of a lack of mastication in birds, it was thought that chemicals are not readily exposed in birds' mouths and hence taste buds occurred in relatively low numbers (Berkhoudt 1985). However, when the number of taste buds is considered in relation to the size of food that is taken into the mouth in one bite, the number of taste receptors in birds (at least in Chickens) is not lower compared with mammals. This suggests that the capacity to taste food in birds and mammals is, in fact, likely to be similar (Roura et al. 2013). Genetic evidence which suggests the general assumption that birds share the same primary tastes as mammals (Berkhoudt 1985) should be reviewed. While birds may detect some of the 'classical' taste categories birds may well have other chemical sensitivities (Roura et al. 2013).

4.2.4 Taste Categories in Birds

Sweet

The response of birds to 'sweet' compounds (i.e. chemicals labelled as sweet by humans) varies between bird species. When faced with a choice, omnivorous and granivorous birds (e.g. domestic fowl) do not respond positively to sugars, but nectivorous birds (e.g. hummingbirds Trochilidae (Baldwin et al. 2014)), and frugivorous birds (e.g. Cockatiels *Nymphicus hollandicus*) do respond positively to sugars (Roura et al. 2013). These examples provide evidence of an ecology for this particular taste, an idea which has been discussed a number of times (Kare and Mason 1986; Klasing 1998; Matson et al. 2001). Some bird species (e.g. Common Starlings) show a preference for fructose and glucose and this may be related to their preference for fruits that are ripe (Schuler 1983).

Umami

In humans, the umami taste is generally identified as the taste of monosodium glutamate or glutamic acid. It is interpreted as a taste whose function is the detection of protein or amino acids. In choice experiments, Chicken chicks showed a preference for a diet containing amino acids (Roura et al. 2013). The actual substances detected by umami or amino acid receptors require a good deal of further investigation, but they include 1-alanine which is detected by both Common Starlings and Red-Winged Blackbirds *Agelaius phoeniceus* (Espaillat and Mason 1990; Werner et al. 2008), while in Chickens umami receptors are thought to be involved in the co-ordination of post-ingestive and metabolic events since they occur in hypothalamus, liver, and abdominal fat tissues. It is suggested that umami receptors in Chickens may be capable of detecting a wide range of amino acids and that they are driving protein-specific appetites (Roura et al. 2013).

Bitter

Bitter tastes are usually associated with naturally occurring toxic compounds from plant or animal origin, especially insects. Bitter-tasting substances, such as tannins and phenylpropanoids, seem to be aversive to birds (e.g. Common Starlings and Red-winged Blackbirds), and there have been attempts to exploit this in the development of bird repellents. Chickens and Cockatiels find quinine (a substance known particularly for its bitter taste in humans) aversive. Quinine has also been shown to be sufficiently unpleasant to Chickens that they learn to avoid it after just a few exposures (Skelhorn and Rowe 2005).

Calcium

Calcium requirements in the diets of birds are particularly high during egg laying. Dietary preferences of many bird species for calcium-rich items such as shells, bones, and calcareous grit have been well documented (Reynolds and Perrins 2010). Young Chickens are among birds which have been shown to have a well-defined appetite for calcium at the time of skeletal development. A similar appetite also occurs in adult female Chickens, presumably because of the high demands for calcium for the formation of eggs over extended periods (Wood-Gush and Kare 1966). However, just how important taste is in the intake of calcium is unclear; it could be that calcium intake is regulated by post-ingestive effects but recent work suggests that the sense of taste is important with 'calcium-like' sensing perhaps representing a distinct taste (Tordoff et al. 2008), but conclusive evidence is required.

Salt

The taste of salt (sodium chloride, NaCl) in birds seems to be well defined and can result in two quite different behavioural responses depending upon the concentration of NaCl presented. In Chickens, high concentrations elicit aversion but low concentrations usually produce an attraction response, particularly after birds have been fed a diet deplete in sodium (Kare and Mason 1986). This suggests a fine discrimination of salt concentration and that this is continuously calibrated against the internal state of the bird. High concentrations of salt (>2% solution) are toxic to birds, except those that have salt glands (Bradley 2009), but birds must at times ingest NaCl at low concentration to ensure good health (Mason and Clark 2000).

Sour

Sourness is related to the acidity of food. The presence of acids is often caused by bacterial fermentation, and it is a source of tastes that typically evokes a rejection response in mammals. However, like salt, a graded response to acid/sour is shown depending upon the degree of acidity (pH) of the stimulus. Thus Chickens are tolerant of medium acidic or alkaline solutions (either side of the pH value of 7), but they avoid extreme pH values (Fuerst and Kare 1962).

The actual taste sensitive cells for both salt and sour have yet to be identified in birds although they have been identified in mammals, and the orthologous genes (i.e. inherited from a shared ancestor) for these receptors have been shown to be present in the Chicken genome, suggesting that these two tastes are similarly detected in birds and mammals (Roura et al. 2013).

Fat

Recent evidence suggests that in mammals a fat-sensing taste receptor exists and that this plays an important role in regulating the caloric content of the diet (Cartoni et al. 2010). However, little is known of the receptor and its function and to date no avian homologue gene for this receptor has been found in the Chicken genome database.

4.2.5 Taste and Foraging in Shorebirds

Evidence that shorebirds of the genus *Calidris* can use taste cues to determine the presence of prey comes from a series of intriguing experiments involving four species: Sanderling *C. alba*, Dunlin *C. alpina*, Purple Sandpiper *C. maritima*, and Red Knot *C. canutus* (Gerritsen et al. 1983; Gerritsen and Sevenster 1985; van Heezik et al. 1983). In these experiments, birds were first trained to forage for natural foods placed in jars of sand. Birds were then presented with pairs of jars: one contained 'taste', the other 'no-taste'. Taste came from polychaete worms or bivalve molluscs which were introduced to washed sand but removed before the experiments; no-taste was simply the washed sand. It was found that all birds spent longer and probed more frequently at the jar containing 'taste' compared with jars containing 'no-taste'. It was concluded that in all species, taste sensitivity is a cue that guides foraging but that its importance probably varied between species, with it being of most importance in Dunlin and least in Red Knot.

Thus it seems that these shorebird species can determine which areas of apparently uniform substrata will be more profitable for foraging using cues based solely on the presence of certain food-related chemical substances, and these taste receptors may play a key role in their regular foraging behaviour. This use of chemical information for finding locations in which foraging may be more profitable parallels the examples discussed in Chapter 3 (3.2.4) in which seabirds and passerines are able to determine profitable foraging locations, but not individual food items, by using olfactory cues.

4.2.6 Conclusion: Taste in Birds

The sensory system underlying taste in birds is clearly complex. The taste system can be viewed as a group of nutrient sensors which have evolved to evaluate the nutritional quality and content of foods primarily at the point of ingestion but also at various points post-ingestion. To achieve this, there exists an array of different receptor types, and their frequencies and distributions, particularly within the mouth, but also in other parts of the body, are complex. These arrays of taste receptors in the mouths of birds constitute a sense organ which may be considered as important and complex as any of the other sense organs found in the heads of birds. Furthermore, these taste systems seem to show distinct relationships with the ecology of the species with respect to receptor types and their positioning within the mouth, and possibly in other tissues. Thus the exact array of taste receptors in any one species may be closely linked to the diet and the means of ingestion. However, more comparative work is required to fully develop an understanding of the sensory ecology of taste in birds.

The fact that birds are continually having to regulate their food intake, minute by minute or hour by hour, suggests that natural selection is highly likely to be exacting and continuous with respect to taste, resulting in the fine tuning of taste reception. Taste has a role that is as exacting in controlling the foraging of birds as other aspects of food acquisition.

4.3 Magnetoreception

The first definitive experiments indicating that magnetic fields can influence the behaviour of animals were not published until the 1960s. Since then, a large body of work has demonstrated the central importance of information about the Earth's magnetic field (the geomagnetic field) in many animal taxa, especially birds (Wiltschko and Wiltschko 2006). The ability to detect the geomagnetic field probably evolved because it is omnipresent and can provide a source of information to guide navigation and orientation.

In birds, detecting the geomagnetic field and extracting information from it can be thought of as analogous to a Global Positioning System. It seems that the geomagnetic field could be used by a bird to provide information about its current position and the direction to travel from it towards a particular goal. There is a strong body of evidence that birds do use the geomagnetic field to determine at least their initial orientation when departing upon migratory flights and also by homing pigeons when returning to their lofts. Although important, magnetoreception is, however, just one cue that may be used by homing birds. Homing pigeons seem to employ a number of different cues based upon vision, olfaction, and magnetoreception and which cues are used in particular instances depends upon weather, location, and the specific experience of individuals (Beason and Wiltschko 2015; Deutschlander and Beason 2014; Guilford and Biro 2014; Guilford and Taylor 2014). Evidence that homing birds can determine their location using the geomagnetic field alone is less convincing (Wiltschko and Wiltschko 1995; Wiltschko and Wiltschko 2006).

Direction can be derived from the horizontal component of the Earth's magnetic field (the geomagnetic vector), and this is the information that humans have learnt to detect through the invention of the technical compass. Location on the globe can theoretically be derived from the total intensity and/or inclination of the geomagnetic field vector. This aspect of the Earth's magnetic field exhibits gradients between the Earth's magnetic poles and the magnetic equator. Although humans have technology to determine this, it has not been commonly employed in human navigation devices (Skiles 1985).

4.3.1 Animals that Detect the Geomagnetic Field

Animals have been shown to be able to detect one or both types of geomagnetic information (direction and location) and the uses to which they are able to put this information are becoming well understood (Wiltschko and Wiltschko 1995). Animals which can use a magnetic compass include members of the phyla Mollusca and Arthropoda, plus all major vertebrate groups including cartilaginous and bony fish, amphibians, reptiles, mammals, and birds. Magnetic cues are known to be used for orientation at different spatial scales, for example, within a home range by birds, for the orientation of buildings by bees and termites, and in the extended migrations of eels, salmons, marine turtles, and birds.

Systematic analysis of magnetoreception is still a work in progress and much is yet to be learnt about the mechanisms that allow magnetic field information to be detected and used. Among birds, magnetoreception is likely to be a widespread ability since it has been demonstrated in a broad range of avian orders including homing pigeons (Columbiformes), domestic Chickens (Galliformes), and a number of species of Passeriformes. However, the actual number of species in which it has been demonstrated is small (Freire et al. 2008; Munro et al. 2014).

4.3.2 Magnetic Compass Mechanisms

Two types of magnetic compass mechanisms have been described and there are fundamental differences in their functional characteristics (Wiltschko and Wiltschko 2006). The first type is a 'polarity compass'; this works in a similar way to a technical compass, using polarity of the magnetic field to distinguish between magnetic 'north' and 'south'. The second type is an 'inclination compass'; this is the type of compass used by birds. It relies upon the axial course of the geomagnetic field lines, with directional information obtained by interpreting the inclination of the field lines with respect to up and down (derived from gravity). This mechanism distinguishes between 'polewards', where the field lines run downwards, and 'equatorwards', where they run upwards.

4.3.3 Detection of the Geomagnetic Field

Although the presence of a compass mechanism is well established in birds, the way in which birds detect the Earth's magnetic field is still unclear. What is, or are, the magnetoreceptors and how information is transmitted from them to the nervous system is still debated. However, a consensus now seems to be established that there are two detector mechanisms which operate in birds, with both mechanisms probably occurring in the same individual. These mechanisms are referred to as the 'magnetite model' and the 'radical pair model'.

The magnetite model proposes a primary process involving tiny crystals of permanently magnetic material located in the head, while the radical pair model proposes a so-called 'chemical compass' based on 'singlet-triplet transitions' in special photopigments in the retina and is thus detected in the eye.

The magnetite model of magnetic field detection

Magnetite crystals were first found to function in magnetic field orientation in bacteria (Blakemore 1975), and similar crystals were subsequently found in birds (Wiltschko and Wiltschko 2013). For example, there are descriptions of deposits of iron oxide (probably magnetite) that lie in sheaths of tissues around the olfactory nerve and olfactory bulb and between the eyes, and also in bristles which project into the nasal cavity, in Bobolinks *Dolichonyx oryzivorus*, a migratory passerine species of North America (Beason and Nichols 1984) (Figure 4.4). Very small magnetite crystals have also been found in the skin of the upper bill in Rock Doves (Fleissner et al. 2003). There is evidence that the nerve fibres which run

Figure 4.4 Magnetoreception in birds. European Robins *Erithacus rubecula* are a species which have been shown to employ the light-dependent mechanism of magnetoreception that is based upon specialized photopigments (cryptochromes) in the retina of their right eye. Bobolinks *Dolichonyx oryzivorus* have been shown to have a mechanism based upon deposits of iron oxide (probably magnetite) in sheaths of tissues around the olfactory nerve and bulb, and between the eyes. Each species may have both mechanisms with photopigments used for the detection of directional information, and magnetite-based mechanism used to detect magnetic field intensity. Photographs: European Robin, Graham Martin; Bobolink, Public Domain, Flickr, JanetandPhil.

from these regions are responsive to changes in earth-strength magnetic fields in Bobolinks (Semm and Beason 1990). This anatomical and physiological evidence has been supported by various behavioural experiments in which birds are placed in high-intensity magnetic fields which overwhelm the low-strength geomagnetic field. This treatment disrupts the orientation behaviour of birds which, however, is reinstated when the high-strength magnetic field is removed (Wiltschko and Wiltschko 2006; Wiltschko and Wiltschko 2013).

The radical pair model

Evidence in support of the radical pair model in the eye was first found through demonstrations that magnetoreception in migratory birds is light dependent. Species in which this has been demonstrated are passerines which have migratory populations; these species are from three families, European Robins *Erithacus rubecula* (Turdidae) (Figure 4.4), Australian Silver Eyes *Zosterops lateralis* (Zosteropidae), and Garden Warblers *Sylvia borin* (Sylviidae). The radical pair model proposes that magnetic fields are detected by specialized photopigments in the retina with the most likely candidates being cryptochromes (Heyers et al. 2007; Müller and Ahmad 2011). They are flavoproteins that are sensitive to blue light and are found widely in plants and animals where their primary function is in the control of circadian rhythms. Although they occur in the retina, they are not located within the rod or cone photoreceptors. However, they do seem to be associated with the visual pathway, and there is increasing evidence for their role in magnetoreception and the control of circadian rhythms in birds (Heyers et al. 2007).

Magnetoreception in birds is not only light dependent, but depends also on the wavelength of light and on the eye that is used. Evidence suggests that radical-pair-based magnetoreception is found only in the right eye of birds (Rogers et al. 2008; Stapput et al. 2010; Wiltschko et al. 2002; Wiltschko et al. 2003). Furthermore, experiments with Rock Doves and Chickens have shown the presence of a magnetic compass that works only under blue, turquoise, and green light (up to a wavelength of 565 nm), but normal migratory orientations are disrupted under yellow light (light with wavelengths longer than 582 nm) (Freire et al. 2008; Wiltschko et al. 2010). The link to blue light is another indication of the involvement of cryptochromes.

4.4 Conclusion: Magnetoreception in Birds

The above picture of magnetoreception is complex and has generated some controversy, and a full consensus is yet to emerge on the functions of magnetoreception and its mechanisms in birds. This is perhaps partly because humans have difficulty in comprehending what the perception of magnetic fields could be like. There is extra confusion, perhaps because it seems that magnetoreception in birds is in some way related to the eye and the visual system, and there has been the temptation to imagine that birds must 'see' the geomagnetic field. However, there is no reason that this should be the case any more than any other sense should be perceived like any other. Each sense provides its own unique suite of information and sensations, and these are integrated to provide the overall perception of a scene or situation.

There is now a strong body of evidence that birds, along with many other animals, are able to extract information from the Earth's magnetic field; humans are one group of animals which seem to be unable to extract this information from their environment. It is well established that birds are able to employ information derived from the Earth's magnetic field to determine migratory orientation at the time of departure, and they probably use this within a suite of other vision-based information (landmarks, star- and sun-based compasses) and olfactory information (Gagliardo 2013) to guide themselves between locations that may be just a few, or many thousands, of kilometres apart. However, this information does not seem to be very precise in that there is typically much variation between individual birds in their chosen orientation at the time of departure on flights (Berthold et al. 2003). Nevertheless, for individual birds, the direction may be precise and the apparent lack of precision may reflect variation of preferred direction within a population.

It also seems that magnetoreception can play a part in the behaviour of bird movements at much smaller scales, perhaps finding their way back to sites within a home range or territory. As such, magnetoreception is a sense that may be employed in birds continually, not just referred to at key times during a year and not only by migratory populations. After all, it seems that other senses potentially provide information continually; they are not switched on and off for the execution of particular tasks. While many fundamentals remain to be determined, it does seem clear that more than one type of magnetic information is used by birds, and that two types of detector mechanisms may be involved. Perhaps cryptochrome photopigments in the right eye function for recording magnetic directions, and magnetite-based receptors in the upper bill for recording differences in magnetic intensity. It would seem that birds have a compass in their right eye and a magnetometer in their bill.

5 From Senses to Sensory Ecology

5.1 Making Sense of the Diversity of Bird Senses

The previous chapters have provided a broad overview of what sense do and what limits them. I have also discussed the flexibility in the structure of different sensory systems and, importantly, how this flexibility has provided the potential for widely differing sensory capacities between different species. It is now time to discuss how senses have been shaped in different bird species and how they facilitate life in different habitats and underpin the execution of different tasks.

There is clear evidence that in any individual bird, its senses provide a great diversity of information about the environment that surrounds them. Some senses provide information about objects that are quite remote, others provide information about what is nearby, and some provide information about what is directly in contact with the bird's body and what is inside the body. The main telereceptive senses of vision, hearing, and olfaction can provide information about environmental features that may be many kilometres away, although they will also provide information about objects that are very close. Magnetoreception may help birds to determine their direction of travel and their position relative to a goal at both large and small scales. Thus, at any instant a bird has available to it a unique suite of information about its environment, both nearby and remote.

While we do have some understanding of what it means to receive information from vision, hearing, touch, taste, and smell, it is clear that birds must live in very different sensory worlds to our own. Our human-eye view of the world is clearly very different to a bird-eye view, and the same can be said about bird-ear soundscapes, bird-olfactory landscapes, etc. The surveys of the previous chapters also make it clear that bird-eye views not only differ from human-eye views but they differ markedly from one bird species to another. Birds of different species sitting beside each other may have available to them quite different information to guide their next actions.

While we can 'know about' the vision of a bird by using information derived through behavioural studies, or from anatomy and physiology, we cannot 'know' what a bird sees. All of us are prisoners of our unique way of extracting information from the world around us and it is very difficult to imagine just what information

The Sensory Ecology of Birds. Graham R. Martin: Oxford University Press (2017).
 DOI 10.1093/acprof:oso/9780199694532.001.0001

is available to a particular bird that might be the focus of our interest. It is clearly erroneous to think that as we watch a bird and try to interpret its behaviour that it is living in the same world as us. We might at a particular instant be sharing the same environment, but the world of each species is unique. It may, in fact, be quite misleading to assume that the world through the eyes of a bird is anything like the world through our eyes; it may be, but we cannot be certain. It is always worth reminding ourselves that birds probably have constantly available to them quite detailed information about the Earth's magnetic field, but we have no direct access to that information. We cannot imagine how that information is perceived or experienced. In fact, it is probably best to rely upon experiments which show what birds might be able to use magnetic information for rather than try to imagine what it might be like to experience geomagnetic field information.

We experience the world through multifaceted information received in a constant stream. Whether awake and active, or asleep and at rest, the sensory systems are providing the brain with streams of information, and it is the brain's function to order and prioritize that information for the control of actions. Each sense in each species is, however, highly selective, detecting only certain information about the environment in which each animal sits. No eye sees everything; no ear hears everything; no olfactory system, or suite of taste receptors, detects all of the chemical compounds in the environment. This selectivity arises because of the flexibility both in the capacities of the individual receptors and in the ways in which receptors are brought together. Receptors may be brought together in an ordered fashion within a sense organ which serves to select and arrange stimuli before they are detected and analysed by the receptors. The receptors of other senses, such as those underpinning touch or taste, may be distributed widely and are not placed within a sense organ.

This selectivity of sensory systems is exemplified very clearly in eyes, but it occurs in all senses. In eyes, the image of the world produced by the optical system is a simulacrum of the world. The visual system does not have direct access to the real world in the way that a taste bud or an olfactory receptor can access directly the molecules of a compound in the air or in the mouth. The retinal image is an interpretation of the real world and when comparisons are made of the images in the eyes of different species, it is found that they may vary markedly with respect to brightness, contrast, precision of focus, and field of view. Furthermore, these images can be analysed by the retina in an almost infinite number of different ways depending upon the populations of receptors employed to extract information from different parts of the image. In a similar manner, the sense of touch is mediated by somatic receptors of many different types, each type detecting different aspects of a mechanical stimulus, and these receptors can be distributed and clustered at different densities in many different locations on the body surface. In olfaction or taste, many different sorts of receptors, each capable of detecting specific compounds, can be grouped together at different concentrations and so provide priority for the detection of particular compounds present in the air or inside the mouth.

Even though the senses of only a relative handful of bird species have been analysed in any detail, it seems clear that the ways of extracting information from the world are close to infinite. This raises profound questions about the nature of reality and truth, something first identified by Sextus Empiricus 2000 years ago and which led to the development of Scepticism and the empirical method, a position and a method of enquiry which underpin scientific enquiry.

Today we know so much more than Sextus Empiricus could have imagined about the diverse ways in which animals gain information about the world. We now have clear evidence that our world view is no more unique or privileged than that of any other species. Each represents a way of gaining information about the world for the guidance of behaviours in particular circumstances. Unlike Sextus, we have the luxury of clear ideas about how the diversity of information gained through different senses came about. We can now view this diversity through the lens of evolution driven by natural selection. This perspective should enable us to bring order to understanding the factors which have driven this diversity of senses and the information that they provide.

We are all familiar with interpreting the diversity of bill shapes and sizes as being the products of natural selection, selection driven by the mechanical requirements for manipulating objects and materials of particular shapes, sizes, and mechanical properties. In light of what is now known about how the senses of birds differ between species, it seems highly likely that for every bill shape and size there is also a diverse suite of sensory information that is used, for among other things, the control of those different bills.

The remainder of this book aims to explore these questions by considering some key examples of how birds use sensory information in particular environments and in the conduct of particular tasks.

6

Birds in the Dark: Complementary and Partial Information

To be 'in the dark' means that we are short of information, so much so, that we cannot make a wise decision. It is a metaphor born of the observation that after the sun has gone down, human vision is not adequate to control everyday behaviour. Modern humans go to great lengths not to be in the dark, to always have light that is bright enough to provide high spatial resolution usually enhanced by colour vision. There are guidelines supported by legislation describing which kinds of work can be conducted safely at different light levels (in the UK, for example, see http://www.hse.gov.uk/humanfactors/topics/lighting.htm#lighting). In general, the more detailed the task, the greater the light requirement.

Many people, at least the majority who live modern city-based lives, are so unsure about receiving sufficient visual information about their environment that they rarely expose themselves to light levels that can be called 'darkness'. At least they rarely expose themselves to natural darkness outside the predictable environments of their own homes. Even at home most people prefer to have bright lights until they close their eyes for sleep. Venturing outdoors at night without an artificial light is regarded as definitely problematic, possibly eccentric, and by many, definitely dangerous. This is mainly because under such circumstances little visual information can be retrieved about objects at a distance. This is seen as a problem principally because we wish to continue at night all those activities that we do during the day; we perhaps regard what we do in the day as our real and purposeful lives. Whatever activities that we can do without artificial light after the sun has gone down are regarded as somehow inferior. In effect, we want to extend our daytime into the night-time. We resent having to modify our behaviour between day and night. We have only to look at a cityscape at night, or the night-time images of Earth from a satellite, to appreciate that night is an inconvenience for modern economic man, but an inconvenience that can be overcome by pouring light on the problem.

The upshot of all this is that life at night is seen as particularly challenging, not just for humans but for life in general. Life at night is portrayed as strange or bizarre since we project the challenges that humans seem to face at night onto other animals, and we do this especially with birds. The desire of humans to carry

The Sensory Ecology of Birds. Graham R. Martin: Oxford University Press (2017).
 DOI 10.1093/acprof:oso/9780199694532.001.0001

on life as usual at night is projected on to nocturnal animals and we tend to focus upon understanding how nocturnal animals conduct behaviours that we deem as essential in daytime active animals.

If the flight behaviour of birds depends upon visual information, it would seem little wonder that the large majority of birds go to roost as soon as light levels start to fall when the sun declines below the horizon. So few birds seem to be active at night that those able to live a nocturnal existence, the birds which fulfil the majority of their life cycle activities at night, have long been regarded as having special powers and they have also become symbols of wisdom. The Greek Goddess Athene, and her successor the Roman Goddess Minerva, were both symbolized by, or depicted with an owl.

Even today, owls are treated popularly with an element of awe or reverence. It seems likely that, because of their nocturnal activity, these birds were imputed to have wisdom that could be a model for humans. Indeed that idea extends through to the present day, captured in the nursery rhyme, 'A wise old owl sat in an oak/The more he heard the less he spoke/The less he spoke the more he heard/Why can't we all be like that wise old bird?' As a lesson from nature this rhyme is in fact quite instructive from the perspective of sensory ecology. While the key message of the rhyme is that wisdom is gained mainly through quiet and concentrated thought, rather than activity, it also implicitly refers to a trade-off in information between two different senses, vision and hearing. Seeing less at night leads to a greater reliance on information from hearing.

It was shown in Chapter 2 that as light levels decrease, spatial resolution and contrast sensitivity inevitably fall (Figures 2.16 and 2.17), and that colour vision, which is a means of enhancing spatial resolution, functions only at high light levels. As explained, these changes in vision are because of fundamental constraints which apply to the vision of all vertebrates, and they arise because of the physical nature of light and because of some fundamental limitations on the physiology of the eye's image-analysing system, the retina.

These decreases in spatial resolution with light levels mean that there is a gradual reduction in the information that can be gained through vision about objects and surfaces that surround an animal. As light levels decrease only larger objects, and objects which contrast highly with their backgrounds, can be detected with certainty. An important consequence of this is that the decrease in resolution, captured in Figure 2.16, translates for an individual animal into decrease in the distance at which objects of particular size or contrast can be detected. The relationship between acuity and distance is straightforward. For example, if some objects are just detectable at a particular distance at a high light level, then when acuity decreases ten-fold those same objects will have to be ten times closer if they are to be detected. This means that the great utility of vision over other senses, its telereceptive abilities, reduces progressively as light levels fall.

Certainly an eye can still see to infinity at night since an eye can detect the stars of the night sky, but the all-important spatial volume about an animal from which detailed information can be usefully extracted to control behaviour decreases dramatically at night. In effect, the radius of the visual world around an animal that can be used to guide behaviours shrinks as light levels fall. This effect is manifest by expressions which describe the night as 'closing' or 'pressing' in. If there is wisdom to be had from owls, it lies in recognition of this reduction in the telereceptive ability of vision at lowered light levels and, importantly, the consequences of this for the control of specific behaviours. It may, perhaps, be concluded that owls and other nocturnal birds have the 'wisdom' to rely upon vision for the control of a more restricted suite of behaviours compared with daytime (diurnally active) birds and to exploit information available through non-visual senses.

Despite these restrictions, many animal species, including a number of bird species, are strictly nocturnal, that is they complete all aspects of their life cycles between dusk and dawn. Furthermore, many bird species are able to complete specific, but nevertheless essential, activities at night. This chapter explores nocturnally active bird species and those which may intermittently be active at night. The focus is, of course, upon the sensory information used to guide their behaviours. Because of the ubiquity of artificial lighting, very few people actually experience what information their senses, especially vision, can provide at night. Yet to be active in natural night-time, guided by the information that our senses can provide, is part of human experience. It is experience that is no less rich than obtains on the brightest and sunniest day.

6.1 The Problem of Night-time

What is the problem of night-time? It has two aspects: the absolute levels of light and the variability of light levels. Both pose problems for the extraction of information from the environment.

Natural light levels experienced at the Earth's surface vary over a very wide range. Naturally occurring light levels from all sources, sun, moon, stars, and airglow, have been reviewed (Martin 1990a) based upon comprehensive measurements of naturally occurring light levels across the globe (Anon 1952) and predictions of astronomical events (Anon 1988). These data show clearly that in any one place daily maximum and minimum light levels are not fixed but vary with latitude and time of year. Furthermore, the absolute rate at which light levels change each day is also a function of time of year and latitude (Figure 6.1). At the poles light levels hardly change within a day; at the equator light levels go through a huge cycle every day. At latitudes between these extremes, light levels change from day to day by different degrees and on cycles of different lengths as the year progresses. This degree of variability means it is difficult to make casual predictions of just what light levels an animal may experience on the Earth's surface from time to time.

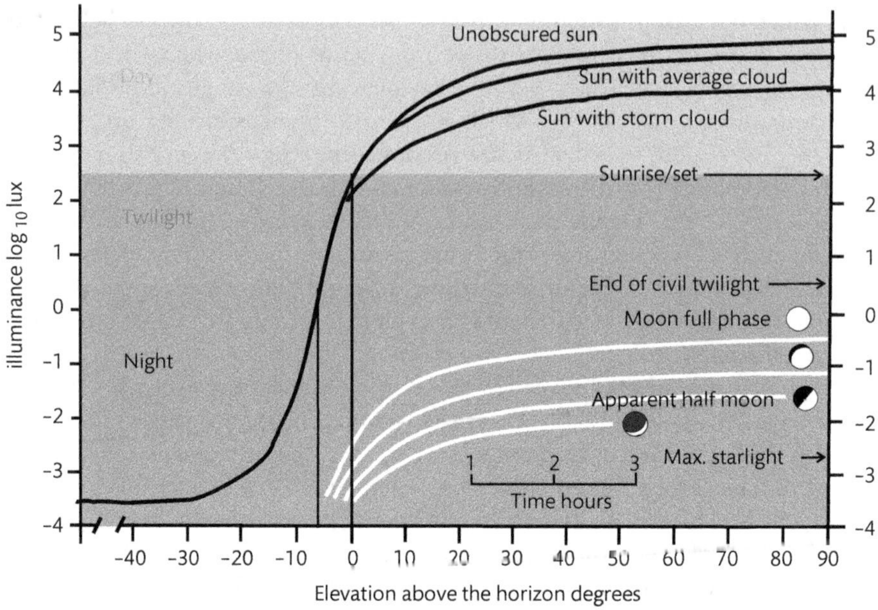

Figure 6.1 Natural illumination sources and light levels. Natural illumination of the Earth's surface comes only from sunlight and starlight. This includes sunlight reflected from the moon. The amount of light from these sources changes constantly over very large ranges, which are determined by the elevations of the sun and moon above the horizon and the amount of the moon's surface from which light is reflected (the moon's phase). All of these change with latitude and time of year, and with the presence of cloud cover. This diagram captures the total range of light levels due to the sun as a function of altitude above the horizon and the range of illumination produced by the moon in different phases and at different altitudes above the horizon. The basic unobscured sun curve is for latitude 50° at the time of the summer solstice. At lower latitudes, the rate of change of illumination levels is more rapid and at higher latitudes it is slower, which means that the periods designated as twilight (the period of transition between daytime and night-time) are shorter or longer than shown here. Note that the scale of illumination is logarithmic. During daytime (sunrise to sunset), light levels may vary over a range of approximately 1000-fold, with average cloud cover usually making just a 10-fold reduction in light levels. Between sunset and sunrise, light levels can vary over a much larger range, up to 1 million-fold. However, the presence of the moon can reduce this to 1000-fold depending upon phase and moon elevation. The diagram is based upon data published in Natural Illumination Charts (Anon 1952).

One particularly important aspect of this variability of naturally occurring light levels is that variation is always larger during the night compared with the day. In essence, daytime is a relatively stable light environment; night-time is not.

The total difference between the light available under clear skies on the brightest days and on the darkest nights at mid latitudes is enormous, just over 100 million-fold ($\times 10^8$) (Figure 6.1). Furthermore, the length of the night also changes

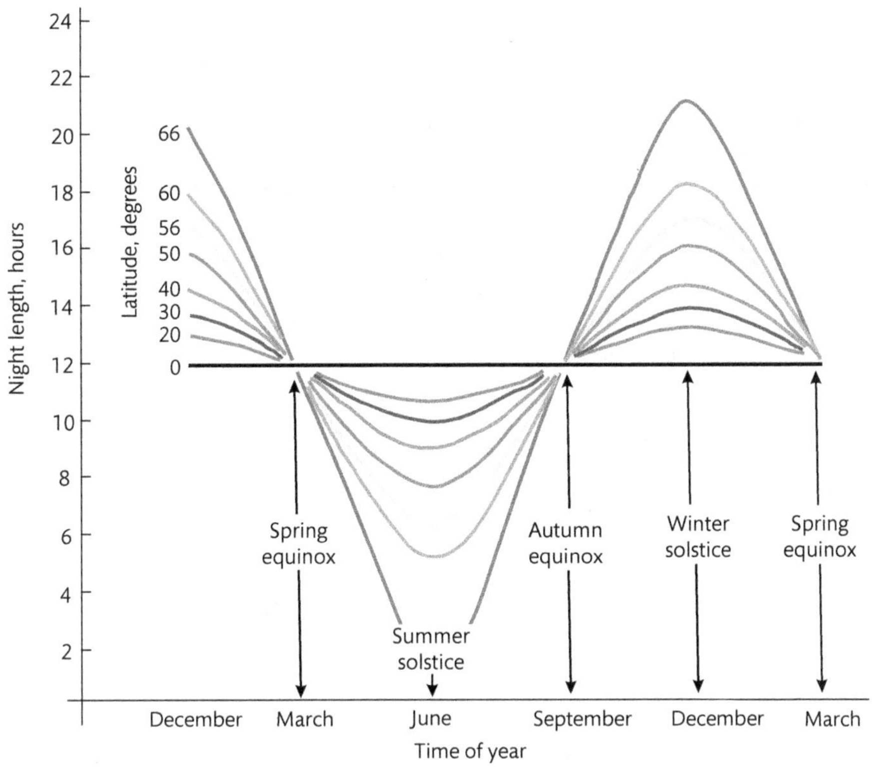

Figure 6.2 Variation in the length of the night. The extreme variability in night length at different locations on the Earth's surface is captured in this diagram. Night length is defined as the period between sunset and sunrise (Figure 6.1); it changes dramatically with both latitude and time of year. Each curve indicates how night length changes at a particular latitude over a period of 14 months. Night length is always within a few minutes of 12 hours at the equator; at mid-latitudes it varies between less than 8 hours (about the time of the summer solstice) and 16 hours (about the time of the winter solstice). Above latitude 66° around the time of the summer solstice, there is no night-time. Based upon data in *The Astronomical Almanac* (Anon 1988).

dramatically with time of year (Figure 6.2). At the equator, night and day are always equal and 12 hours long, but even at mid latitudes (e.g. 50°) night length can be as short as 7 hours in mid-summer and as long as 16 hours in mid-winter.

During the day (when the sun is at or above the horizon), under clear skies light levels vary by about 100-fold from sunrise to full overhead noonday sun. The coming and going of cloud cover can extend this range to 1000-fold. During the night, however, light levels can be much more variable. This is because the main sources of light, sunlight reflected by the moon onto the Earth, varies not just with the elevation of the moon, but is also affected by the moon's phase which changes on a monthly cycle. This results in night-time light levels that can vary by 1 million-fold between sunset and starlight on a moonless night. On top of this, there is a further potential 10-fold in variability brought by cloud cover.

The variability of light levels at the Earth's surface is made even greater by the presence of vegetation. A tree canopy can reduce light levels by up to 100-fold. Thus, entering under a woodland canopy from an open habitat can instantly decrease light levels 100-fold regardless of whether the light source is the sun, moon, or stars, while the presence of cloud cover, coming and going, extends this variability a further 10-fold. The ranges over which light levels can vary in open habitats and beneath a closed canopy woodland due to the full range of natural light sources are shown in Figure 6.3.

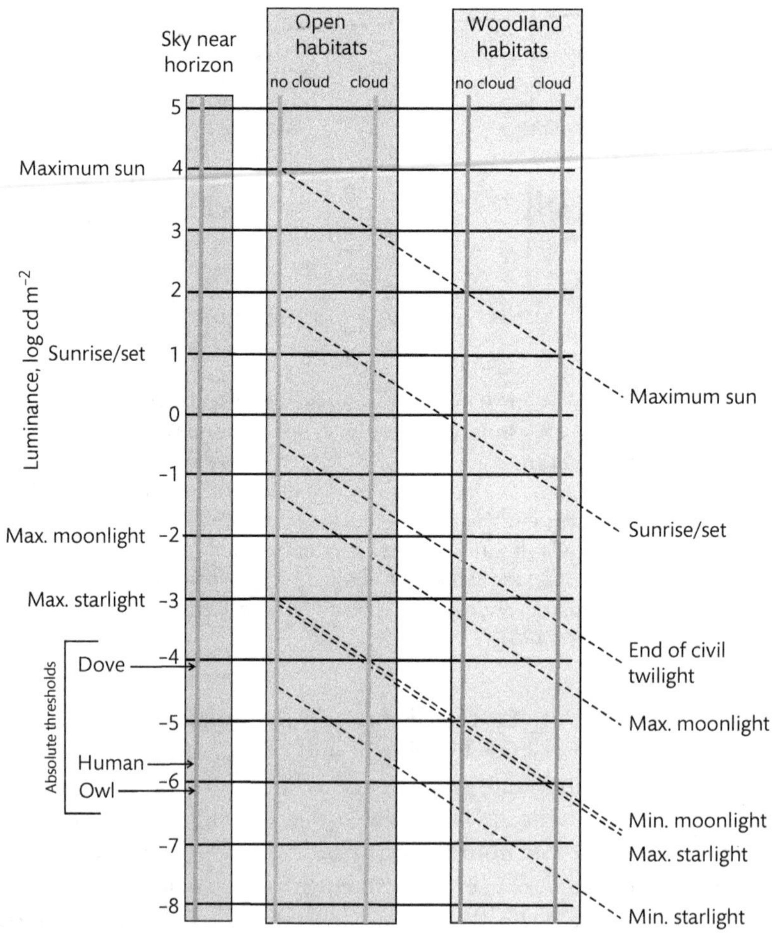

Figure 6.3 Light levels in and out of the woods. Variability in the luminance of a surface due to natural sources of illumination, in open habitats and beneath a woodland canopy, is complex and highly variable, especially at night. Along each vertical scale, the full ranges

of luminance levels due to different natural light sources are shown. The diagonal dashed lines can be read across the habitats types. For example, when the sun is at its maximum altitude in the sky (maximum sun) in the open habitats, the luminance of the ground surface will be about 4 log cd m^{-2}. Beneath an average woodland canopy, this is reduced 100-fold to 2 log cd m^{-2}. The presence of average cloud cover reduces light levels by a further 10-fold regardless of habitat type. Reading the diagram vertically, it can be seen that the total range of light levels that an animal could experience between maximum sunlight and minimum starlight (plus cloud), is about 8.5 log units (300 million-fold). At night, the full range of luminance levels is about 1 million-fold. Note that the luminance of the sky near the horizon is always brighter than within a woodland. Also indicated are the absolute visual thresholds of doves, humans, and owls. The absolute threshold is the minimum luminance of a surface that an animal can see. It is clear that a dove would often experience light levels that are below its threshold under starlit conditions even in open habitats. Furthermore, under maximum moonlight beneath a woodland canopy vision would not be possible for a dove. Even owls and humans would experience light levels that are below their visual threshold in woodlands, for example, when there is only minimum moonlight. However, owls and humans would always be able to see the sky even under the darkest conditions; hence larger objects could be detected in silhouette against the darkest night sky. Data derived from Figure 6.1, the average attenuation of light produced by a broad-leaved tree canopy in full leaf (Federer and Tanner 1966), and the luminance of the sky (Middleton 1958). Data on absolute visual thresholds; dove (Blough 1956), human (Pirenne et al. 1957), owl (Martin 1977).

The importance of all this variability in light levels for the gathering of information from the environment cannot be overstated. Also, the effect of entering beneath a woodland canopy has a dramatic effect on information gathering which is crucial in understanding the sensory ecology of nocturnality.

The important effect of light level variability during the night at any one location is that the spatial resolution of an animal's vision must be changing continuously by non-trivial amounts. Consequently the spatial information, and hence the detectability of objects and surfaces, also changes continuously. Furthermore, as Figure 2.16 shows, the way in which spatial resolution changes with light level is not linear. Over the 1000-fold range that the high light levels of daytime can vary, spatial resolution changes relatively little, but over the million-fold range of light levels that can occur at night-time, resolution changes are precipitous in humans, although they are less so in owls, principally because daytime (maximum) resolution is lower. These changes in acuity with light level have the important consequence that a diurnally active animal will experience relatively stable spatial resolution from day to day and hence relative stability in its potential to gather visual information from its environment. On the other hand, a nocturnally active animal, such as a Great Horned Owl *Bubo virginianus* or a Barn Owl (Figure 2.16), at the same location, will experience much greater, and less predictable, changes in its ability to retrieve spatial information from night to night.

One way to envisage the effects of these differences is to think of information retrieval in terms of the radius of the world about an animal from which it can extract information at the limit of its visual resolution. The radii of these limits will not only be smaller but they will change far more at night than during the day.

These differences can be readily verified from our own experience. For example, if we set up a target stimulus, say a black and white grating pattern, we can readily determine the distance at which the pattern is just detectable. If we did this repeatedly under natural light regimes between sunrise and sunset, we would find that the detectable target is further away during the middle of the day than just after dawn, and we could also see that as clouds obscured the sun, the distance would become less, but the distance of target detection would lie within clear bounds. If we really persevered we might find that the daytime detection distances would become on average less during the winter months when ambient light levels are lower.

By carrying out this simple exercise, we would see how the world about us, the world from which we could reliably extract visual information, is not fixed but depends crucially upon ambient light conditions. More telling would be the comparison between the night and day detection distances. If we repeated the measurements with the same target stimulus at night, we would find immediately that the detection distances are much less, and this is a direct result of our acuity decreasing with light levels as depicted in Figure 2.16. But, we would also find that because of the differences between moonlit and starlit nights, moon phase and cloud cover, the range of target detection distances would vary even more dramatically at night than during the day, especially at the very low light levels when acuity drops more steeply with light levels.

Such an exercise demonstrates graphically that information extraction during natural daytime is relatively stable and predictable. The distance at which a given target could be detected would vary relatively little. This translates into there being a predictable and relatively narrow range of distances at which, for example, a branch of a certain size, or a known predator, are likely to be detected. This reinforces the value of vision as a telereceptive sense during daytime.

Gaining information, however, from vision during night-time is far less predictable. There will be circumstances in which vision would yield such little information about the target at any significant distance that the telereceptive properties of vision would have almost disappeared. Certainly the distance at which a known target could be detected would be less, not only in absolute distance but also in predictability. Furthermore, the lower the light levels the greater the problem, and detection distances are not only smaller but less predictable inside a wood than outside it, because of the shading effects of the canopy.

This hypothetical series of measurements of the radii of target detection neatly captures the problem that any nocturnal animal faces in using vision as a source of information for the guidance of its behaviour. For daytime active birds, using vision as a primary means of extracting information from its environment is relatively reliable. This is not the case, however, for nocturnal birds—not only are detection distances small but they can vary in relatively unpredictable ways from night to night.

6.2 Absolute Visual Sensitivity and the Challenges of the Nocturnal Environment

One obvious and simple solution to explain how animals can be active at low light levels is to assume that natural selection will have driven eye structure and physiology to achieve higher visual sensitivity. Put simply, the sensitivity of eyes in nocturnally active species should be higher than those of strictly diurnal species.

There are two primary ways in which visual sensitivity can be increased and these are well understood (Land and Nilsson 2012; Warrant 2008): first, increasing the light-gathering capacity of the eye's optical system, and second, increasing the sensitivity of the retinal photoreceptors that analyse the image. The former will increase the brightness of the image so that more photons will be available for image analysis, while the latter will increase the probability that light photons in the image are detected and signalled to the brain. As explained in Chapter 2, the gaining of high retinal sensitivity is, however, usually at the expense of resolution. This is because it is achieved by either increasing photoreceptor dimensions, or by pooling photoreceptor outputs so that many photoreceptors feed into a single ganglion cell which sends information to the brain (Warrant 2008). Sensitivity and resolution are, in effect, traded-off, so an eye that has very high sensitivity will have relatively low resolution, and this is indeed found to be the case in many animals, both vertebrate and invertebrate, and underlies the general relationship between spatial resolution and light levels shown in Figure 2.16 (Warrant 2008).

One very obvious prediction of the above is that eyes which have evolved to maximize light capture should be absolutely large, since this allows the entrance pupil to be large. However, the focal length of the eye should not increase in proportion to the overall eye size. This is because the important determinant of image brightness is the f-number, familiar to photographers who worry about the 'speed' of their lenses. It is the ratio of the focal length of the optical system to the entrance pupil diameter; the lower the f-number the brighter the image. Both large eye size and low f-number are indeed found in the eyes of owls (Figure 2.13) suggesting that natural selection has operated to maximize sensitivity in the eyes of these birds (Martin 1982).

In owls, the eyes are so large, that unlike other bird species they do not sit within the protection of the skull in a deep orbit, but instead they are attached almost to the surface of the skull (Harmening and Wagner 2011) (Figure 2.2). Owl eyes cannot be readily moved since the muscles which usually control the movements of the eyes have to function primarily to keep the eyes in place (Steinbach and Money 1973). These features are readily seen in Figure 2.2. In addition to being very large, owl eyes are tubular in shape. This shape may not, however, have any particular optical function, rather it is probably an adaptation which allows an absolutely large eye to be fitted within, or in the case of owls upon, a small skull. Such tubular shapes are a feature of other eyes which have evolved to function at low light levels, especially some deep sea fish (Locket 1977; Warrant 2008). In owls, the tubular shape may have the particular advantage of reducing the weight of the eyes.

Analysis of the optical system of Tawny Owl eyes shows that, indeed, the absolute size is disproportionately large compared to the total mass of the bird and the optical system has a relatively low f-number. Thus, although the overall size and focal lengths of the eyes of owls and humans are very similar, they differ in the maximum brightness of the image which they produce. The f-numbers of human and owls' eyes are 2.13 and 1.30 respectively with the result that, when viewing the same scene, the image in owls' eyes is approximately 2.7 times brighter than in human eyes (it is the square of the f-numbers which must be compared) (Martin 1977).

The suggestion that this difference is functionally significant comes from studies of the absolute visual sensitivity of Tawny Owls determined using behavioural training techniques in which the birds indicated the lowest light levels that they can detect under controlled conditions. When tested using the same apparatus, the absolute sensitivity of Tawny Owls is on average 2.2 higher than humans (Martin 1977). Compared to Rock Doves, which are strictly diurnal species, owls are 100 times more sensitive (Blough 1956; Martin 1977). What these results show is that the difference between humans and owls with respect to absolute visual sensitivity can be accounted for by the greater light-gathering capacity of the owls' eyes, suggesting that the sensitivity of the retinas is very similar in both species. However, the differences in sensitivity between owls and doves cannot be explained by differences in image brightness (the f-number of dove eyes is very similar to that of human eyes) and so it must be attributable to differences in retinal sensitivity between these nocturnal and diurnally active birds.

6.3 Visual Sensitivity in Context: In and Out of the Woods

The finding that natural selection has indeed maximized the visual sensitivity of owls' eyes compared with other birds, and that this is achieved by both maximizing image brightness and retinal sensitivity, is not particularly surprising.

In a broader context, however, the measures of absolute sensitivity are very intriguing. The three figures at the start of this chapter (Figures 6.1–6.3) showed just how variable light levels at night are and also just how low they can be, especially beneath a woodland canopy. If comparison is made between these absolute minimum light levels and the absolute maximum sensitivity of owls' eyes (Figure 6.3), it is clear that there will be occasions at night within a wood when light levels fall below the threshold of owls, i.e. they would be truly in the dark, they would be unable to discern at close range even gross spatial details of objects around them. On the other hand, it is also clear from Figure 6.3 that in open habitats, visual sensitivity is sufficient to give owls (and humans for that matter) some kind of vision at all naturally occurring night-time light levels. Furthermore, under all conditions the luminance of the sky will always be above the absolute threshold for vision. Even under conditions of just starlight and thick cloud, owls can probably see something in silhouette against the sky and they could use this to orient themselves visually.

Two general conclusions are warranted. First, although absolute visual sensitivity is close to the theoretical limit for a vertebrate eye, and is considerably superior to that of diurnal birds, it is not adequate to ensure vision throughout the range of light levels which occur naturally beneath a woodland canopy. In open habitats, however, vision of some sort is always possible. Second, although the spatial resolution of owls at low light levels is superior to that of diurnal birds, the actual degree of detail which can be resolved is not high. Indeed, at light levels close to the absolute threshold, spatial resolution is very poor and darkness must 'close in' on owls and severely limit the distance at which objects can be detected—only the grossest details within the scene can be discerned. But, some details can always be discerned if objects are viewed in silhouette against the sky.

Thus, there appears to be a paradox. An owl hunting outside a wood should always have sufficient light for seeing something. It is only beneath a vegetation canopy that gaining any information through vision may sometimes be impossible. So why do some owls prefer life in woodland habitats and how do they cope? What general lessons can be learnt from this analysis for understanding the nocturnal activity in other bird species?

6.4 Nocturnal Birds

Few bird species are strictly nocturnal, completing all aspects of their life cycle between sunset and sunrise (Martin 1990a). True nocturnality is restricted to non-passerine species, notably some species of owls (Strigiformes), nightjars, Oilbirds, potoos and frogmouths (Caprimulgiformes), kiwi (Apterygiformes), and some species of parrots (Psittaciformes). Many birds, however, show occasional nocturnal activity, but this is also primarily found among non-passerine species.

Nocturnal behaviours do occur among passerines but are very restricted in scope. Among these are the notable vocal displays of species such as Common and Thrush Nightingales *Luscinia megarhynchos* and *L. luscinia* in Europe and Central Asia, and of Northern Mockingbirds *Mimus polyglottos* in North America and the Caribbean, although in none of these species is singing restricted to night-time. Many species of passerines are known to undertake migratory flights at night, a phenomenon first confirmed through the use of the early radar systems which detected the movements of birds flocks at night at high altitudes (Eastwood 1967), and since confirmed many times by the use of sophisticated radar and radio-telemetry systems capable of detecting and following individual birds (e.g. Bruderer and Boldt 2001; Sjoberg et al. 2015). The sensory problems faced by birds migrating at night are relatively minor since flights take place at high altitudes well away from obstacles (Martin 1990b).

The occasional nocturnal activities of other bird species are usually associated with foraging. They are found in some species, or even just some populations, of the following taxa: shorebirds, skimmers, auks and gulls (Charadriiformes), wildfowl (Anseriformes), cormorants (Suliformes), and penguins (Sphenisciformes). Many species of the procellariiform seabirds are active at night, primarily attending nest burrows under the cover of darkness. Some non-passerines species may give vocal displays at night and, as in the passerine nocturnal singers, these displays are given for just a few weeks each year and are not exclusive to night-time—most notable among these are the nocturnal vocalizations of crakes (Gruiformes) and cuckoos (Cuculiformes).

Details of the nocturnal behaviours and the possible sensory information that makes them possible are available for only a few of the above species. Those that have been investigated, however, provide valuable insight into how information from different senses is complementary or is traded-off, and how behavioural adaptations play an important role.

6.5 The Owls' Solutions to Nocturnality

Attractive as it may be to impute owls with great wisdom, this cannot, of course, be their sole solution. Even a wise bird needs information about their environment and, as has been explained above, visual information is likely to be severely constrained in an owl's nocturnal environment. However, it does seem that knowledge of the environment plays a key role in nocturnality. Wisdom, of a particular kind, can indeed be considered to have a role in explaining nocturnality in owls.

The general conclusion from the preceding section showed an intriguing split between what might be possible at night outside a wood, and what might be possible under a woodland canopy. Owl species which operate in open habitats may have sufficient information to guide their movements under practically all night-time conditions, especially as open habitats have very few small obstacles to avoid. Barn Owls, Short-eared Owls *Asio flammeus,* and Snowy Owls *Bubo scandiacus* are examples of owls that live predominantly in open habitats. Each species specializes in taking a narrow range of small mammal prey from the ground in open habitats, and they may hunt during the day and in twilight, as well as occasionally at night (Mikkola 1983). The principal problem which they face is not guiding their own mobility, but in detecting their prey which are typically hidden beneath grass and leaf litter, sometimes under snow. There is very good evidence that such prey detection is achieved primarily not through visual information but through the detection of sounds that are emitted as the prey disturbs litter. Aspects of the hearing of owls was described in some detail in Chapter 3, and the most notable feature of owl hearing, which sets these birds apart from all other bird species, is the high accuracy of sound localization. It has been shown that in a number of owl species, but most convincingly in Barn Owls, that sound cues are sufficient to allow the birds to capture moving prey guided by the sounds of leaf litter rustles (Payne 1971). Clearly auditory information is sufficient to guide prey capture, and this information, coupled with high visual sensitivity, must go a long way in accounting for the habits of owls, which forage in open habitats.

It is clearly impressive that owls can catch prey in total darkness using information from sound cues. Fully nocturnal species, however, must do far more than just catch prey in darkness. Tawny Owls and Great Horned Owls, for example, are highly nocturnal and rarely are abroad in twilight or during the day. They must complete all aspects of their life cycle at night and furthermore they must be mobile in complex woodland habitats at the lowest light levels; light levels at which their spatial resolution, and hence radii of obstacle detection, is low.

Two adaptations of the flight of owls may make this mobility possible. The first is that the flight speeds of owls are low and the birds can stall readily. These features can allow avoidance of obstacles detected at short range and also if the birds do collide with an obstacle impact speed is low. This slow buoyant flight of owls thus plays a part in collision avoidance or, at least, when impacts do occur, damage may not be serious. In addition, the slow flight of owls is also very quiet. Not only is air displaced less vigorously by slow wing movements and low air speeds, owls also have particular adaptations to their feather structures. These reduce turbulence of the air flowing across the wing surface and result in near silent flight (Jaworski and Peake 2013; Sarradj et al. 2011), and this presumably functions to reduce the probability that an owl will be detected as it flies towards its prey.

The open habitat owls, such as Barn Owls and Short-eared Owls, have a relatively easy task. Small rigid obstacles are few within their usual flight spaces. Furthermore, they are able to hunt on the wing using slow flight and occasional hovering, effectively detecting their prey by sound from an 'aerial perch'. The woodland owls, on the other hand, must face many small rigid obstacles in their usual flight space, and it may be that for this reason they do not hunt on the wing. They use a perch-and-pounce technique and drop onto prey from fixed perches. This is the kind of task that owls have been shown capable of doing in total darkness. However, as discussed in Chapter 3, 3.1, sound ranging can be performed in all birds and mammals only after extensive familiarization with specific sounds and with the ways that they are degraded in a range of environmental conditions.

In experimental studies of sound location by owls, accurate ranging is possible only when the birds have had sufficient time to familiarize themselves with the experimental situation at low light levels. During this familiarization period, which may be a number of weeks long (Payne 1971), the birds are presumably learning both the positions of perches and the ranges of sounds heard from those perches. Thus, as in the case of passerines, they become able to determine the range of sounds through repeated exposure during the early trials. This need for familiarity with both sounds and perch positions introduces a particular behavioural adaptation which may be a vital component to nocturnality in woodland owl species.

As argued above (6.1), night-time is characterized by light levels which vary over a very wide range. At the higher night-time light levels (e.g. under full moonlight), relatively fine details can be detected, and this could allow an owl in a woodland to become familiar with the fine spatial details of smaller branches, as well as with the structure of the larger trees that provide the overall framework of obstacles. This information, however, is quite specific to particular locations and does not readily generalize. Thus the structure of a particular patch may become well known to an individual bird. This knowledge would allow the bird to orient itself to avoid smaller obstacles at low light levels because it could still detect the larger structures.

This introduces a clear hypothesis that there is a cognitive component to nocturnality. Knowledge of the spatial structure and sound degradation characteristics within a particular location are built up when light levels are high, and this knowledge allows mobility and accurate sound ranging when light levels are low.

There is good evidence to support this hypothesis. It comes from the high degree of territoriality exhibited by woodland owls and which is not found in owls of open habitats. Tawny Owls are, in fact, among the most sedentary of all birds species studied to date. Basically Tawny Owls do not go anywhere. Once settled, all of their lives are lived out in relatively small patches of woodland habitat which may be no more than 5 ha (Hirons 1985; Southern 1970; Wendland 1984). Possession of a territory is crucial for annual survival and the birds have an average longevity of 4 + years and many live for more than 20 years (http://blx1.bto.org/birdfacts/results/bob7610.htm). Once a young bird has established a territory in the first few months of its life, annual survival can be as high as 85% but failure to establish a territory results in an equally high probability of not surviving (Wendland 1984). Both sexes defend their own individual territories but they overlap to some extent in breeding pairs. Interestingly territory boundaries out-survive their owners (Wendland 1984), suggesting that occupancy on either side of the boundary can change due to the demise of a holder but an adjacent holder does not expand into the territory. It seems likely that the crucial reason for this is that a neighbouring bird cannot readily move in and exploit the resources of an adjacent vacant patch because it lacks sufficient knowledge of the structure of the patch. Thus new birds may take over a vacant territory but they learn the existing boundaries by being driven away by adjacent territory owners who stick rigidly within the patch that they know (Wendland 1984).

Such a high degree of territoriality is not a feature of owl species which live in open habitats. These birds may hold a territory for an extended period but it is not so essential for survival as it is in the woodland owl species (Bunn et al. 1982; Mikkola 1983; Thoms 2014). Open habitat owls will defend the resources of the territory during breeding but outside of the breeding period they may wander widely, and some species, such as Short-eared Owls and Barn Owls, periodically range widely, even crossing stretches of sea, for example between continental Europe and the British Isles. These birds do not have regular migratory patterns but seem to undertake movements away from breeding areas primarily in response to a shortage of their preferred prey types, which are small mammals that are prone to population cycles (Thoms 2014). Woodland owls do not follow such patterns: they are far less specialized in their diet and will take a very wide range of prey some of which would seem of suboptimal size compared with the small mammals which typically make up the core of their diets. The interpretation of this is that dietary breadth results from the fact that woodland owls simply cannot move away from their familiar patch (territory) in pursuit of more optimal prey because they would not have sufficient knowledge of the spatial structure and sound attenuation patterns at novel locations to be able to catch prey under low light levels. In effect, woodland owls simply have to sit tight and broaden their diet rather than move to seek out more optimally sized prey in new situations. The open habitat owls do not experience this constraint since their prey capture and mobility does not depend upon such detailed knowledge of a specific location.

The nocturnal habits of owls seem, therefore, to be based upon two different strategies which are driven by habitat type and the range of light levels which occur in them. First, the open habitat owls rarely, perhaps never, experience the lowest range of naturally occurring light levels. There is always sufficient light to give them the benefit of visual guidance, albeit based upon relatively low spatial resolution and the fact that their habitats are spatially uncluttered so large obstacles can be detected at a distance sufficient to avoid collisions when flying at their characteristic low speeds. Second, owls which live under a woodland canopy face a more exacting task in that light levels of the forest floor may frequently fall below their threshold for vision. Under many circumstances, fine spatial details cannot be detected even at close range. The solution for these birds is to build knowledge of the spatial structure of a particular location which becomes the place in which they live the whole of their lives. This knowledge, their 'wisdom', allows resident birds to interpret the spatial information that is available to them under the lowest light levels. It enables them to be mobile and able to forage for prey items.

In open habitat and in woodland species, however, it is clear that vision alone is insufficient to mediate prey capture. Nocturnality in both groups of species is possible only because they depend primarily upon prey which by their movements through vegetation cause sounds that can be used by the owls to locate prey position accurately. Thus in both types of owls, the senses of vision and hearing work in tandem, or complement each other, to provide sufficient information to allow prey capture and mobility at night-time light levels. To exploit the resources of a woodland habitat requires the additional behavioural adaptation of a highly sedentary life and this facilitates the accumulation of knowledge that allows minimal spatial cues to be successfully interpreted. Thus it can be seen that 'The wise old owl' has little option than to 'sit in an oak'.

6.6 The Oilbirds' Solution to Nocturnality

Oilbirds are among the most nocturnal of all bird species. The majority of individuals probably never experience daylight in their entire lives. They roost and breed in caves usually at depths where no light penetrates and emerge from their caves at dusk and return before dawn. They live close to the equator and therefore night and day length varies little over the year. They emerge from caves at dusk to feed on fruits in the tropical rainforest canopy and the majority of birds return to the cave before dawn, although some may roost in trees during the day (Holland et al. 2009; Roca 1994; Snow 1961; Thomas 1999).

Oilbirds (Figure 6.4) have eyes which are both relatively and absolutely large and their retinas are dominated by rod receptor cells (Martin et al. 2004b). There is evidence that the rod photoreceptors in some parts of the retina are arranged in multiple layers, a mechanism that enhances sensitivity by increasing the probability that photons in the retinal image will be intercepted. Efficient as this mechanism is for enhancing sensitivity, it is highly likely to decrease spatial resolution. It is a

Figure 6.4 Nocturnal birds. Both North Island Brown Kiwi *Apteryx mantelli* (left) and Oilbirds *Steatornis caripensis* (right) are highly nocturnal, but they differ in the suit of senses used to guide their behaviour in the very low light levels of their preferred habitats.

mechanism found in some deep sea fish which live permanently at extremely low light levels (Locket 1977; Warrant 2008). The minimum f-number of Oilbirds' eyes is low (1.07), and it has been argued that this, combined with the structure of the retina, means that their eyes are pushing at the theoretical limits of visual sensitivity for vertebrate eyes (Martin et al. 2004b; Rojas et al. 2004).

From the perspective of vision, Oilbirds appear well adapted to the demands of life at low light levels. Like the owls, however, vision cannot be the sole answer to their mobility and for their foraging at night. Deep inside caves where Oilbirds nest and roost, no photons of light penetrate, vision is simply not possible. Unlike an owl under a woodland canopy at night, there is no residual vision providing gross spatial cues which can provide orientation. As described above, owls use hearing to complement their vision in the location of prey. Oilbirds, on the other hand, use hearing to guide their mobility. This is used, however, only within the spatially simple and static interiors of caves.

As described in Chapter 3, hearing in Oilbirds has evolved to enable spatial information to be detected using active SONAR. However, the spatial resolution of this system allows the detection of only large objects at close range (Konishi and Knudsen 1979). The Oilbirds' sound-based spatial resolution is very poor compared with the echolocatory mammals, but it is clearly sufficient to enable birds to locate obstacles, cave walls, other birds, and possibly nest and roosting ledges, within the caves' totally dark interiors. Like the owls, the flight of Oilbirds is slow and includes momentary hovering, and this must reduce the probability of damage should collisions occur. Thus as an Oilbird enters deeper into its cave, its spatial guidance must switch from visual to auditory-based information and the latter probably includes both active and passive sonar. It does seem clear that once outside the cave, the birds do not echolocate because they fly in silence. At this point visual guidance presumably takes over.

While Oilbirds attempt to locate food resources in their rainforest habitats, they usually fly above the canopy. In effect, they fly in an open habitat devoid of fine spatial obstacles. As in the owls of open habitats, light levels encountered above the tree canopy will always exceed the visual threshold. This will allow visual orientation to large obstacles on the darkest of nights and probably allow the detection of relatively fine spatial detail under moonlit conditions. There is, however, no experimental evidence on the visual spatial resolution of Oilbirds and data would be very valuable.

To guide their foraging as they fly over the canopy, Oilbirds seem to switch to a third telereceptive sense, olfaction. Again, no detailed experimental work on olfaction in Oilbirds has been reported but Snow (1961) provided good observational evidence that the birds are using olfaction to at least detect locations where ripe fruits are concentrated. This would allow the birds to enter the upper layer of the canopy and perhaps seek out individual fruits using olfactory and possibly visual cues. It should also be noted that Oilbirds, like other nightjars (Caprimulgidae), have prominent whisker-like feathers (rictal bristles) around the mouth (Figure 6.4). It seems likely that these have a tactile function and could guide Oilbirds' mouths towards individual items. Such tactile information would also facilitate close contact between individuals when on roosting and nesting ledges in caves.

While there is clearly more to be learnt about the sensory capacities of Oilbirds, it is clear that they must rely upon rather gross spatial information to guide their behaviour at all times. Fine-grained spatial information, of the kind that we might assume to be essential for bird behaviour, is not available to them. Vision and hearing both play key roles in providing spatial information about the birds' environment but they do not work together. SONAR takes over when vision is no longer viable and vision seems to be the default sense when there is natural light available outside the cave, presumably because echolocation is effective only at short range. Effective as these two information sources are, a third telereceptive sense, olfaction, is employed in the guidance of foraging.

Oilbirds can be viewed from the perspective of sensory ecology as unique. They are certainly extraordinary birds and challenge commonly held ideas of what birds do and how they do it, and certainly bring a different perspective to the idea of a 'bird's eye view'. They provide a particularly clear illustration of how it is not always necessary for birds to have available to them highly detailed spatial information of the world about them. Whether detailed spatial information is necessary clearly depends upon specific environmental conditions and behaviours. Oilbirds also show that the senses of vision, hearing, and olfaction can provide spatial information at different scales and can complement each other in providing information to guide the key tasks of mobility and foraging.

6.7 The Kiwi's Solution to Nocturnality

Five species of kiwi are currently recognized (Gill and Donsker 2015; Wilson 2004). They are endemic to New Zealand and are part of a unique avifauna that evolved in the absence of terrestrial mammals over a period of 80 million years (Wilson 2004). (Note, the plural of kiwi is kiwi, it is a Maori word which does not take an 's' to indicate a plural in the same way that an 's' is not added to indicate plural in certain nouns in English, such as sheep.)

The biology and behaviours of kiwi are unique among birds. Like the Oilbirds and owls, they challenge many popular assumptions about what birds are. They certainly cannot be regarded as a wing guided by an eye. Kiwi are both flightless and nocturnal. They exploit the resources of forest-floor habitats where they forage for invertebrates living in the soil or on the surface among litter, and they use their bill as a tool to probe and fossick through material (Marchant and Higgins 1990). Structural differences among kiwi species are relatively minor: species differ mainly in their body mass, relative length of bones, bill length, etc. The biology of the different species is rather similar having diverged as populations became geographically isolated from each other. One species, Okarito Kiwi or Rowi *Apteryx rowi*, was proposed in 2003 (Tennyson et al. 2003) and its status was accepted only recently (Robertson 2013).

From the sensory perspective, kiwi present a paradox. As explained in Chapter 2 and in the above sections of this chapter, evolution of a nocturnal habit would be predicted to result in the evolution of eyes of absolutely large dimensions and low f-number in order to maximize the extraction of information from the retinal image. Kiwi, however, have small eyes (Figure 6.4). Their axial length is approximately 7 mm, no larger than those of small passerine species. Common Starlings' eyes have a length of about 8 mm and in Rock Doves the eyes are about 12 mm long (Martin 1986a) (Figure 2.11). This means that a kiwi eye is only a quarter the length of a Tawny Owl's. Furthermore, one of the important constraints upon eye size in birds would seem to be removed in this flightless species. Kiwi would seem to be released from constraints imposed by the need for both reduced body mass and the distribution of mass towards the body core (King and King 1980). Freed from these mass constraints, it would be predicted that flightlessness should favour the evolution of large eyes and possibly large brains. Among all terrestrial and aquatic vertebrates, the eyes of the flightless ostriches, rheas, cassowaries, and emus (Struthioniformes, Rheiformes, Casuariiformes), and penguins (Sphenisciformes) are among the largest (Brooke et al. 1999; Martin et al. 2001), suggesting that flightlessness removes an important constraint upon eye size in birds. In kiwi, the eyes are exceptionally small with respect to body mass, there being a general positive correlation across all birds between eye size and body mass (Brooke et al. 1999). However, the brain size of kiwi, like those of other ratites, are large (Figure 6.5). Furthermore the flightless moas (Dinornithiformes), some species of which were massive (estimated body mass in excess of 200 kg in some species) that lived until recently alongside kiwi, appear from their skull structure to have had large eyes (Worthy and Holdaway 2002).

Clearly, kiwi cannot be relying upon vision to guide their nocturnal behaviour, and there is good evidence that they gain most information to guide their foraging from olfaction and tactile cues. In effect, they move around their forest-floor habitat, within the low light levels that occur beneath a canopy, probably without any visual information or at best, guided by information at very low spatial resolution. There is evidence that kiwi eyes are functional in that they have an array of rod receptors (Corfield et al. 2015b) and a functional optical system (Howland et al. 1992). Recent comparative analysis of the pathways and brain structures associated with the processing of visual information, however, shows a marked reduction in size compared with diurnally active birds (Martin et al. 2007b).

The use of olfactory information in the foraging behaviour of kiwi was first demonstrated experimentally more than half a century ago (Wenzel 1968), and this was backed up by work on the relatively large size of the olfactory bulbs (Bang

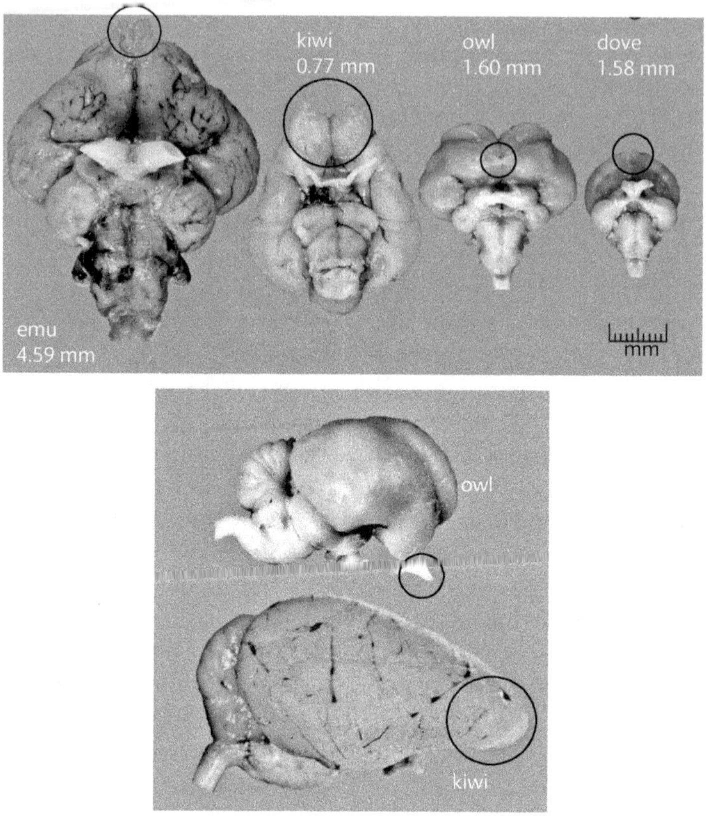

Figure 6.5 Comparative brain structure in nocturnal and diurnal birds. The brains of two mainly diurnally active species, Emu *Dromaius novaehollandiae* and Rock Dove *Columba livia*, are shown alongside those of two nocturnally active birds, North Island Brown Kiwi *Apteryx mantelli* and Common Barn Owl *Tyto alba*. The olfactory bulbs are highlighted by circles and the dimensions beside each brain refer to the diameters of the optic nerve in each species. Photographs courtesy of Jeremy Corfield.

and Wenzel 1985). These observations have also recently been confirmed using more advanced anatomical techniques (Corfield et al. 2015a; Corfield et al. 2014; Martin et al. 2007b). The nostrils of kiwi open, uniquely among birds, at the bill tip (Chapter 4, section 4.1.2 'Bill tip organs in shorebirds, kiwi, and ibises'), and this can clearly function as a means for olfactory sampling of the substrate at a relatively fine spatial scale or allow specific olfactory sampling below ground when the long bill is used as a probe (Figure 4.3).

The behaviour of foraging kiwi is described as 'walking slowly along tapping the ground and when prey is detected they probe their bill into the leaf litter or a rotten log; occasionally plunge their bill deep into the ground' (Robertson 2013).

It seems likely that the 'tapping' behaviour could be involved in the collection of olfactory information, but it is equally likely to be involved in the collection of tactile information.

Recent work has shown that detailed tactile information can also be obtained from the bill by virtue of their bill tip organ (Chapter 4, 'Bill tip organs in shorebirds, kiwi, and ibises') (Cunningham et al. 2009; Cunningham et al. 2007; Cunningham and Castro 2011; Martin et al. 2007b). Furthermore, it has been shown that this bill tip organ can possibly function as a 'remote touch' device, capable of detecting objects buried some distance away from the bill when it is probed into soft substrates.

Around the base of the bill, there is an array of long whisker-like feathers (rictal bristles) (Figure 6.4) and, as in Oilbirds, these may function as tactile receptors at close proximity and allow objects to be detected and social interactions to be mediated without visual cues, perhaps between individual birds in the complete darkness of nest and roosting burrows. Kiwi are also vocal animals and use auditory signalling extensively as a means of controlling social interactions between pairs and declaration of their presence in territorial disputes. There is evidence of some specialization in the anatomy of the auditory system of the brain, although to date there is no experimental evidence on their function, nor is there any evidence on the ability of kiwi to determine the direction and range of sounds (Corfield et al. 2011a; Corfield et al. 2012).

Thus the picture which builds of the sensory world of kiwis is really very different to those of other nocturnal bird species. The flightlessness and relatively slow locomotion through a dimly lit forest-floor habitat do not pose the kinds of problems faced by the owls and Oilbirds. The kiwi diet of relatively immobile invertebrates, allows the detection and identification of individual items using a combination of tactile and olfactory cues. Furthermore, the detecting mechanisms are incorporated into the very tool, the bill, which is used to explore and manipulate the environment, and to seize and ingest the food. In effect, this is a neat combination of a tool with the sensory detectors which gather information for its control.

The control of general mobility with respect to structures remote from the body does however, seem to rely upon vision. The small eyes probably provide rather gross spatial cues, which presumably even on the darkest nights are able to detect large objects in silhouette against the sky. At higher night-time light levels, more detailed spatial information should be available, but clearly experimental determination of spatial resolution is needed to settle just what spatial information is available to a kiwi through vision.

It seems likely that, as in the case of the woodland owls, a cognitive component may also be vital to explain the nocturnal mobility of kiwi. Although kiwi effectively move only in two dimensions compared to the three in which a woodland owl moves, the birds probably still need to build up spatial knowledge of a familiar patch of woodland. This can be achieved through these birds' long-term residence in the same patch of habitat which is defended against neighbours (Taborsky and Taborsky 1992; Taborsky and Taborsky 1995). This territorial system may parallel that described in Tawny Owls, in which once established in a territory the birds never leave that area for the rest of their lives, and so are able to build detailed knowledge of its structure that can be interpreted using minimal spatial cues when they are available. Like owls, kiwi are long-lived birds; in fact, they probably live longer than owls on average. Brown Kiwi *Apteryx australis* have been recorded as living up to 35 years in captivity and at least 20 years in the wild (Carey and Judge 2000). Furthermore, it has been shown that their populations are extremely subdivided with many cryptic species, similar to the situation in small mammals, and this suggests that kiwi are highly sedentary with very low dispersal (Baker et al. 1995). In other words, kiwi stay put and live long, and this may be the key behavioural component to their nocturnal existence in the same way that it plays a key role in the woodland owls.

The above analysis can account well for the different kinds of information that kiwi employ in a complementary manner to control their behaviour. It also provides another good example of how the use of different sensory information is tuned to the sensory challenges of particular environments and the conduct of tasks within them.

An intriguing questions remains. Why is vision of such little importance to kiwi? Owl-sized eyes could have evolved in kiwi. Are the small eyes just a legacy from the time when kiwi were diurnally active and their ancestors arrived on New Zealand, or are they an indication of the regression of vision? The latter seems the most probable, not just because the eyes are so small, certainly very much smaller than would be expected on body mass alone (Brooke et al. 1999), but also because while kiwi brains are relatively large only a small proportion of the kiwi brain is associated with the analysis of visual information. The diameter of the optic nerve which carries information from the eyes to the brain is small and the thickness of the optic tectum is also small. For example, the optic nerves of Barn Owls and Rock Doves have the same diameter while in kiwi it is half that (Figure 6.5) (Martin et al. 2007b). In Emus, the optic nerve is nearly six times the diameter of those in kiwi.

It seems safe to conclude that reduced reliance upon visual information is a derived characteristic in kiwi, that is the kiwi's early ancestors were diurnal and capable of flight (Phillips et al. 2010), and presumably had similar reliance upon vision as the flying diurnal birds of the present day. The downgrading of vision in kiwi is probably an example of adaptive regressive evolution which has been noted to have occurred in the vision of other animals (Jeffery 2005). At some point in the evolution of kiwi, natural selection favoured foregoing visual information in favour of other sensory information, while vision has remained important in other descendants (emus and cassowaries) of their common ancestor (Hackett et al. 2008; Harshman et al. 2008).

The ecological circumstances favouring the downgrading of vision in kiwi are unclear. However, a similar reliance upon tactile and olfactory information over visual information is found in nocturnal mammals, especially rodents, with which kiwi are often compared ecologically. Such rodents do not occur naturally in New Zealand, it being considered that kiwi fulfil a similar niche to that exploited by small rodents in other parts of the world (Wilson 2004). This suggests the evolution in kiwi and in these mammals, of similar sensory performances tuned to a common set of perceptual challenges presented by the forest-floor environment at night, which cannot be met by vision. While regressive evolution of visual systems have been described in both vertebrate and invertebrate animals (Jeffery 2005; Leys et al. 2005), these examples have involved a complete loss of vision following colonization of subterranean habitats devoid of all light. In kiwi, complete regression of the eye and parts of the brain associated with visual information processing has not occurred. However, while kiwi roost and nest in the complete darkness of burrows, their foraging habitats are not completely devoid of light; they must experience light regimes similar to those experienced by the woodland owls, as described in the early sections of this chapter.

The final, and probably a key factor that allowed the regression of vision has been the absence of mammalian predators throughout the whole of kiwi evolution in New Zealand. Avian predators (Haast's Eagle, Eyles's Harrier) were present during this time, and certainly Haast's Eagles were large enough to take birds the size of modern kiwi. It could well be that it was predator pressure from these birds' daytime hunting that set in train the evolution of nocturnality in kiwi (Wilson 2004). Also there were nocturnal predators (Laughing Owls) present, but they took a range of prey smaller in size than kiwi. Not only are the eyes of kiwi small, their visual fields are also very small, each eye encompasses a field of only 120° and is the narrowest so far described in any bird. Furthermore, the total visual field of the two eyes combined is the smallest of all birds determined to date. There is a very narrow (11° wide) area of binocular overlap and a 120° wide blind area behind the head (Martin et al. 2007b) (Figure 6.6). In addition, kiwi can only just see their own bill tip, it falls at the very edge of their binocular field and it seems unlikely that the bill tip could be placed accurately under visual guidance (Martin 2009). In those other bird species in which the bill falls at the very edge of the visual field (some ducks and shorebirds), bill position and timing does not appear to be under visual guidance. Unlike the kiwi, however, the eyes in these birds have evolved to a position high up in the skull, providing comprehensive vision of the world above and around the birds, as an adaptation to maximize predator detection (Martin 2007). Clearly, this has not occurred in kiwi and so it can be concluded that kiwi are simply unable to detect predators using vision, but there is no pressure to do so.

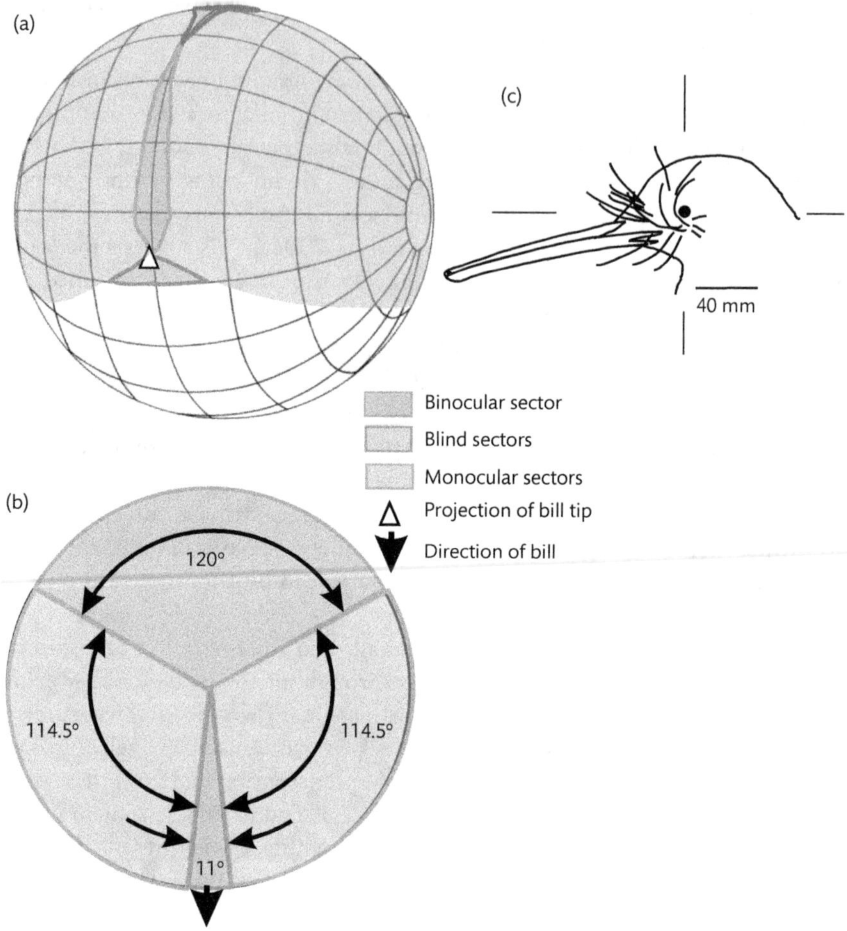

Figure 6.6 The visual fields of kiwi. The diagrams follow the same conventions as used in Figure 2.14. The visual field of each eye is narrow (125°) and they are combined to give a very limited region of binocular overlap and a very large blind area behind the head. The bill tip falls at the very periphery of the visual field suggesting that the birds cannot use vision to guide the position of the their bill tip. This is quite unlike the situation in birds which use vision to guide their bill towards targets. In such birds, the bill falls centrally, or just below the centre, in the binocular region; see, for example, the visual fields of bustards and storks shown in Figure 2.14.

Thus an absence of nocturnal predators is likely to have been a key factor in driving the regression of kiwi vision.

It seems likely that the regressive evolution of kiwi vision is the result of the trade-off between the requirement for an eye large enough to gain useful spatial information at low light levels, and the metabolic costs of maintaining such large eyes and the costs of extracting and processing information from them (Laughlin 2001a; Laughlin 2001b). At some point, insufficient spatial information can be gained about the environment from vision to balance the metabolic costs of vision against the information and metabolic costs of other sensory systems which provide comparable or complementary information. It seems that there is a natural ambient light level below which the costs of maintaining a large eye and associated visual centres are not balanced by the rate at which information can be gained, and that this occurs in forest-floor habitats at night. This principle finds its expression in the sensory systems of kiwi.

6.8 Nocturnality in Other Birds

The three examples discussed in the preceding sections show that a fully nocturnal habit in birds depends upon a suite of sensory and behavioural information, which complement each other in rather complex ways. There are few other bird species which are totally nocturnal. There are, however, many bird species which may be active at night for the conduct of specific behaviours and/or at particular times in their annual life cycle.

6.8.1 Nocturnal Parrots

Just over 400 species of parrots (Psittaciformes) are currently recognized but only two, Kakapo *Strigops habroptilus* and Night Parrots *Pezoporus occidentalis*, are nocturnal. Night Parrots are found in a distinctive type of grassland, *Spinifex*, in dry areas of Australia. They were once thought to be extinct but a live specimen was caught in April 2015. Their ecology, behaviour, and anatomy are hardly known (BirdLife International 2015). Kakapo are better known and their breeding biology has been intensively studied because of their high conservation status and intense management (Powesland et al. 2006). Both species are regarded by BirdLife as critically endangered.

Kakapo (Figure 6.7) are found in New Zealand and as well as being nocturnal they are also flightless and are the largest, by weight, of all parrots. Their habitat preference is for mossy forests (primarily *Nothofagus)*, particularly where forest is adjacent to open ground along river flood plains or the subalpine scrub belt bordering tussock 'meadows'. Kakapo are described as 'versatile and opportunistic vegetarians' with different seasonal diets: fruits, berries, nuts, seeds, green shoots, leaf buds, roots, rhizomes, tubers, bark, stems, even moss, fungi, and ferns. Some of these foods are dug out from below ground using the large bill and feet. In part of their range, the main food plants are fern fronds *Blechnum*, followed by *Dracophyllum* shrubs but also cushion-forming plants, *Gahnia* (a sedge) and *Astelia* which are rhizomatous tufted perennials, plus club mosses, for example

Figure 6.7 A Kakapo *Strigops habroptila*. One of 11 young birds photographed when being transferred to Codfish Island, New Zealand in 2002. The birds had been hand raised as part of a conservation management programme aimed at preventing the species' extinction. Photo courtesy of Dianne Manson, Nationwideimages.co.nz

Lycopodium. All of these food sources have generally relatively low nutrient value and large quantities have to be eaten. However, the birds do seems to target their foraging on more nutritious parts of these plants; for example, new growth and developing foliage are targeted in spring and summer, and subterranean parts of plants from autumn through to early spring. Fruits of *Fuchsia* species are also thought to be important.

Kakapo are described as 'living life in the slow lane' in that they do everything more slowly than most birds, but they live on average for 58 years and potentially to almost 90 years, reproduction is intermittent occurring only in years when certain foods are abundant (Powesland et al. 2006). From a conservation management perspective, Kakapo present a particular challenge because of their intermittent breeding and elaborate reproductive system which involves lek-type behaviour. This is controlled primarily by vocalizations in which males compete to attract females using low frequency booming calls that can propagate long distances, in excess of 1 km (Merton et al. 1984).

From the sensory ecology perspective, the life of Kakapo may not be as challenging as those of other nocturnal bird species. Like the kiwi, they evolved in an environment free of natural nocturnal predators capable of taking birds of their size. Although they may live under a forest canopy where light levels can be very low at night, they also forage outside of the forest where light levels are usually higher (Figure 6.3). Thus the visual challenges which they face are at times more similar to the open habitat owls rather than the woodland owl species. Kakapo are also highly sedentary and live long and well-regulated lives within their territory, so much so that they make defined tracks which they use to get around their territories when foraging. This clearly suggests that Kakapo parallel the kiwi and the woodland owls in having the opportunity to gain detailed knowledge of particular locations, and this knowledge can be built up through very long-term residence and regular movement patterns under the full range of naturally occurring night-time light levels. As in these other nocturnal species, this knowledge could be used to interpret information of low spatial resolution that is available when light levels are low. Thus there may be an important cognitive component to their nocturnality. Furthermore, because these birds move slowly about their habitat there is little danger of collisions with obstacles likely to cause injury.

There is some evidence that both the vision and olfaction of Kakapo show adaptations to nocturnal conditions, at least compared with other parrots that are diurnally active; however, differences may be subtle. While the eyes of Kakapo are relatively large in terms of axial length and corneal diameter they are, however, well within the range of other parrots and scale with head size and body mass (Corfield et al. 2011b). Similarly their retinas contain both rod and cone receptors of similar length and diameter to those of diurnally active parrots and of other birds. Nothing is known, however, about photoreceptor density and distributions within the retina. Kakapo do, however, show evidence of a decreased reliance upon vision in that the portion of the brain involved in analysis of visual information is reduced compared with diurnally active parrot species. Thus, Kakapo have a significantly smaller optic nerve and tectofugal visual pathway, and the optic tectum, nucleus rotundus, and entopallium are significantly reduced in relative size compared to other parrots (Corfield et al. 2011b).

This reduction of the visual pathways suggests that spatial resolution may be lower than in diurnal parrots of similar skull size and body mass. The only suggestion of a difference between the eyes of Kakapo and other parrots is that the orbits of the eyes appear to be more forward facing in Kakapo. In some parrot species, the eyes diverge very widely (Demery et al. 2011), and in Kakapo evidence from orbit position suggests they may be less diverging, and this could provide increased binocular overlap, although, direct measurements have not been made (Corfield et al. 2011b). Even if binocular overlap is larger, its function may not be associated with the nocturnal habit since the greatest binocular overlap is found in diurnally active passerine birds, not owls, as commonly supposed (Troscianko et al. 2012) (and see Chapter 8). Therefore, a slightly greater degree of binocular overlap may be correlated with inspection of food items including items held in the foot or bill. The lower degree of binocular overlap in other parrots may be a result of the need for more comprehensive vision for the detection of predators, a task which is not required of Kakapo in their natural habitats.

Kakapo clearly use auditory signals to regulate their social behaviour, especially during breeding (Merton et al. 1984), but hearing is unlikely to play any role in foraging because the diet is entirely vegetable. Like other parrots, it can be assumed that Kakapo have a bill tip organ (Chapter 4, 'Bill tip organs in parrots'), although this has never been investigated. Such a clustering of tactile receptors would function in the seizure and manipulation of objects held in the bill. This may facilitate foraging and manipulative behaviours, including climbing using the bill as a third limb, in the absence of visual cues (Demery et al. 2011), behaviour which Kakapo readily exhibit. In other parrots, it seems that the bill tip organ has particular importance because the birds cannot see their own bill tip or what is held in the bill. The use of tactile cues from a bill tip organ could, of course, help to predispose Kakapo to feeding at low light levels as well as aid other uses of the bill, as are found in other parrots.

There is good evidence that olfactory information may play an important role in Kakapo nocturnal behaviour. Kakapo not only have a large olfactory bulb, they also produce a distinctively sweet feather odour (Butler 1989). Hagelin (2004) has suggested that feather scents could signal the presence of other individuals, competitors, or possibly mates; that is, Kakapo may employ semiochemical signals (Chapter 3, 3.2.5). Hagelin (2004) has also presented evidence that Kakapo use olfactory information when foraging, at least under experimental conditions. Thus, it was demonstrated that a tame male bird that had been trained to come to a food hopper (food hidden from sight) could be guided to the hopper by the smell of the food that it contained. Interestingly, it was found that spatial learning dominated in these experimental conditions with the bird always going to the hopper where it last found food and going to others containing food after it had discovered that food was not available at the previous location. Further information on the use of chemical signals by Kakapo could provide insights that are applicable to the conservation management of this endangered species.

The sensory ecology of nocturnality in Kakapo thus seems to throw up both parallels and differences with the sensory ecology of the other obligate nocturnal birds. Kakapo do not seem to show evidence of particular sensory specializations of the kind found in kiwi which combine information from olfaction and tactile sensitivity of the bill, or in the hearing specializations, and the eye structures found in owls and Oilbirds. However, in all of these nocturnal species, it is seen that their nocturnality depends upon a unique combination of sensory information and specific behaviours that operates within the environments preferred by each species. Perhaps the behavioural key for Kakapo is their slow pace of life, territoriality, and longevity; these allow the birds to gain knowledge of their specific environment at close proximity. Kakapo may explore their environment using spatial information of low resolution from vision and tactile information from the bill tip, and probably olfaction. Information about objects at a greater distance may be retrieved from olfaction and hearing, but the spatial resolution of these information sources will be low.

6.8.2 Nightjars, Frogmouths, and Potoos

The avian order Caprimulgiformes is usually referred to as the 'Nightjars and their Allies' (Holyoak 2001). It is currently regarded as containing 122 species (Gill and Donsker 2015), all of which are active at night and/or around dawn and dusk. With the exception of Oilbirds, their senses have not been investigated in any great detail, but it is possible to piece together some interesting general ideas about their sensory ecology. Unravelling the taxonomy of this avian order has been problematic. It has already been noted that Oilbirds have only relatively recently become regarded as a caprimulgiform (Holyoak 2001). A more recent surprise has been the proposals, based upon genomic analyses, that the order Apodiformes (swifts and hummingbirds) are, in fact, embedded within the Caprimulgiformes and that the more diurnal habits of swifts and hummingbirds evolved from nocturnal/crepuscular ancestors (Hackett et al. 2008), a surprising conclusion since it has long been assumed that it is nocturnality which is secondary to a diurnal habit (Walls 1942). From a sensory perspective, however, this may not be so surprising as regards the swifts and swiftlets because, as explained in Chapter 3, some species of swiftlets exploit the dark interiors of caves for roosting and breeding sites, and some species of swifts are thought to fly continuously through night and day for many months each year, landing only to breed, and so they must be regarded as regularly nocturnal although they may do little more than fly in open airspace at night.

Nightjars

The nightjars, although foraging at night, do so in mainly open habitats, above a tree canopy or above low ground cover in habitats that contain scattered trees (Holyoak 2001). As seen in the discussion on open and close habitats at night (6.3), it seems that even on the darkest of nights, there is likely to be sufficient light to see objects in silhouette against the sky, although spatial resolution will be low.

All species of nightjars tend to spend the day roosting on the ground, or on large tree branches, and become active around dusk when they begin foraging for insects, which are their exclusive diet. Nightjars tend to take the larger flying insects, especially the moths and beetles which start to fly as the evening progresses. However, smaller insects and flightless forms are not uncommon in their diet. In Europe, the habitats preferred by nightjars for breeding and foraging include heathland, broad woodland rides, grassland, tree plantations in their early stages, and regenerating burnt areas. For some species, such as the Common Nighthawk *Chordeiles minor* of North America, the airspace above towns may be used for foraging. The one common feature is that none of the nightjars forage amongst complex vegetation. Where trees are present, the birds tend to forage above the vegetation or around it, not within it.

The actual foraging techniques and the specialized anatomy of the mouth have been the subject of a number of investigations. Data on the foraging techniques even for well-studied species are, however, somewhat contradictory. It does seem clear that nightjars capture prey by 'open-billed trawling' through dense clouds of insects, but they are also recorded pursuing larger individual prey items (Cramp 1985; Holyoak 2001). There are conflicting opinions as to whether nightjars take prey when on the ground or only in flight. As shown in Figure 6.1, however, ambient light levels at night vary over a very wide range, and so it is not clear to what extent light levels affect the particular foraging technique employed. Nightjars will take prey of a relatively wide size-range, from mosquitoes and micro-moths (Microlepidoptera) to large moths, beetles, and cockroaches. While the larger prey may be taken by individual pursuit, it is difficult to believe that the smaller prey are captured in this way. Indeed, the specialized mouth structure of nightjars suggests that they may be exceptionally well adapted for the capture of insects by trawling—in effect they may be foraging almost blind, guided by only gross spatial cues that are used for general orientation.

The first aspect of their anatomy that may play a crucial role in the foraging of nightjars are the rictal bristles growing from around the margin of the upper mandible. The bristles are so arranged that they can trap or direct insects towards the open mouth. As in other bird species, the possible sensory properties of the bristles are unknown but it seems likely that they could have sensory function, perhaps indicating when insects have been struck.

The second anatomical feature is that the open gape of nightjars is very wide, much wider than might be supposed from inspection of the closed mouth (Figures 6.8 and 6.9). This wide gape is achieved by virtue of a unique structure of the skull and the arrangement of joints, flexible bones, and musculature involving the articulation of the mandible (Buhler 1970; Buhler 1981). This unique anatomy spreads the sides of the mandible wide apart and results in a nearly circular cross section to the open mouth.

Third, the roof of the mouth (palate) of nightjars has been reported to contain a unique sensory structure. In the majority of bird species, the palate is lined by a horny sheath which is relatively tough and apparently insensitive to touch. This presumably functions to protect the palate from damage when eating. In the Caprimulgidae, the horny sheath is absent and in its place is a highly vascularized membrane, which is bright red in colour because of the high number of blood vessels close to its surface. Cowles (1967) who first described this structure concluded that, 'it would be very sensitive and easily stimulated by an insect striking the surface, enabling the bird to react quickly to the prey while in flight'. Cowles was further supported in the idea that this sensitive palate is associated with nocturnal insect feeding in that while he was able to show that this structure is present in the 12 species of insect-eating Caprimulgidae which he examined, it was not present in other Caprimulgiform species including the fruit-eating Oilbirds (6.6), the ground-feeding frogmouths, or owlet-nightjars (see 'Frogmouths and Potoos', this chapter).

Figure 6.8 The wide gape of nightjars. Two photographs of European Nightjars *Caprimulgus europaeus* showing features associated with its foraging technique of taking insect on the wing at low light levels. In the left hand photograph, the highly vascularized structure of the palate can be seen. The palate embodies a touch-sensitive mechanism that triggers bill closure when an insect is encountered during 'blind trawling'. The rictal bristles around the edge of the upper bill may also serve a sensory function but they also increase the area trawled for insects and help to funnel them into the mouth. The right hand photograph shows the full extent of opened bill. The near circular cross section of the open gape is achieved by virtue of joints in the mandible which allows the bones to spread outwards as the mouth opens. The approximate position of this joint in the mandibles is indicated by an arrow.

Thus, it seems that nightjars have three special features which facilitate the capture of insect prey on the wing. The picture emerges of birds able to trawl on the wing with their mouths wide open, whose capture areas are further enlarged by rictal bristles, and endowed with a touch-sensitive mechanism inside the mouth that enables insects that strike the palate to be detected, and the mouth snapped shut on them. It is, of course, extremely difficult to verify this in the field, and field observers are much more likely to witness other types of prey-catching activity, such as sorties from a fixed perch towards a large insect, presumably seen in silhouette against the sky under the higher night-time light levels.

Little is known about the visual capabilities of any species of Caprimulgidae but what evidence there is suggests that vision is unlikely to provide sufficient spatial resolution to indicate a precise role of vision in foraging.

First, measurements of visual fields show that their eyes are laterally placed in the skull and the degree of binocular overlap is relatively small and vertically elongated, as in many other bird species (Martin et al. 2004a) (Figure 6.9). Secondly, the eyes of some, but not all nightjars, contain a tapetum (Nicol and Arnott 1974). This is a reflective layer behind the retina. Tapeta give rise to the bright eye-shine easily seen in many nocturnally active mammals (such as dogs, cats, cattle, and deer) when a beam of light is shone into the eyes at night. The function of a tapetum is to increase the chance that light will be intercepted by the photoreceptors, and

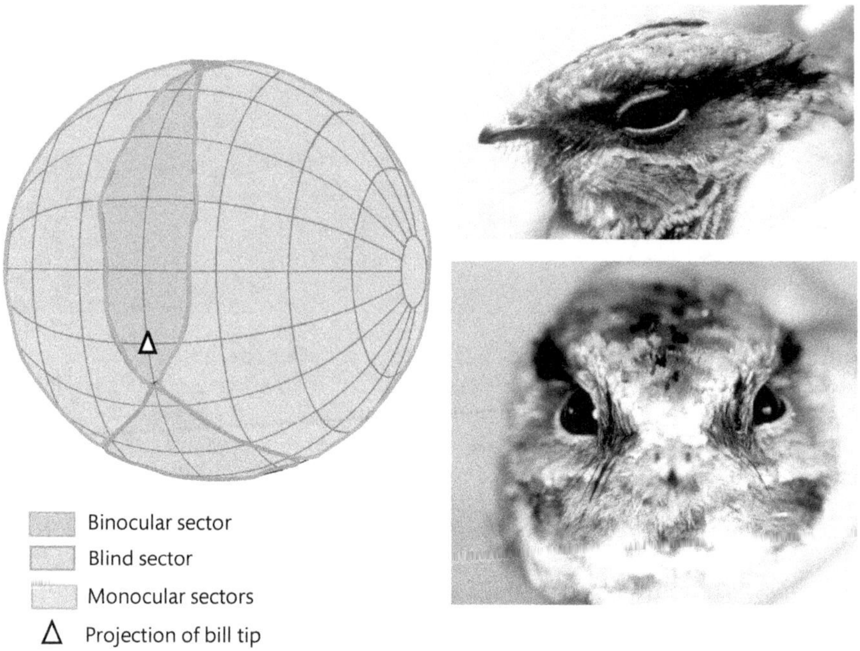

Figure 6.9 The visual fields of a nightjar. The diagram follows the same conventions as used in Figures 2.14 and 6.6. The visual field is that of Pauraques *Nyctidromus albicollis* and shows that when the mouth is closed there is a relatively narrow, but vertically elongated, binocular region. The bill is placed as the lower limit of the visual field suggesting that visual information does not play a key role in guiding bill position even when the bill is closed. In level flight, the bird has a relatively broad (20° wide) binocular area in the direction of travel. However, when the mouth is fully open as in Figure 6.8, it is likely that the bird does not have a forward view.

this results in an increase in the absolute sensitivity of the eye (Warrant 2008). The tapetum lies behind the photoreceptors of the retina and reflects back to the photoreceptors light photons that were not captured as they passed through the retina the first time. A tapetum thus serves to increase the probability that photons will be detected. When we see eye-shine, we see light which, having entered the animal's eye, has been reflected from the tapetum back out of the eye towards us. Throughout the animal kingdom, tapeta are found only in species which are either nocturnally active or live in environments where there is little natural light, for example, deep sea fish (Rodieck 1973; Warrant 2008). For animals that live mainly in bright daylight conditions, the increased sensitivity which a tapetum can provide is of no consequence and may be a disadvantage because the light reflected from behind the retina is likely to reduce spatial resolution. This may not be disadvantageous to an animal functioning at low light levels because spatial resolution is already likely to be low. Increasing the chance of detecting any of the scarce photons that are around at night can clearly be to the advantage of a nocturnal animal. It is not surprising, therefore, that tapeta are absent from the eyes of occasionally nocturnally active birds or from birds which need to retain as high as possible spatial resolution at night. The fact that tapeta are present in at least some nightjars (but they are absent from all other bird species) suggests that evolution has favoured sensitivity over resolution.

A second piece of evidence that nightjars may use vision of low spatial resolution in their foraging comes from field observations of the foraging tactics which birds have been observed to employ. Thus nightjars may normally approach large flying insect prey from below. Only occasionally do birds swoop down to flying insects (Cramp 1985). This could be because they are better able to detect insects in silhouette against the sky rather than against the darker background of vegetation or soil. If vision was not important and some other cues were sufficient to guide nightjars to their prey, it might be expected that they would be approached as often from above as from below.

Thus the nocturnal foraging of nightjars would seem to be possible because of three key elements: 1. The possibility of relatively high visual sensitivity enabling the larger objects of their environment to be seen in silhouette against the night sky; 2. Dependence on prey that fly in open airspace that are either relatively large and can be seen at least in silhouette against the night sky at short range, or prey that are small but highly abundant that can be trawled blindly from the airspace; 3. The exceptionally wide gape, rictal bristles, and a specialized tactile surface in the palate. Together these can provide a plausible explanation for nightjars' foraging ability. However, these adaptations cannot account for the social behaviour, especially breeding behaviour, of these birds. All nightjar species are highly vocal, and it seems that social contact is maintained through vocalization and hearing, although nothing is known of auditory sensitivity or the ability to locate and range sound sources. Contact between birds on the ground at short range, especially at nest sites, may be via vocalization and tactile cues, as well as vision if there is sufficient light. The plumage of nightjars typically incorporate reflective and relatively large white patches and these may provide coarse spatial cues which enable birds to more readily locate each other (Holyoak 2001).

Frogmouths and Potoos

Frogmouths (Podargidae), potoos (Nyctibiidae), and nightjars present a similar division among the Caprimulgiformes as is found between the woodland and open habitat owls, and this division may be key to understanding their sensory ecology. While the nightjars primarily take insects from the airspaces above structurally simple habitats, the frogmouths and potoos are birds which forage primarily beneath woodland canopies for insects. All of the species live in the tropical and sub-tropical climate zone. Frogmouths are distributed around Southeast Asia, Indonesia, and Australia, while potoos are found only in the American tropics. Another family, the owlet-nightjars (Aegothelidae) are also nocturnal and forage for insects in more wooded habitats. They were placed until recently in the Caprimulgiformes (Holyoak 2001) but are now placed within the Apodiformes alongside the swifts and hummingbirds (Gill and Donsker 2016), attesting to the close affinity between these two avian orders (Hackett et al. 2008).

All of these species feed primarily upon large insects and other invertebrates, although the larger species of frogmouths, for example, Tawny Frogmouths *Podargus strigoides*, also take small vertebrates such as frogs and mice. A striking feature of all frogmouths is their large gape which in some species is surrounded by rictal bristles. The large gapes of these birds are not, however, used for trawling insects from the air. They feed primarily by taking prey from the ground after pouncing from a low perch, in a similar manner to the woodland owls, but while the owls primarily take prey in their feet, frogmouths strike or capture prey directly with their large and relatively heavy bills (Serventy 1936). The owlet-nightjars also take prey from the ground but they tend not to use an elevated perch for the approach, preferring to pursue prey along the ground. Again prey is taken directly with the bill, and the diet consists mainly of non-flying invertebrates such as millipedes, spiders, and ants. The potoos do, however, take prey from the air, but rather than trawl them, these are taken 'flycatcher fashion' by sorties from a fixed perch to which the bird returns. Potoos generally live in more open woodlands than frogmouths which prefer more densely forested areas.

Little is known about the sensory capacities of these birds, although the eyes are relatively large, suggesting that vision is of importance to them. The influence of light levels on their foraging technique and foraging success has not been investigated. The possible role of hearing as a means of locating prey, especially on the ground, may be particularly important in the prey capture techniques of the frogmouths and owlet-nightjars. Similarly, there is no evidence on the possible role of the rictal bristles in prey capture.

Unfortunately, only a rather sketchy picture emerges of the possible sensory ecology of these species, but it is worth noting that their prey capture techniques do not seem to involve flight, or if flight is involved, it is of short range from a fixed perch down to prey and back again. It is possible that hearing plays a role in the same way that it does in owls, possibly gaining information through the detection of litter rustle sounds as prey moves on the ground. However, nothing is known of these birds' abilities to determine either the direction or range of sounds. Clearly, there may be some parallels between the sensory bases of their prey capture technique and general mobility, and that described for the woodland owls. It is worth noting that all of these species seem to be solitary and sedentary at all times of the year and that they do not live in strongly seasonal habitats. As argued for the woodland owls, this sedentary habit may be a key behavioural adaptation that allows these birds to build up familiar landscapes when nocturnal light levels are high, and this knowledge can be used to successfully interpret lower spatial resolution cues when light levels are lowest. This is different to the situation in nightjars. Many nightjar species breed in seasonal habitats in which suitable prey occurs in high abundance but also fluctuate markedly during the annual cycle. In consequence, many nightjar species are migratory and would therefore be unable to build detailed knowledge of particular locations, but because they exploit relatively open habitats they do not need such detailed knowledge of the spatial structure of a particular place in order to be able to forage for insect prey in open airspaces.

6.9 Occasional Nocturnality

Activity at night can occasionally occur in a wide range of bird species which are otherwise thought of as diurnal. Some species sing at night to declare their presence to potential mates; some species undertake migratory journeys at night; other birds may come ashore from the sea and enter nest burrows at night; and some species occasionally forage at night. The first three activities listed are not particularly challenging from an informational perspective in that the birds are either more or less stationary, or when flight does occur it is in open habitats well clear of obstacles. There are, however, interesting questions concerning how night migrating birds may actually guide themselves, and also interesting questions about how seabirds can locate their burrows when coming ashore at night. Occasional nocturnal foraging would, however, seem to be a challenging task.

6.9.1 Occasional Nocturnal Foraging

Two groups of birds are noted for their occasional nocturnal foraging; shorebirds, particularly species in the Scolopacidae (sandpipers and snipes) and waterfowl (Anseriformes). In a number of these species foraging can occur at more of less any time of day or night and in some species the pattern of feeding is determined primarily by the state of the tide on coastal and estuarine feeding locations (Robert et al. 1989; Sitters 2000). While there is evidence that occasional nocturnal feeding is widespread, there is also evidence that night-time feeding may not be preferred over daytime feeding in a range of species including, Eurasian Oystercatchers *Haematopus ostralegus*, Bar-tailed Godwits *Limosa lapponica*, Grey Plovers *Pluvialis squatarola*, and Common Redshank *Tringa totanus* (Cramp and Simmons 1983; Goss-Custard 1969; Sitters 2000), suggesting that feeding at night is less efficient. However, it has also been suggested that Grey Plovers prefer to feed at night because of an increase in the availability and activity of invertebrate prey suggesting that the occurrence of night feeding is sensitive to local conditions. At one study site, it was shown that the feeding rate at night of Oystercatchers was on average about half that recorded on the same foraging area the following day (Goss-Custard and dit Durell 1987).

Nocturnal feeding may occur on a seasonal basis even in shorebird species whose activities are not governed by tidal cycles. Thus radio tracking of Eurasian Woodcocks *Scolopax rusticola* showed that they foraged 90% of the time in spring at night outside woodlands but foraged 80% of the time by day inside woodlands in summer (Hoodless and Hirons 2007).

Among shorebirds, night feeding occurs principally among the longer-billed species which forage by probing, rather than the short-billed species which feed by picking individual items from the surface.

A survey of the occurrence of occasional nocturnal feeding among waterfowl in the Western Palearctic showed some systematic patterns associated with particular taxonomic groups (Martin 1990a):

Swans and geese *(Anserini)*: These species generally roost at night and feed by day with night feeding rarely recorded. One notable exception are Brant Geese *Branta bernicla* which may feed regularly at night, and this is the only species of this group that habitually feed in tidal areas during the non-breeding season. Hence, their feeding on submerged vegetation may be under the influence of the tidal cycle, but there is evidence that when these birds do feed at night they appear to do so primarily on moonlit nights.

Sheldgeese and shelducks (*Tadornini*): Egyptian Geese *Alopochen aegyptiaca* appear to feed predominantly at night in freshwater habitats while Common Shelducks *Tadorna tadorna*, tend to be coastal feeders at night according to the tides.

Dabbling ducks (*Anatini*): All dabbling ducks may feed at night, especially if they exploit tidal locations, but birds which feed in non-tidal locations may feed at night primarily if they are persecuted or disturbed by man or natural predators.

Diving ducks (*Aythini*): All species will feed at night even in non-tidal habitats and when not disturbed by day. Their food is primarily vegetable matter and small invertebrates, so it seems highly unlikely that food becomes more easily available at night-time. Common Pochards *Aythya ferina* and Ferruginous Ducks *A. nyroca* seem to feed mainly at night whatever the conditions.

Eiders (Somateriini): These birds usually feed by day but in some locations tidal cycles seem to result in feeding at night.

Scoters and sawbills (Mergini): All of these are diving ducks which feed on molluscs and fish and they are primarily daytime feeders and there is no evidence that coastal populations are influenced by tidal cycles in when they feed.

It can be seen from this survey that feeding at night for most species of ducks and geese is a flexible response to local situations (e.g. tides, daytime disturbance, winter food shortages), rather than a preferred strategy. It may be best to consider that in these birds, and in the shorebird species, it is due to the combination of their feeding techniques, their sensory bases of food location, and the habitats in which they feed, that feeding at night is possible should that be necessary due to energetic requirements. That is, these birds have an option to forage at night, although night foraging may be less efficient than foraging during daytime. It does seem, however, that in Common Pochards and Egyptian Geese night feeding is a preferred strategy. What information do all of these birds have at their disposal to guide their nocturnal foraging?

In both wildfowl and shorebirds, the key factor is that foraging can rely upon tactile and taste cues for the location and identification of food items. There is also some intriguing evidence that auditory cues may also play a role in the foraging of some shorebirds. It seems that the ability to exploit non-visual information frees these species to forage, perhaps with lowered efficiency, but somewhat independently of ambient light levels. Furthermore, the foraging of all these birds takes place in open habitats largely devoid of obstacles. This means that the birds may not need long-term familiarity with particular locations in order to be able to move relatively safely when toing and froing between feeding and roosting locations.

Both shorebirds and wildfowl are noted for the variety of feeding techniques that a particular species may use even in the same location (Cramp and Perrins 1994; Cramp and Simmons 1977; Hale 1980). Field observations can be interpreted as suggesting that these different techniques depend to various degrees on hearing, and visual and tactile information, although these have not been subject to experimental investigation.

This possible use of different information underlying different foraging techniques is exemplified well by observations of Eurasian Curlews *Numenius arquata* which can use three main feeding techniques: 1. Pecking, the bill just touches the surface; 2. Jabbing, the bill is rapidly inserted and withdrawn but only up to about half its full length; 3. Probing, which is a more prolonged movement of the bill which is partly or fully inserted into the soft substrate. Presumably the first technique employs visual and/or auditory information while the latter two depend upon tactile information gained through remote touch from the bill tip organ (Chapter 4, 'Bill tip organs in shorebirds, kiwi, and ibises'). Prey identification may involve the use of tactile and possible taste information since the birds can seize prey and ingest it without the need to bring it to the surface for visual identification.

A similar use of tactile and taste information from the bill may underlie the nocturnal foraging of wildfowl. The bill tip organs of shorebirds may primarily help in the location of buried objects using their 'remote touch' capability. The bill tip organs of wildfowl (Chapter 4, 'Bill tip organs in waterfowl'), however, do not provide remote touch but they probably permit fine-grained spatial discrimination of objects held in the bill tip probably allowing birds to identify buried objects (Zweers and Wouterlood 1973) using tactile cues alone, and for their palatability to be verified by taste cues (Berkhoudt 1985) (Chapter 4, 4.2).

Experimental evidence has shown that shorebirds of the genus *Calidris* can use taste cues to detect the presence of chemical information left by polychaete worms or bivalve molluscs in an apparently uniform substrate (Gerritsen et al. 1983; Gerritsen and Sevenster 1985; van Heezik et al. 1983) (Chapter 4, 4.2.5). This indicates that at least some species of shorebirds can determine, in the absence of visual cues at the surface or tactile cues from below the surface, which areas of apparently uniform substrate will be more profitable for foraging. Clearly, further work on the use of chemical (taste) cues in the foraging in other shorebirds species would be very valuable and may help explain their ability to locate suitable foraging locations at night.

There is one further piece of information which suggests that shorebirds and waterfowl can forage successfully without the need of visual cues to guide their bills. This comes from studies of visual fields. In Mallards, Northern Shovelers *Anas clypeata*, and Blue Ducks *Hymenolaimus malacorhynchos*, which are all species known to employ tactile cues from the bill when foraging, the eyes are placed so high in the skull that the birds cannot see their own bill tip but the birds gain comprehensive visual coverage of the hemisphere around and above their heads (Guillemain et al. 2002; Martin 1986c; Martin et al. 2007a). These birds are therefore well equipped to detect predators but are probably unable to place their bill accurately using vision with respect to an object or surface. The same applies to some shorebirds exemplified by Eurasian Woodcocks (Figure 6.10). These birds which are known to habitually forage at night at certain times of the year (Hoodless and Hirons 2007) cannot see their bill tip and have compressive visual coverage of the hemisphere above their head (Martin 1994). This clearly predisposes the birds to be able to forage at night using tactile cues from their bill as the primary source of information.

Finally, mention should be made of evidence which suggests that hearing could play a role in the foraging of at least some shorebird species, notably the plovers (Charadriidae). All species of this family seem to feed occasionally at night, but their foraging typically involves taking prey from the surface, or by probing just a couple of centimetres at most into the surface. Although field observations have been interpreted as suggesting that their foraging is guided primarily by visual information, there are observational studies which suggest that sound may also be used (Fallet 1962; Lange 1968). The sound information would be derived from the sounds made by invertebrates as they move on or below the surface. Fallet proposed that Golden Plovers *Pluvialis apricaria* and Northern Lapwings *Vanellus vanellus* can use hearing to detect prey which is then searched for using the bill as a short tactile probe. The evidence is rather tenuous and nothing is known of

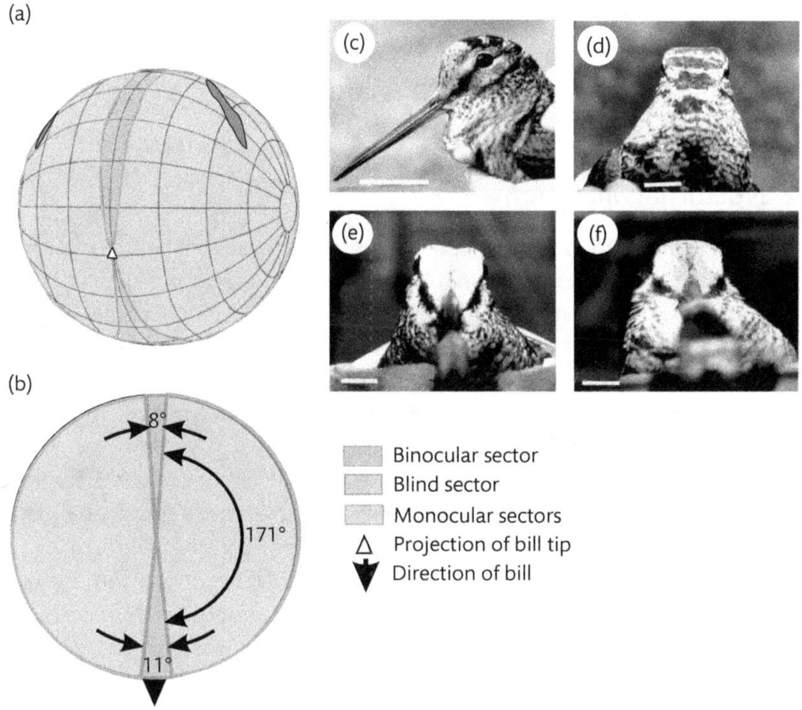

Figure 6.10 Visual fields in woodcocks. Woodcocks are an example of a small number of birds which have comprehensive visual coverage of the hemisphere about the head and extensive visual coverage to the sides. In fact, the only location from which the birds may not be able to retrieve visual information is from the portion of space occupied by its own body. The diagrams show the visual fields of Eurasian Woodcocks *Scolopax rusticola*. The visual field of an individual eye exceeds 180° in the horizontal plane and the fields of the two eyes overlap from directly in front to directly behind the head (b); this gives a narrow binocular field which stretches through 180° above the head. The projection of the bill tip falls at the very edge of visual field and the binocular field has a maximum width of only 11°, approximately in the horizontal plane which is in the direction of travel when a bird is in flight (a). In flight, the bill is held approximately at the angle shown in (c). Photographs (d) and (e) show views of the bird from directly behind and directly in front in the horizontal plane when the bird is positioned as in photograph (c). The eyes can be seen in both views emphasizing the comprehensive visual field depicted in the horizontal section of diagram (b). Photograph (f) is taken looking directly along the line of the bill; the eyes cannot be seen.

hearing or sound location in these birds. If hearing is used, it would function only at close range.

Regardless of whether auditory information is available it seems clear that tactile and taste information can play an important role in the foraging behaviour of many, if not all, wildfowl and shorebirds that feed at night. It would seem that this information is sufficient to allow feeding in the absence of visual information. Because these birds forage in open habitats, spatial information from vision can be rather poor and still allow the birds to move around feeding locations and to move safely to and fro between them.

6.9.2 Nocturnal Migration

There is a wealth of evidence that many birds undertake at least part of their migratory journeys at night (Berthold et al. 2003) and the sensory problems that night migration poses have been reviewed (Martin 1990b). Most nocturnal migration involves bird species which are diurnally active during the rest of their life cycle and night migration is particularly common among passerine species.

Studies of circadian rhythms in birds (mainly passerine species held in cages) have shown that at the time of year when migration usually takes place, instead of roosting, the migrating birds begin to show high levels of activity after the end of the normal daylight period. This activity is known as 'Zugunruhe' (migratory restlessness) with the German word preferred in recognition of the pioneering work into the phenomenon undertaken by German ornithologists (see, e.g., Gwinner (1975)). This restless behaviour is not simply random activity but shows orientation towards a particular direction which is correlated approximately with the direction in which a bird would travel if they were free to do so. Also, the time of year and the number of nights during which a bird shows Zugunruhe correlates approximately with the time period over which the bird migrates.

A great deal of research has been devoted to determining the hormonal changes and triggers which elicit this behaviour (Berthold 2001; Berthold et al. 2003) and to how the birds determine direction when showing Zugunruhe. This has involved much work on the role of the geomagnetic field and how it is detected (Chapter 4, 4.3), as well as evidence that birds employ a compass based upon the stars, which was first demonstrated in Eurasian Blackcaps *Sylvia atricapilla* 60 years ago (Sauer and Sauer 1955). See Berthold (2001) for a summary of work in this area. This work has revealed that birds appear to have available to them a system of redundant information that can orient them when on nocturnal migratory flight, or at the time of departure. Alternatively, geomagnetic information may be used to calibrate another compass based upon star patterns or the position of the setting sun, with the calibration occurring around the usual dusk period when the birds depart on migratory flights.

There does not, however, appear to be particular sensory information associated with these mechanisms. Beyond the detection of the geomagnetic field, or the imaging of the evening sky or night-time star patterns, it is inherited predispositions to respond to these relatively simple stimuli in a particular way, which seems to be key. There is a developmental or cognitive component in that birds do seem to need exposure in early life, at the nestling and fledging period, to the star patterns or setting sun positions for the orientation of Zugunruhe to become refined later in life. Species and populations probably do not differ one from another in their ability to detect these features of the environment. It is the interpretation and use to which that information is put that is key to understanding how birds on different parts of the globe, or the same birds at different times of the year, respond to what might be similar stimuli. Thus, beyond the ability to detect the star patterns or the geomagnetic field it is the use that the brain makes of this information which is key.

There is evidence that while nocturnal migrants may be using star patterns and geomagnetic information for at least their initial orientation, vision does play a role in their general orientation during flight. Thus, under most night-time conditions, birds seem to be able to maintain their orientation even when flying below a cloud ceiling but the birds do become disoriented if they fly within cloud or fog. This suggests that some kinds of visual cues, derived either from the stars and/or moon above them, or from the land/sea below, are necessary to maintain orientation. Furthermore, migrating birds are attracted, often fatally, to isolated illuminated structures under low visibility conditions suggesting strongly that at night these birds can be dominated by visual cues from below them and that they can be easily confused.

Whatever the basis of the perceptual confusions they do suggest that, as predicted from knowledge of how resolution decreases with lowered light levels, these passerines' ability to make visual discriminations at night is rather limited. Thus it seems that although night migrating birds have at their disposal a range of mechanisms whereby they can determine their migratory direction, they are still reliant upon visual cues for some aspects of their guidance during flight. Although the Earth's magnetic field may provide important information concerning the compass direction for migratory orientation, it seems that this is used to calibrate visually based compasses (stars and position of the setting sun) rather than referred to directly. While the attraction of night migrating birds to man-illuminated structures needs explanation (Martin 1990b), the phenomenon of attraction to illuminated buildings would seem to suggest that these birds can be subject to considerable perceptual confusion at night if flying in the lower airspace. However, it is important to note that night migrating birds are flying at considerable altitude, often thousands of metres above land (Berthold et al. 2003; Bruderer and Boldt 2001), well away from obstacles and are thus not being called upon to conduct fine spatial discriminations. They may, in effect, be flying almost blind and it is only under adverse weather conditions that birds fly lower and then this blind flight poses problems which can lead to confusion and ultimately to collisions with illuminated structures that protrude into the birds' air space (Huppop et al. 2016).

6.9.3 Night Attendance at Nests

Most prominent among the birds that come ashore to visit their nest at night are the smaller species in the orders Procellariiformes (shearwaters, petrels, and storm-petrels) and some species of penguins, for example Little Penguins *Eudyptula minor*.

These birds breed colonially with nests underground in a burrow or in a natural rock crevice. Entering and leaving these sites only at night is regarded as a strategy to avoid daytime active aerial predators, such as gulls and skuas (Brooke 2004). These seabirds are vulnerable to such attacks, especially the smaller species of penguins and the smaller Procellariiformes, because adaptions for an aquatic existence (short legs set well back on the body) have resulted in their movements on land being slow and cumbersome. Some of these birds are reported as coming ashore to the nest site only under the darkest conditions. For example, Manx Shearwaters are less likely to visit their nest site at night if a moon is present (Brooke 2004).

The sensory capacities associated with the Procellariiform species' ability to locate their nest sites at night have received considerable investigation. The studies, however, have been without definitive conclusion since no one experiment has been able to control all of the possible cues (olfaction, audition, and vision) in a systematic way, but it may well be that these cues can be used in a redundant way and there is no single sensory answer to nest burrow location.

The fact that Procellariiform birds have a well-developed sense of smell is established (Chapter 3) and the possibility that they could use this sense to locate their own nest burrow has been demonstrated (Bonadonna et al. 2003; Bonadonna and Bretagnolle 2002). Some of these species are able to use olfaction to locate food sources in mid-ocean (Nevitt 2008), and there are a number of studies which show that different species can discriminate between their own nest material and similar litter on the basis of olfaction. These studies have led to the conclusion that olfaction is perhaps the primary means by which these birds are able to locate their own nest burrows within a relatively crowded colony (Bonadonna et al. 2003). However, studies of nest burrow location in some species of Procellariiformes have led to conclusion that visual cues are employed, although no one study has definitively demonstrated that visual cues alone are used to find burrows. There is evidence of visual cues being used for nest location in Manx Shearwaters (Brooke 1978; James 1986), Wedge-tailed Shearwaters, and Short-tailed Shearwaters *Puffinus tenuirostris* (Brooke 2004), but the experimentally introduced visual markers that these birds appeared to attend to were large white objects placed near the burrow entrance that did not mimic the kinds of stimuli that could be used by birds as markers in natural situations.

Hearing could also play a role in nest burrow location through the recognition of the call of the mate inside the burrow. The fact that such a cue is available to

Manx Shearwaters was demonstrated by Brooke (2010) who showed that individuals could recognize each other's calls. However, this cue is not essential for burrow location since birds can find the correct burrow when it is occupied only by a silent chick.

Thus it can be seen that Procellariiform seabirds have available to them a range of information based upon vision, audition, and olfaction which could perhaps be used in a hierarchical or redundant manner to locate their nest burrows at night.

6.10 Conclusion: Birds in the Dark–Complementary and Partial Information

This chapter has covered a wide range of activities conducted by birds at low light levels. Many activities are concerned with foraging and involve the detection of a wide range of different foods including small mammals, reptiles, amphibians, invertebrates (buried in substrata, on the surface, and in the air), and fruits. In Chapter 7, the taking of fish and invertebrates underwater at night-time light levels will also be discussed. In addition to foraging, activities at night also include reproductive behaviours and migratory movements. Clearly, birds do a lot at night. Activity during night-time is an essential part of the life of many birds. In this chapter, it has also been shown that the sources of information that may be used to guide these activities include the telereceptive senses of vision, audition, and olfaction, and also the close-quarter senses that provide tactile and taste information. In all instances of nocturnal activity, vision plays only a limited role, mainly because spatial resolution is low.

It is clear that the information from a single sense is insufficient to account for the behaviour described in every instance of nocturnal activity. The use of information from multiple senses must, of course, apply to many behaviours during daytime. However, considering in detail what birds do at night has demonstrated explicitly that the sensory ecology of birds involves trade-offs and complementarity between different types of information. It has also been shown how cognition often plays a key role in the interpretation of minimal or partial information so that sophisticated and complex tasks can be accomplished. This cognition enables minimal cues, especially spatial information about objects in a specific environment, to be interpreted sufficiently well to allow mobility. Information derived from other senses allows specific targets, usually food items, to be located and identified.

Clearly, these 'nocturnal birds' eye views' are far removed from a general notion of all-seeing eyes that are able to take in information about objects both near and far and integrate it into a whole scene. In many birds that are active at night, the information available is perhaps best considered to be a patchwork occurring at quite different spatial scales, a patchwork of information that is partial or lacking in detail. It is the integration of this complementary and partial information that allows birds to complete complex behaviours at night. There are no simple 'super-sense' explanations of nocturnal activity. It is true that senses in nocturnal birds are often close to the theoretical limits of their sensitivity, but it is also true that the information that they provide is often insufficient to meet all the challenges of nocturnal environments. Far more fascinating is how partial information from different senses is integrated, and often used only in specific locations, to enable nocturnal birds to complete their lives.

7

Birds Underwater: A Paucity of Information

A large number of bird species feed exclusively by diving for food. Many others feed on food taken at or just below the water surface. Some of these birds could have been considered in Chapter 6 because they forage both underwater and occasionally at night, or they dive to such depths that even during the day they experience night-time light levels when foraging. Despite the large number and diversity of species that forage underwater very little is actually known about their foraging behaviour or how they find their prey. Some birds habitually forage at depths which cannot be reached by humans without the use of elaborate diving technology, and bird dive-depths and their possible foraging behaviours are deduced from automatic recording devices attached to the birds, rather than through direct observation.

All of these birds are amphibious; they forage exclusively underwater but all other aspects of their life cycles are carried out in air—in the terrestrial environment. This means that they have to gain sensory information in two quite different media which impose different limitations on the information that is available. Going through the water surface poses a number of significant sensory problems for a terrestrial animal.

First, going through the water surface changes the optical properties of the eye, and it has been assumed that an ideal amphibious eye should be able to function as well in water as it does in air. Whether this is the case is unclear. Second, light levels immediately fall and continue to do so with increasing depth, and the spectral composition of the ambient light also changes with depth. Third, because birds are air breathers, the majority of divers stay down for only short periods, just a few minutes at most. This means that they transition very rapidly between different light levels both as they dive and as they travel back to the surface. Such rapid transformation in light levels is unlikely to allow sufficient time for the process of retinal dark adaptation to occur (Warrant 2008). Fourth, many diving species forage during night-time, compounding the loss of light.

Each of these sensory challenges will be discussed. Possible solutions to them will be illustrated by case studies. It is clear there is no one solution to foraging

The Sensory Ecology of Birds. Graham R. Martin: Oxford University Press (2017).
 DOI 10.1093/acprof:oso/9780199694532.001.0001

underwater. Before considering these solutions, a brief survey is presented of the species which forage underwater and the prey that they take.

7.1 The Underwater Foragers

Diving birds can be subdivided into five main groups based upon the prey that they seek (Martin and Crawford 2015).

1. Surface-diving species that take sessile prey. Prey are typically hard-shelled molluscs which have to be located and removed from substrata at depths down to about 50 m. All birds in this group are ducks (Anatidae) and include Greater Scaup *Aythya marila*, Long-tailed duck *Clangula hyemalis*, and Eiders *Somateria* spp. The range of species and their foods are summarized in Cramp and Simmons (1977).
2. Surface-diving species that take evasive prey (fish) from the water column at a range of depths down to 300 m, mainly in open pelagic waters. These include the penguins (Spheniscidae) and auks (Alcidae) (see e.g. Hedd et al. (2009) and Williams (1995)).
3. Species which dive from the surface, to prey upon evasive fish, in predominantly shallow depths, usually less than 10 m, and typically in coastal waters but sometimes in estuaries. The most prominent group that feed in this way are the divers (loons) (Gaviidae) (Carboneras 1992; Cramp and Simmons 1977).
4. Species which take evasive prey at shallow to mid-depths from surface dives with the prey often disturbed from substrata or hiding places. This group includes all cormorants and shags (Phalacrocoracidae) and grebes (Podicipedidae). This foraging may take place both in relatively clear and in turbid coastal and inland waters.
5. Species which take either evasive prey (primarily fish) and/or slow moving prey (such as crustaceans and molluscs such as squid) following plunge dives to shallow depth, although sometimes individuals may also pick prey from the surface. Among these species are the pelicans (Pelecanidae), gannets and boobies (Sulidae), albatrosses (Diomedeidae), petrels and shearwaters (Procellariidae), northern storm petrels (Hydrobatidae), tropicbirds (Phaethontidae), and diving petrels (Pelecanoididae).

This summary shows that feeding underwater involves a wide range of bird species, different prey types, and different foraging techniques. Foraging in each of these different ways is likely to pose different sensory challenges for the detection and capture of food items. Thus underwater foraging is far from a uniform task. Some foraging would seem to be more exacting in that it involves the detection of mobile prey which may have evasive strategies to avoid capture. Other foraging poses less exacting sensory challenges in that it involves sessile foods but they have to be detected and then detached from a surface. Although detection of sessile items may be challenging, once found they do not have to be pursued and caught. The foraging of the diving petrels differs from all the other groups listed in that prey is often, perhaps always, detected from above the water surface, as the birds fly over or sit upon the water surface, whereas in all other types of foraging the actual search for food does not begin until the birds are underwater. Also, for this last group, prey items may be detected either as individual items or as concentrations. Indeed, as described in Chapter 3, among the petrels, shearwaters, and storm petrels (Procellariidae, Hydrobatidae), olfactory cues can play a key role in the detection of profitable feeding areas in the open ocean.

7.2 Optical Challenges of Foraging Underwater

It was seen in Chapter 2 that all vertebrate eyes, including the eyes of all birds which forage underwater, employ the same optical structure (Land and Nilsson 2012; Martin 1983). With this same basic structure, eyes have evolved to function primarily in air (terrestrial eyes) or to function primarily in water (e.g. the aquatic eyes of fish and octopuses). A particular problem arises when an eye is required to function in both media. Thus a terrestrial eye immersed in water will lose the refractive power of the cornea and will become long sighted (Figure 7.1). The eye ball is now effectively too short and the image no longer falls on the retina but is in effect focused behind the eye, and so the image on the retina is blurred. We experience this loss of the cornea's refractive power and the blurred image that results when we open our eyes underwater. It can be overcome by the use of a face mask which is simply a device for maintaining the air–water interface at the cornea. The reverse happens when an aquatic eye enters air; now the cornea comes into play and the image is focused in front of the retina, again resulting in a blurred image. The size of the refractive changes and hence the degree of defocus when a terrestrial eye enters water will depend upon the relative contribution of the cornea to the total refractive power of the eye. A flatter cornea is less powerful in air than one that is more highly curved and so its loss upon immersion will be less severe.

The loss of corneal refractive power on entering water has two other important consequences: the brightness of the image is reduced because the effective size of the entrance pupil becomes smaller (in air the cornea magnifies the size of the real pupil), and the limits of the visual field of each eye will decrease. The extent of these changes upon immersion is a function of the refractive power of the cornea in air. The more powerful (more highly curved) the cornea, the greater will be these changes upon immersion.

One adaptation which may have evolved to reduce the magnitude of these changes in the eyes of amphibious species has been relatively flat (hence low refractive power) corneas.

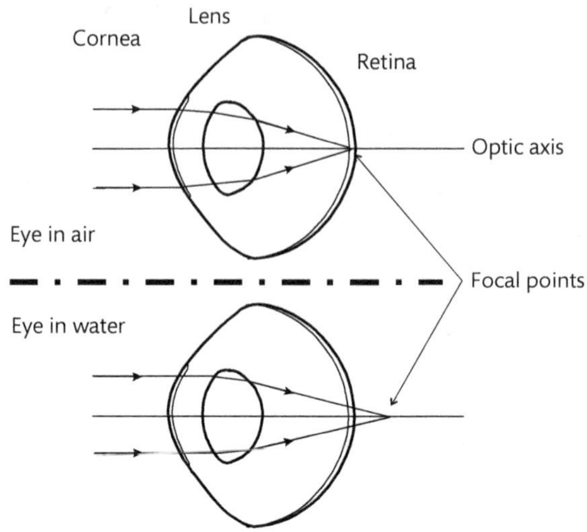

Figure 7.1 The effects of immersion on focus. An eye which is well focused in air produces a sharp image of a distant object on the surface of the retina. This focusing is achieved by the combined optical powers of the cornea and the lens (Figures 2.1 and 2.11). The cornea gets its optical power because it is a curved surface that separates media of different refractive indices, in this case air and the fluid which fills the chamber of the eye. The refractive index of the fluid that fills the eye is almost the same as water. Thus, upon immersion, the cornea no longer separates media of different refractive indices and so it loses its refractive power. The result is that the focal length of the immersed eye is determined by the lens alone and the image becomes focused behind the back of the eye, producing a blurred image on the retina.

Sivak (1976) pointed out that penguins seemed to have flatter corneas and argued that this could be viewed as an adaptation to reduce the detrimental effects of immersion. However, this observation did not take account of the scaling effect of eye size on the power of the cornea (Martin 1985). In effect, in a larger eye, the power of the cornea needs to be lower. If eye size is taken into account, penguins' eyes may not have corneas flatter than might be predicted based upon their size alone. This has been noted in some albatross species (Martin 1998) and in some penguins (Martin 1999a; Martin and Young 1984). It is clear, however, that for some of the smaller-eyed amphibious bird species, for example, ducks, auks, grebes, and cormorants, the corneas are relatively highly curved and the loss of their power upon immersion will have marked effects upon focus, image brightness, and visual field size. In the case of the latter, it has been shown that in Great Cormorants and Humboldt Penguins *Spheniscus humboldti*, binocular vision may be abolished when the eyes are immersed; this is a consequence of the narrowing of the fields in each eye so that they no longer overlap (Martin and Young 1984; Martin et al. 2008) (Figure 7.2).

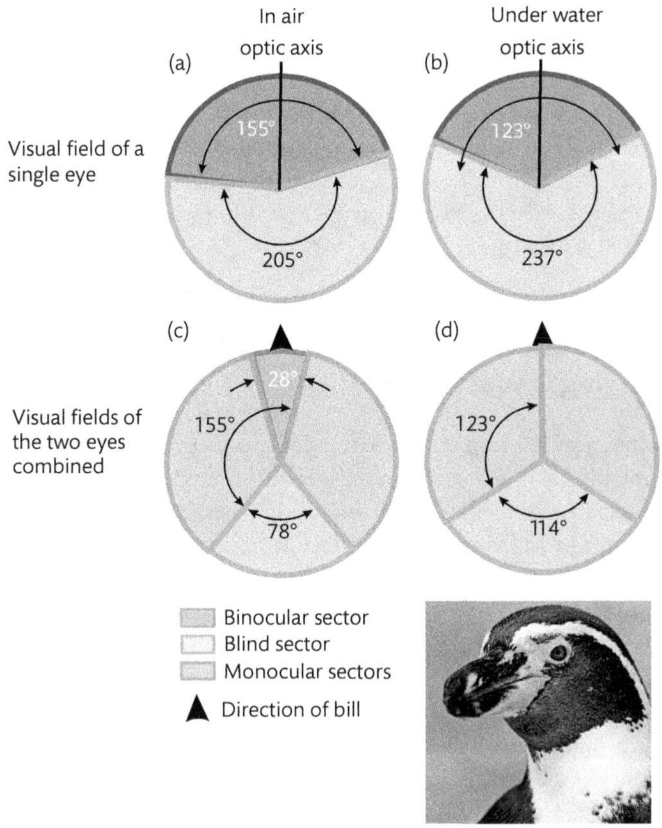

Figure 7.2 The effect of immersion on visual fields. A further effect of the loss of the refractive power of the cornea upon immersion is that the visual field of an eye narrows. In the case of the eye of a Humboldt Penguin *Spheniscus humboldti*, analysis of its optics showed that in air the visual field of each eye is 155° (a) but upon immersion the field shrinks to 123° (b) (Martin and Young 1984). In air, the visual fields of the two eyes combine to produce a binocular field that is 28° wide (c). The effect of immersion is to abolish the binocular sector, although the margins of the eye will just meet, so visual coverage of the frontal field is maintained (d). However, the width of the blind sector behind the head increases by nearly 40°. Eye movements that might achieve some degree of binocular overlap underwater would result in an even larger increase in the size of the blind sector behind the head. Photograph of Humboldt Penguin courtesy of Santa Barbara Zoo/David Orias.

Evidence has been presented which suggests that some amphibious birds can compensate for the loss of the cornea by dramatically increasing the power of the eye lens by significantly increasingly the curvature of its front surface (Sivak 1978). However, this effect could be an artefact of the drugs used to induce changes in the eyes of the species (Hooded Merganser *Lophodytes cucullatus*) in which this was shown (Levy and Sivak 1980) and may not occur in nature. Such a mechanism was first hypothesized by Hess (1909) in Great Cormorants and there is evidence that their vision is well corrected underwater (Katzir and Howland 2003), suggesting that they can in some way compensate for the loss of corneal power upon immersion. However, as discussed in 7.7.1, there is good evidence that spatial resolution underwater in Great Cormorants is relatively low and so while they may have the ability to compensate for the loss of corneal refraction when diving it does not support high spatial resolution.

7.3 Light Levels and Spectral Challenges of Foraging Underwater

Entering water results in an immediate decrease in the levels of ambient light. Natural light falling on the water surface is reduced immediately because of reflection. We appreciate this readily by the high brightness of light reflected back from any water surface. In addition to this initial loss of light, water absorbs and scatters light such that at the depths at which many bird species forage, the attenuation of light is significant. This effect is seen in the clearest of natural waters (Tyler and Smith 1970) with the result that at a depth of 200 m (a depth frequently reached by penguins and auks), even in pure waters, irradiance would be decreased about 100-fold (Figure 7.3).

Scattering of light within the water column also results in significant changes to the distribution of light. Below water, but close to the surface, light appears to come from directly above and the direction of the sun can be clearly determined. However, with increasing depth, the distribution of light becomes less clearly delineated as coming from above because of light scattering, and by 40 m below the surface, the position of the sun may not be discernible.

In addition to these reductions in irradiance, there is a very marked differential spectral absorption of light with increasing depth. Again, this is a property of the clearest natural waters, such that by 200 m deep, the available light becomes noticeably blue to the human eye. This indicates that there has been selective absorption of light at longer wavelengths and also at shorter (violet and ultraviolet) wavelengths.

Thus, the ambient light becomes increasingly narrow in its spectral distribution and eventually is centred at wavelengths 420–430 nm (Figure 7.3). These light conditions in themselves pose a significant perceptual challenge for any animals which forage below the water surface. They are particularly challenging for amphibious species as they inevitably face quite different light environments above and below the water surface and, because they tend to dive rapidly, they typically will experience rapid changes in the perceptual challenges that they face.

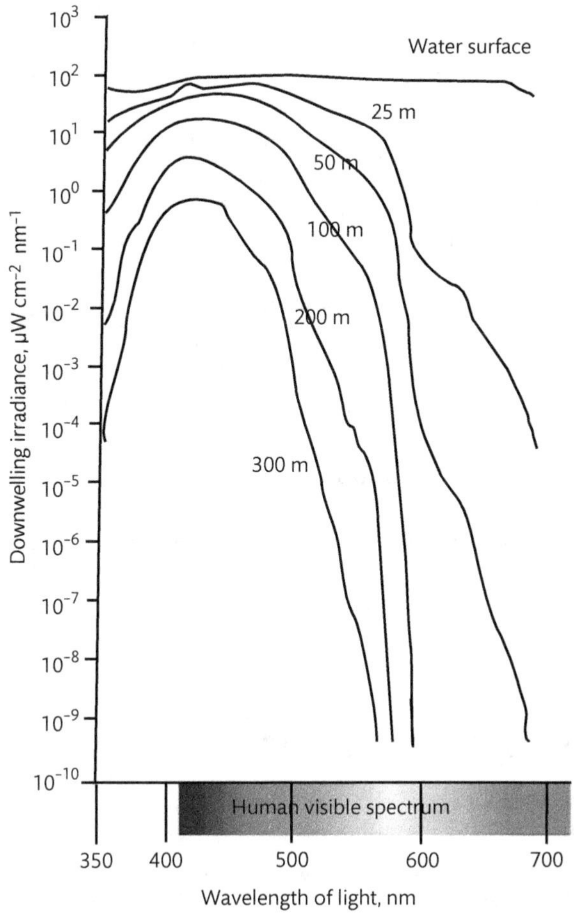

Figure 7.3 Light levels and spectral changes underwater. Water absorbs light selectively in the spectrum. This selective absorption has the effect that the broad spectrum of sunlight falling upon the surface is progressively narrowed, and the level of the downwelling irradiance decreases with depth. The diagram shows the effect of this based upon field measurements made in the deep pure waters of Crater Lake, Oregon. With progressive increase in water depth, the spectrum of the downwelling light becomes increasingly narrow and its peak shifts from the blues to the deep violets of the human visible spectrum. Many seabirds such as auks dive to depths of about 100 m (Regular et al. 2010) and some penguins regularly dive to depths close to 300 m (Kooyman et al. 1992). Redrawn and modified from Tyler and Smith (1970).

The effects described above are a property of pure water but this rarely occurs in nature. Pure waters tend not to be foraging environments because they are nutrient poor and would not support significant number of organisms that can be foraged for. In most waterbodies, and certainly in most foraging waters, the light environment is more complex due to the presence of suspended matter (living organisms, products of decay, dissolved and suspended minerals) which serve to scatter light even more and add other selective spectral filtering effects. The cumulative effects of these are that light will be attenuated at depths of 100–200 m by much more than the 100-fold of pure water, and the spectrum of light may have its peak shifted slightly to longer wavelengths, between 420 and 500 nm (blue-green part of the visible spectrum) with very little, or no light, at shorter wavelengths and also none at wavelengths beyond about 550 nm (Lythgoe 1979).

There are extensive data on the light environments of naturally occurring freshwater types, and these can be highly variable in colour due to dissolved materials and the presence of simple organisms (Lythgoe 1979). A comprehensive description and classification of natural marine water types was produced by Jerlov (1976). He showed that, on a global scale, natural water types vary significantly in spectral absorption and attenuation, especially between pelagic and coastal situations. The result is that the light environments at different depths, and at the same depth in different locations, vary markedly in both irradiance and spectral distribution.

7.4 The Challenge of Rapidly Changing Light Levels

In the nocturnally active terrestrial species mentioned in Chapter 6, the retinas should have sufficient time to adapt to the lowering ambient light levels as twilight progresses. This is because light levels change at a relatively slow and steady rate as the sun descends below the horizon (Figure 6.1). Furthermore, many nocturnal species remain in the dark or in well-shaded locations all day and so are already partially dark adapted when they emerge at dusk. The time course of the process of dark adaptation in vertebrate retinas is slow, taking up to 40 minutes for full adaptation (Warrant 2008), and it is clear that such a time course occurs in birds (Blough 1956). This slow rate of adaptation is sufficient to keep pace with the rate at which natural light levels decrease anywhere on the globe. The result is that although spatial resolution decreases as twilight progresses (Figure 2.13), the retina is able to maintain optimal sensitivity and resolution as light levels fall. Entering rapidly into a dark environment, however, does not allow the retina to maintain optimal adaptation. This has recently been explored in the case of cavity-nesting birds (Maziarz and Wesolowski 2014; Wesolowski and Maziarz 2012). It has been shown that when entering a natural cavity a bird may transition from bright sunlight to dark cavity interiors in less than a second and that light levels can fall by 10,000-fold. There is no information as to how these birds' eyes can cope with these rapid changes, and it may be that on entering a cavity their vision is simply not well adapted to the darkness. The birds may be able to detect only very low spatial detail, and they rely primarily upon information from hearing and tactile cues to feed young or attend to their eggs.

A bird that dives rapidly from the water surface may face a similar, though less dramatic, problem to the cavity-nesting birds. It is, in fact, a problem that we all face in the modern world in which artificial light sources are ubiquitous. These allow us to transition rapidly between areas with very different ambient light levels. For example, spaces are often flooded at night with the light of powerful 'security' lighting. Although we may feel secure when in that light, we are taken aback when peering beyond the pool of light since we find that little can be seen. In effect, security lights can create 'menacing dark areas' adjacent to them, menacing because as we enter the dark areas very little spatial information can be retrieved. This effect is, in part, the result of inadequate dark adaptation. If we allowed our eyes to follow their natural pattern of dark adaptation, we would find that the menacing dark spaces are not so devoid of information—objects can be detected. Although resolution would not be as high as in the pool of bright light, we could see adequately to guide many of our actions, but it takes time to dark adapt. Diving birds, however, do not have the opportunity to take their time. As air breathers, their dives must be of short duration and descent can be rapid to regions in which light levels are low. An example of this problem is illustrated by King Penguins *Aptenodytes patagonicus.*

King Penguins dive during daytime and night-time. During daytime, they reach depths of between 100 m and 300 m in 1–4 minutes (Kooyman et al. 1992), and at these depths light levels fall to within the nocturnal range. Thus the birds are, from a sensory perspective, nocturnal foragers at all times: either they forage at night at relatively shallow depths (30–40 m) or they forage during the day at night-time light levels at greater depths. King Penguins would thus seem vulnerable to a similar problem faced by the cavity-nesting birds or faced by ourselves when we dodge between lit and unlit areas at night. However, whenever they forage, King Penguins are able to catch small fish (Olsson and North 1997). How they can achieve this is discussed in 7.7.2.

7.5 Aquatic Foraging and Nocturnal Foraging

Many of the species listed in 7.1 forage during daytime, but many also forage regularly during twilight or night-time (see e.g. Camphuysen 1998; Cramp and Simmons 1977; Grémillet et al. 2005; Nilsson 1969; Piersma et al. 1988; Regular et al. 2010; Regular et al. 2011; Systad and Bustnes 2001; Wanless et al. 1999). Furthermore, as explained above, even when foraging in daytime a bird that dives to in excess of about 50 m will almost certainly experience twilight or night-time light levels, depending upon the turbidity and amount of dissolved material. Thus, some taxa, especially penguins and auks, may regularly forage during the day at depths at which ambient light levels equal those commonly experienced at the water surface under twilight and moonlit night-time conditions (Hedd et al. 2009; Martin 1990a; Martin 1999a; Regular et al. 2010; Regular et al. 2011; Wilson et al. 1993). The explanation for diving to these different depths during daytime and night-time is that auks and penguins are following the diel vertical migrations of their preferred fish prey species. These rise to shallower depths at night and dive deeper during the day (Regular et al. 2010; Regular et al. 2011; Wilson et al. 1993). Thus, from the sensory ecology perspective, these birds may be regarded as twilight or nocturnal foragers, either foraging at night at shallow depths, or if foraging during the day they do so at depths where night-time light levels occur (Martin 1999a). Thus a diving bird may routinely experience the same general visual challenges faced by terrestrial nocturnal foragers.

As described in Chapter 6, the perceptual challenges of nocturnal foraging in terrestrial species have led to the evolution of a wide range of adaptations of vision, the recruitment of non-visual senses in the location of prey, and the evolution of specific behavioural strategies for the conduct of specific tasks. Aquatic foragers not only have the challenge of having to be able to see in both air and water, but in many instances they also forage at nocturnal light levels while completing most other aspects of their lives in daylight, and they frequently alternate between these two sets of perceptual challenges.

7.6 Tactile Information and Underwater Foraging

The above analysis shows that underwater the use of vision to detect objects faces clear challenges and limitations. A reliance upon high spatial resolution that can be used for the detection of objects at a distance is unlikely to underpin the foraging of most amphibious birds. Tactile cues are an obvious source of information that could supplement vision under low light levels as has been shown in the ducks and shorebirds that forage at night (Chapter 6, 6.9.1).

Some ducks have a bill tip organ that is probably capable of making fine tactile discriminations of objects held at the bill tip (Chapter 4, 'Bill tip organs in waterfowl'). The possibility that tactile information is used to detect specific prey items underwater is exemplified by investigations of foraging in Blue Ducks, an endemic and endangered species of New Zealand. It seems likely that these birds use tactile cues when foraging underwater in certain water conditions or trying to retrieve prey items from particular locations. Blue Ducks presently inhabit forested headwater catchments of rivers with medium to steep gradients, providing well-oxygenated and fast-flowing water, although they probably had much wider distributions and occupied a broader range of river habitats before the arrival of humans and the introduction of mammalian predators (Collier et al. 1993; Kear and Burton 1971; Wilson 2004).

Blue Ducks feed on a wide variety of aquatic invertebrates (Collier 1991; Veltman et al. 1995) and occasionally on small amounts of plant material (Marchant and Higgins 1990). The species specific name *malacorhynchos* means soft bill and this refers to the presence of a specialized structure either side of the tip of the maxilla (Figure 7.4). This consists of a pair of flexible flaps that overhang the mandible when this bill is closed. These flaps are relatively simple structures with a heavily keratinized epidermis containing a small number of Herbst's corpuscles (Chapter 4, 'Herbst corpuscles') mainly confined to the parts nearest the point of attachment to the maxilla (Kear and Burton 1971). These tactile receptors may function to detect when the flaps are deformed as they touch or move over a surface

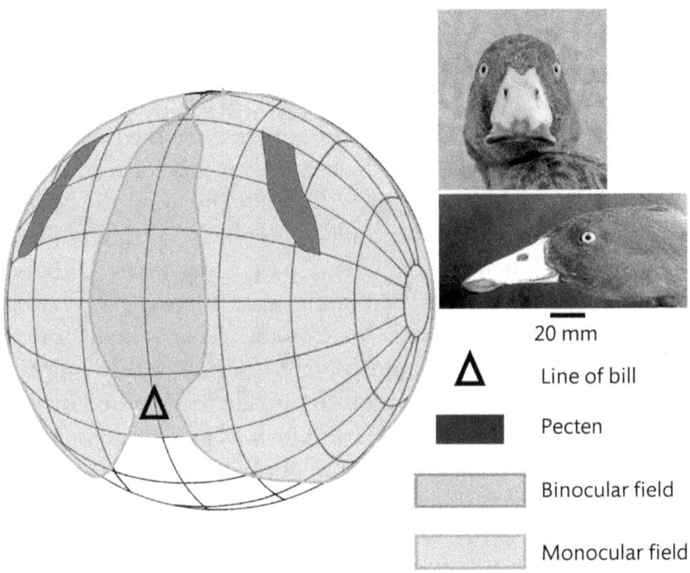

Figure 7.4 Vision in Blue Ducks *Hymenolaimus malacorhynchos*. These birds prefer to feed in fast-flowing rivers where they take invertebrates anchored to rocks surfaces and, at certain times of the year, larvae which are free swimming in the water column. Detection of the anchored prey is probably achieved by tactile cues from receptors situated in the flaps of keratin that characterize these birds' bill tips. Such foraging is probably done without visual cues. However, taking small prey items from the water column is a foraging task that probably requires visual guidance of the bill and the detection of prey items in silhouette against the downwelling light. The broadening of the binocular field above the bill may aid such prey detection. These visual fields were measured in air and so the binocular field may be considerably narrower when the birds are foraging below the water surface (see Figure 7.2). Redrawn with modifications from Martin et al. (2007a).

and probably serve to extend the region of tactile sensitivity from just around the bill tip to along the sides of the bill.

Studies of Blue Ducks' foraging behaviours and diet demonstrate that foraging occurs within specific underwater microhabitats and that their invertebrate prey which are attached to rock surfaces are unevenly spaced, suggesting that the birds cannot forage by simply grazing over rock surfaces but have to detect individual items (Collier 1991; Collier 2004; Veltman et al. 1995; Veltman and Williams 1990). Blue Ducks exploit small, sessile chironomid and cased caddis fly larvae, which can be scraped from rock surfaces and tactile information from the bill is likely to be the primary cue used to guide this behaviour. Hence the birds can 'forage blindly' on, between, and beneath rocks for these prey types, exploring with their bill into crevices that cannot be seen into. Visually guided foraging, however, is presumably required when Blue Ducks exploit their larger and more nutritious prey of swimming mayfly and stonefly larvae. These prey are mobile within the water column.

It has been shown that Blue Ducks have more frontally placed eyes than ducks, such as Mallards, Shovelers, and Pink-eared Ducks, which feed mainly guided by touch and taste cues (Martin et al. 2007). These birds have eyes that are positioned well on top of the head and the birds can only just see their own bill tips, but they do gain comprehensive vision of the hemisphere about the head (Guillemain et al. 2002; Martin et al. 2007) (Figure 7.5). Blue Ducks on the other hand can see their own bill tip and thus vision could be used to guide bill position and bill opening when approaching individual items in the water column. This visual field configuration is similar to the situation in Eurasian Wigeons *Anas penelope* which feed by selective grazing (Guillemain et al. 2002). Thus it seems that the more frontal eyes of Blue Ducks and the resultant increase in binocularity, coupled with their narrow tapering bill, may function to provide the precise visual control of bill position necessary for the capture of certain of its prey from the water column (Martin et al. 2007). However, the presence of Herbst's corpuscles in the flaps of skin attached to the edges of the maxilla suggests a tactile sensory function, and information from these serve to allow the birds to detect sessile prey hidden in crevices, or in turbid waters, or in other situations when prey items cannot be seen.

7.7 Solutions to Underwater Foraging

It cannot be concluded that even if underwater foragers are visually guided, that they are capable of detecting prey at a distance. It seems likely that even at high light levels prey may be detected and pursued at close range only. The main evidence to support this conclusion comes from research into the underwater vision of cormorants and observations of the foraging behaviour of both cormorants and shags Phalacrocoracidae.

Figure 7.5 Vision in Pink-eared Ducks *Malacorhynchus membranaceus*. A number of duck species forage exclusively at the water surface using tactile cues and a filtering mechanism to extract small items from water pumped into the mouth and out through filter structures at the side of the bill (see Figure 1.3, Northern Shoveler). Pink-eared Ducks feed in this way and this is reflected in their visual fields. Unlike the Blue Ducks (Figure 7.4), the bill falls at the very periphery of their visual field (c) and like Woodcocks (Figure 6.10) they gain comprehensive visual coverage of the hemisphere above the head with a narrow binocular field that extends through 180° from just above the bill to directly behind the head. The high position of the eyes in the skull is clearly seen in both front (a) and back (b) views of the head. Modified from (Martin et al. 2007a).

7.7.1 Cormorants

The fish prey of Great Cormorants *Phalacrocorax carbo* (Figure 7.6) is often hidden, highly cryptic, or of low contrast, especially when the birds are foraging in the turbid waters of estuaries and rivers. Foraging does not take place at great depth and dive duration is relatively short at 30–40 seconds. Some populations forage at night in the middle of winter at high latitudes (Grémillet et al. 2005). In spite of these apparently testing conditions, cormorants are credited with being the most efficient aquatic foragers of fish, measured as grams of fish captured per minute underwater (Grémillet et al. 2004), and they can forage efficiently in highly turbid waters (Grémillet et al. 2012). Cormorants usually forage alone although in some locations they may forage co-operatively and apparently force aggregations of fish prey towards the surface (Martyniak et al. 2003; Orta 1992).

The most detailed information on vision underwater in birds is on the underwater acuity and visual fields of Great Cormorants (Martin et al. 2008; White et al. 2007; White et al. 2008). Behavioural measures of acuity showed that resolution is surprisingly low. It is in fact about 40 times lower than an eagle and 3 times lower than a Rock Dove (Table 2.1 and Appendix 1). The highest acuity underwater in cormorants is, in fact, very similar to the highest acuity achieved underwater by young humans without a face mask (Figure 7.6). Using these acuity data, combined with data on how resolution is affected by stimulus contrast (Figure 7.7), it has been possible to model 'a cormorant's eye view' of typical prey. The cormorant eye views include prey at different distances and with different degrees of contrast (Figure 7.8). These show that only at very close distances (< 1 m) can a fish be seen in any detail; beyond this distance a prey fish will appear no more than a faint blur. Furthermore, most typical fish prey will have very low contrast against the background and among the fish known to be taken by cormorants are sculpins which are renowned as examples of cryptic body shape and skin patterns in fish.

The answer to how cormorants can so efficiently detect and take fish seems to lie in their use of a predatory technique which can be guided by low spatial resolution vision. This technique can be characterized as a 'flush-foraging close-capture'. It seems that foraging cormorants force hidden fish to make an escape response—they flush them out. Prey may be hidden among rocks or tree roots, or sheltering on or within different substrate types. This provoking of a fish into making an escape response is achieved by the bird poking with its bill into nooks and crannies, or at the substrate, with darting movements, a technique that has been directly observed in foraging shags, poking with their bills into sandy substrates (Watanuki et al. 2007; Watanuki et al. 2008). Thus, prey is forced to escape, and it is then detected and taken at close range, with its capture probably achieved by rapid neck-extension in a fashion similar to the rapid neck-extension prey capture technique used by herons (Ardeidae) (White et al. 2007). Herons take evasive prey through a water surface or on the ground by stealthy approach, or by sitting and

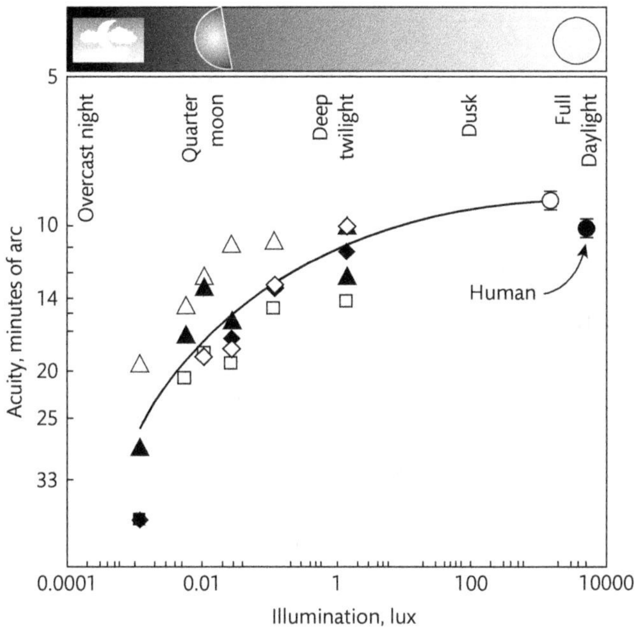

Figure 7.6 Visual acuity in Great Cormorants *Phalacrocorax carbo* underwater. Acuity was measured in five birds each indicated by a different symbol in the diagram. At high light levels, the birds showed very similar acuity of about 10 minutes of arc, but as light levels decreased acuity differed between the birds. For example, the visual acuity of the bird indicated by the open triangle symbol was always higher than that of the bird indicated by the open square symbol. The line is the best fit through all of the data points and shows clearly how acuity decreases with falling light levels. The data point for humans is the average acuity of European children underwater without the aid of face masks; the best performance of cormorants is very similar. However, underwater acuity twice as good as cormorants (5 minutes of arc) has been found in children of the Moken people of Southeast Asia who dive to collect food from the sea-floor without the use of visual aids (Gislen et al. 2003). Illustrations by Craig White and Graham Martin, published in White et al. (2007b).

waiting for prey to come close, followed by very rapid neck-extension to capture what are highly evasive prey. Typically, herons have only one opportunity to capture an individual prey animal or it will, in most instances, have escaped for good. Such single opportunity capture is also likely to be the case in cormorants.

Although cormorants can travel rapidly underwater, they are also highly manoeuvrable and when investigating objects can turn in tight circles, usually holding the neck retracted, but the neck can be extended rapidly to seize the escaping prey. It seems highly likely (given the range of low light conditions, the turbid conditions of some waterbodies, the cryptic nature of prey, and the low acuity of

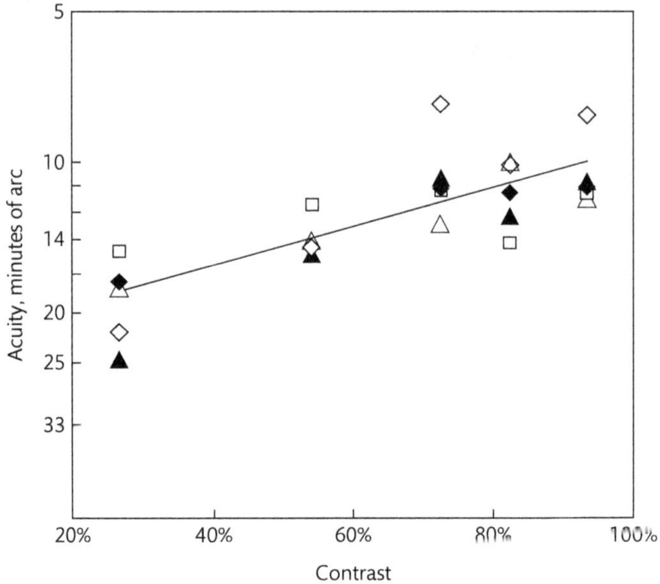

Figure 7.7 The influence of contrast on the underwater vision of Great Cormorants *Phalacrocorax carbo*. This is indicated here by how acuity falls as the contrast of the target changes from high 'black-and-white' contrasts (close to 100% contrast) to low contrast when the target contains only different shades of grey (20–40% contrast). This figure shows that in Great Cormorants their acuity underwater almost halves from a best resolution of about 10 minutes of arc when contrast is high, to 20 minutes of arc when contrast is low, even though the ambient light level has not changed. (Illustrations by Craig White and Graham Martin, published in White et al. (2007b).)

the birds) that cormorants do not identify the prey item—they simply lunge at an 'escaping blur'. Of course, if the blur is escaping it is highly likely to be edible food. Cormorants typically bring their captured prey to the surface in the bill. This may allow identification of the caught object as well as placing it into position for swallowing. The visual fields of cormorants allow the birds to see what is held between the mandibles. Cormorants can see between their opened bill (Figure 7.9), whereas the vision of most birds begins beyond the bill tip (Martin et al. 2008).

The above foraging technique allows the birds to capture prey when none can be seen initially and to catch it at short range as it makes an escape response. This is not to say, however, that the birds will not take prey that is swimming freely in the water column, if it can be detected when waters are clear and light levels relatively high. Even so, their low acuity will only allow the birds to detect relatively large prey at distances of a few metres (Figure 7.8). The habit of co-operative foraging reported in some cormorant populations (Martyniak et al. 2003; Orta 1992) may have the function of concentrating fish close to the surface when they can be seen,

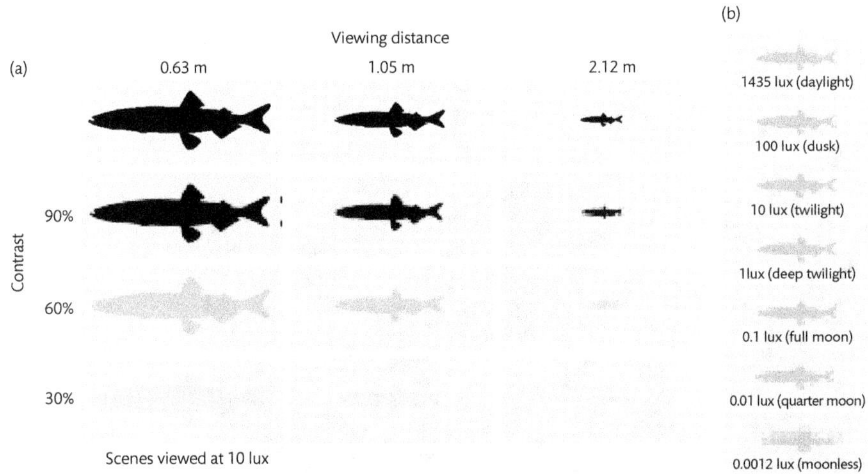

Figure 7.8 Simulations of a Cormorant's eye view. These views are based upon the kind of data shown in Figures 7.6 and 7.7. Images are shown of how a fish of the same size and shape (exemplified here as a silhouette of a 10 cm long Capelin *Mallotus villosus*, a species commonly taken by Great Cormorants in coastal waters of the Atlantic Ocean) will appear to a Cormorant at three different viewing distances and under four different levels of contrast. In (a) the light level is held constant at the lower end of the day light range. The simulated view shows that apart from the high contrast fish viewed at close range, all target fish will appear quite indistinct. In (b) the fish has a realistic mid-range contrast of 60% and is viewed at a distance of 1 m under a range of naturally occurring light levels from daylight to low moonlight. Again the target fish appears as an indistinct blur for many naturally occurring conditions. (Illustrations by Craig White and Graham Martin, published in White et al. (2007b).)

possibly *en masse*, in silhouette against downwelling light. Individual fish do not have to be picked off, rather the birds can lunge into an aggregation of fish.

A further caveat regarding the flush-foraging close-capture technique is that for it to work efficiently (a high rate of grams of fish captured per minute underwater), prey must occur at a reasonably high density. However, what that density is needs investigation. Also, the technique may have particular utility at low water temperatures when the escape response of fish is slowed. This may be part of the key to the successful foraging of cormorants at high latitude coastal waters around Greenland where highly cryptic sculpins form a large proportion of their diet (Grémillet et al. 2004).

The flush-foraging close-capture technique, rather than the pursuit of individual fish through the water column, could well be one employed by other birds that forage underwater. These might include the divers and grebes, since they typically forage at shallow depths where they can disturb prey hiding or resting on substrates. There is evidence that grebes may disturb fish which are driven towards the surface where they could be detected in silhouette from below (Piersma et al. 1988).

Figure 7.9 Seeing between the bill. Analysis of the vision and foraging technique of Great Cormorants suggests that they may often not know what they are catching (Figure 7.8). Holding the item in the bill and bringing it to the surface for visual inspection prior to ingestion is likely to be important. Measurements of the visual fields of cormorants show that their bill falls within the binocular portion of the frontal field and that the eyes are positioned so that a bird can see between its opened mandibles and inspect what it is holding.

7.7.2 Penguins

The close-quarter flush-foraging technique cannot apply to the foraging of species that take fish prey from the water column. The prey has nowhere to hide or settle from which they can be disturbed by the foraging bird. Rather the fish must be detected in the water column. Of course, fish can hold position in the water column and be disturbed from there, but it seems likely that fish either have to be detected visually or encountered more or less at random. This seems to be the kind of foraging task that penguins face.

An indication that vision does play a role in the foraging of penguins comes from investigations of the eye structure, visual fields, and retinal photopigments in Humboldt Penguins (Bowmaker and Martin 1985; Martin and Young 1984) and from studies of the visual fields and eye structure of King Penguins (Martin 1999a).

The optical structure of the eyes of Humboldt Penguins shows features which suggest that its optical design is primarily shaped by the aquatic environment. The lens has a power of 100 dioptres (D) while the cornea has a power of only 29 D. This means that the cornea is relatively flat and a schematic model of the eye's optical system suggests that it is myopic in air but well focused (emmetropic) in water. This indicates that the eye actually has to reduce refractive power when the birds leave water to become well focused or the birds may simply be somewhat short sighted when on land (Martin and Young 1984). This is partly supported by evidence from refractive studies of the eyes of four species of penguins: King Penguins, Gentoo Penguins *Pygoscelis papua*, Southern Rockhopper Penguins *Eudyptes chrysocomes*, and African Penguins *Spheniscus demersus* which suggest that their eyes are emmetropic or myopic in air (Sivak 1976; Sivak and Millodot 1977).

A further clue to the role of vision in the foraging of penguins comes from the observation of the unusual iris and pupil in King Penguins. The eyes of this species are large, and the cornea has a diameter of 18 mm and a curvature of approximately 33 mm. This means that the cornea has very low refractive power in air (10 D). Therefore, its loss upon immersion is unlikely to alter appreciably the focus of the eye, so the eye may be considered relatively well focused in both air and water. The iris is unusual in that it can stop down to a virtual pinhole and also be dilated to almost the full diameter of the cornea. This is a large dynamic range for a pupil and it alters appreciably the brightness of the retinal image. In fact, the image in the eye is 300 times brighter with a fully dilated pupil compared with the image when the pupil is stopped down (Martin 1999a). The 300-fold range compares with the much smaller 16-fold range over which image brightness can vary due to pupil diameter change that occurs in the eyes of Rock Doves, and in humans (Marshall et al. 1973; Woodhouse and Campbell 1975).

The highly dynamic pupil size of King Penguins could serve two functions. First, by being able to close down to such a small diameter, the eye can become a virtual pinhole camera, in effect, by-passing the optics and producing a well-focused retinal image in air, thus overcoming any myopia. Second, the closed-down pinhole aperture would produce a very dim retinal image (a problem of pinhole cameras is that the image is dim and requires long exposure times, and hence they are best for photographing static scenes). This dim image would aid the dark adaptation of the retina when the bird is above the surface during daylight. Pinhole pupil apertures have long been recognized as having such a role in the eyes of nocturnal mammalian predators, such as cats, and in some reptiles (Miller 1979; Walls 1942). These animals typically have a well stopped-down pupil when exposed to the sun. In penguins, opening up the pupil aperture in the low light levels which the birds experience when they reach their foraging depth (100–300 m), would mean that the retina is already dark adapted. On a sunny day, at depths of between 100 and 300 m, light levels are in the range between twilight and minimum moonlight. The birds dive to these twilight levels in a little over 1 minute; therefore, any mechanism that ensures that the eye enters these low light levels already well adapted for them could be essential if vision is of use for foraging.

Thus, the highly dynamic pupil of a King Penguin could help to optimize retinal adaptation to low light levels which, in turn, optimizes the retrieval of spatial information at the low light levels in which the birds seek prey. Throughout a foraging session involving many dives, the retina could remain well adapted to the low light levels experienced at foraging depth by opening the pupil on descent and closing it on each ascent.

Of course, even with an adequately dark-adapted retina, spatial resolution is likely to be low at these twilight-moonlight light levels (Figure 2.16) and the foraging success of King Penguins at these depth may be explained also by the type of prey that is taken. Myctophids (lantern fish) are the principal diet of King Penguins (Kooyman et al. 1992; Olsson and North 1997). These fish are small (46–79 mm in length), but they have many photophores on their body surfaces. The fish are patchily distributed in shoals. To detect these fish, it seems likely that King Penguins must rely not upon detecting fish, but on detecting the photophores which indicate the presence of fish. Photophores are essentially point sources of light against a dark background and for the detection of point light sources it is the absolute size of the pupil aperture, rather than the f-number of the eyes, which determines sensitivity (Land and Nilsson 2012); this may be another reason why King Penguins' eyes have corneas and pupils of large diameter.

The ingestion rate of small prey items that are widely dispersed in the water column is high in King Penguins (Kooymand et al. 1992; Olsoon and North 1997). It seems likely, therefore, that they must search for their prey within a wide sector of the space surrounding them and King Penguins have a broad visual field that correlates well with this requirement (Martin 1999). The foraging task of King Penguins can perhaps best be characterized as requiring the location of pinpoints of light at any position about them. Such foraging is accomplished through the combination of large eyes whose retinas can be maintained permanently dark adapted during a foraging session.

7.7.3 Auks

Like penguins, auks also take fish from the water column and some species, notably Common Guillemots (Common Murres) *Uria aalge*, are known to forage at night and down to a depth of 150 m (Regular et al. 2010; Regular et al. 2011). Therefore, they face similar visual challenges to those described above for penguins, notably foraging at low light levels and the rapid transition during the daytime from bright sunlight to twilight light levels. Practically nothing is known about the eyes and vision of auks apart from a recent study of visual fields in two species, Common Murres and Atlantic Puffins *Fratercula arctica* (Martin and Wanless 2015). This study showed that the visual fields of the two species differ significantly, and it has been suggested that this is because of differences in the diets of the two species during the annual cycle. Thus, while Guillemots feed on small shoaling fish throughout the annual cycle, this prey may be less important to Puffins when away from breeding colonies, and they probably switch to a diet of much smaller prey items detected in silhouette near the surface.

The most intriguing suggestion about the foraging behaviour of auks is that Common Murres may be able to forage blind, relying upon random encounters with fish at depth. Thus, Paul Regular (Regular et al. 2010; Regular et al. 2011) has argued from careful monitoring of ingestion rates and energy expenditure that these birds can achieve sufficient food capture rates using random-encounter foraging, i.e. individual fish are not sought but just encountered randomly. This technique is viable if the fish prey is at a sufficient density in the water column, but the threshold density for successful random-encounter foraging is, in fact, relatively low (Regular et al. 2011). How a bird registers an encounter with a fish could be through vision at very close range (as in cormorants, the prey may appear as a moving blur), and/or simple tactile cues as a fish is struck, most likely by the bill. Clearly, random search and the encountering of prey distributed in the water column could be a technique employed by penguins. In fact, it could be used by any bird species which take fish from the open water column.

This is an intriguing possibility and has parallels with the feeding technique thought to be employed by nightjars as they trawl, perhaps almost blindly, for prey items at a certain density in open airspace (Chapter 6, 'Nightjars'). Such trawling is possible because nightjars forage in open airspaces that are devoid of natural obstacles. The same could be true of a number of underwater foragers which move within the open water column. Clearly direct or remote observation of these birds foraging within the water column, especially at low light levels, would be very valuable.

7.8 Birds Underwater: A Paucity of Information

Foraging underwater often means foraging at nocturnal light levels. Absorption and scattering of light by even clear water means that a bird that dives beyond about 50 m during daytime enters twilight and because of suspended and dissolved substances twilight may be reached at much shallower depths. Furthermore, many aquatic foragers are also nocturnal foragers. These birds must often be foraging in complete darkness. Unlike the situation of terrestrial nocturnal activity, there seems to be no scope underwater for the senses of hearing and olfaction to complement vision, although some birds that feed on sessile prey may rely upon tactile cues gained from the bill.

Spatial information from vision underwater, even at high light levels, is perhaps best considered to be partial, at least it is lacking in detail. From the perspective of a diurnally active terrestrial bird, underwater foragers would seem to suffer from a paucity of information. It is perhaps surprising, therefore, to find that underwater foraging by birds can be highly efficient. It is perhaps even more surprising that birds may rely upon random encounters with prey or be guided towards prey by the minimal cues of small points or spots of light from photophores viewed against a totally dark background. As in the case of nocturnal foragers, the key to aquatic foraging lies in the highly specialized diet and the prey capture technique. These allow prey to be taken guided by minimal or partial information. Birds' eye views below water are clearly very different from those in air. The same bird will experience radically different worlds and experience radically different perceptual challenges, as they oscillate between their life in air and their life below water. Their world view must change from one that is rich in information to one in which there is a paucity of information.

Unlike mammalian amphibious foragers, birds do not seem to have new sources of information that become available when they dive; the birds' world must become less rich in information when they dive. Pinnipeds (sea lions, fur seals, and true seals) possess a sophisticated tactile sense derived from vibrissae on the head, particularly around the mouth, which can be used in water for detailed and subtle tactile discriminations between objects, and also to detect disturbances within water and track the movements of another animal (Dehnhardt et al. 1997; Dehnhardt et al. 2001; Dehnhardt et al. 2003). While this functions only at relatively close range (Dehnhardt et al. 2003), it does give a diving seal or sea lion sources of information that are not available to birds. There is good evidence that echolocation (active SONAR) is the primary sense employed by cetaceans under many conditions, especially by taxa such as porpoises Phocoenidae and dolphins Delphinidae, and they use this particularly in turbid inshore waters (Au and Nachtigall 1997; Philips et al. 2003). Apart from one observation concerning possible SONAR in penguins (Poulter 1969), which has not been replicated or investigated further, there does not seem to be any evidence that birds can echolocate underwater.

It can be concluded that the underwater behaviour of birds is restricted to a small range of relatively simple tasks which are guided by simple information. This combination is sufficient to support huge populations of birds which exploit the resources of both fresh and marine waters. This does mean, however, that these birds can be particularly vulnerable to problems caused by human artefacts, especially fishing nets, placed in their underwater environments. These vulnerabilities will be discussed from a sensory ecology perspective in Chapter 9.

8
What Drives Bird Senses?

The previous chapter about the sensory ecology of birds underwater discussed the information that guides a single task, foraging. This emphasis was inevitable since birds that dive do so for one purpose only, i.e. to find food. All other tasks of their daily and annual cycles take place in air. The chapter demonstrated an important general point: foraging can be highly efficient but guided by what would appear to be a paucity of information. Certainly the information employed in amphibious foraging would appear minimal compared to what might be assumed to be available to a predatory bird in a terrestrial environment, or even to these same diving birds when they are above the water surface.

Chapter 6 on the sensory ecology of nocturnality showed more complex linkages between information sources and behaviour. Diverse species complete all aspects of their life cycle at night-time but they are guided by what can be considered only partial information from any one sense. However, in most cases information from different senses are used in combination, in a complementary way, and this information is sometimes augmented by knowledge of a specific location. This cognitive component enables partial information to be used for the control of a wide range of complex behaviours. Again the emphasis was upon tasks to do with foraging. In light of these general conclusions, it is legitimate to ask: What drives bird senses? Is foraging the key task that shapes the senses of birds?

When discussing the evolutionary pathway that led to the optical design of modern vertebrate eyes, Dan Nilsson (Nilsson 2009) argued that the changes to camera eyes as they first evolved were neither continuous nor incremental. He argued that the evolution of eyes has been subject to periods of rapid change, as new visually guided tasks were hit upon through natural selection, followed by relative stasis. To capture this idea, Nilsson suggested that the evolution of eyes has been the subject of 'task-punctuated evolution', in which there were longer periods of stasis alternating with shorter periods during which rapid structural and physiological changes occurred. But what might be the tasks that drove these changes? Did these tasks operate to keep sensory structures relatively stable once each rapid change had occurred, or have eyes, and other sensory structures, actually been subject to incremental change between the more obvious punctuations of rapid change?

The Sensory Ecology of Birds. Graham R. Martin: Oxford University Press (2017).
 DOI 10.1093/acprof:oso/9780199694532.001.0001

The continuous presence of eyes in most vertebrate lineages suggests that selection in favour of vision as the primary source of information has been very strong and consistently maintained. As argued in the case of kiwi (Chapter 6), however, this may not always be the case, even among birds. Vision may become less important, and eyes and the brain structures used in the analysis of visual information can regress relatively rapidly. This suggests that for any sense to be maintained, there must be constant natural selection based upon its utility in the conduct of specific tasks.

Nilsson and Pelger (1994) argued that the main features of vertebrate eyes may have been arrived at through the process of punctuated evolution over a relatively short period of time. A conservative estimate suggested that a fully functional camera eye could have evolved in as little as 400,000 generations. However, vertebrate eyes have become increasingly differentiated and specialized in many and subtle ways. These are exemplified by the interspecific differences found today in the more readily measured visual capacities, such as resolution, visual fields, and the patterns of image analysis within retinas (Chapter 2).

Natural selection driving these differences may have occurred more or less continuously, and subtle differences have been noted between closely related species. For example, the differences between the visual fields of closely related ducks, shorebirds, and ibises (Guillemain et al. 2002; Martin and Piersma 2009; Martin and Portugal 2011) must have evolved relatively recently on an evolutionary time scale.

In a similar manner, the differences between species in the number and distributions of somatic receptors in the bills of ibises seem finely tuned to differences in foraging tasks (Cunningham et al. 2010). Furthermore, it is now established in birds that significant changes in other key structures associated with foraging can occur at very short time scales. In merely decades, driven by differences in foraging opportunities and subtle changes to the foraging task, it has been shown that bill structure can change (Grant and Grant 2014). However, whether such rapid evolutionary change occurs in sensory structures and capacities is not established, but it is a possibility. Rapid changes in sensory capacity could certainly arise, because the structure of sensory organs is inherently flexible. In Chapter 2, it was made clear that the optical system, the image-analysing system, and the way that eyes are combined in the head, can vary independently and in many ways. This flexibility also applies with respect to the relative numbers and distributions of other types of sensory receptors which gather information concerned with olfaction, taste, and tactile cues.

Numerous examples of interspecific differences in the sensory capacities of birds were summarized in Chapters 2, 3, and 4. Fascinating as these instances of differential sensory capacity are, the sensory challenges of environments and tasks that have provided the general drivers of sensory capacity are not clear. However, all of the previous chapters have provided hypotheses of what these might be in particular instances. In Chapter 1, the idea that a bird is a 'wing guided by an eye' was discussed and it was shown to rest on the assumption that flight was the key driver of vision in birds. That is, Rochon-Duvigneaud assumed that the gaining of information to control flight had been, and still is, the key driver of avian sensory capacities. In view of the many examples discussed here, this now looks less likely. Certainly the examples discussed in Chapters 6 and 7 would not seem to support that assumption.

8.1 Visual Ecology, Trade-offs, and 'Just-so Stories'

The intent of the previous section and of the discussions that now follow has a clear purpose: resolving the question of what drives avian senses. This is somewhat problematic but potentially has considerable utility: problematic because is relies on a diverse suite of information that has to be carefully brought together to make an argument that is more than a 'just so story'. However, understanding what drives the sensory capacities of birds should lead to better understanding of how information limits a wide range of behaviours and hence prove to have applied value, particularly for understanding why birds often have fatal interactions with a variety of large and apparently obvious human artefacts. These include power lines, wind turbines, fences, glass sheets, and fishing nets, even aircraft and motor vehicles. Collisions with all of these structures can have severe impacts on local and sometimes wider populations of particular species. The reasons for such collisions and what might be done to mitigate them is the subject of the final chapter. First, however, a general understanding of the functions and drivers of avian sensory systems, especially vision, needs to be established.

A 'visual ecology' explanatory framework for eye diversity, i.e. a framework which combines measures of an animal's natural environment and identification of key behaviours guided by visual information, is implicit in the works of Walls (1942) and Rochon-Duvigneaud (1943). However, it was not until 1977 that the idea of a trade-off in visual information was theoretically and mathematically analysed (Snyder et al. 1977), and these ideas of trade-offs became a key part of the first explicit discussion of visual ecology by John Lythgoe (Lythgoe 1979). This approach has subsequently been expanded by many authors (e.g. Archer et al. 1999; Cronin 2008; Cronin et al. 2014; Cronin and Douglas 2014).

Snyder, Lythgoe, and others showed that in addition to the need to determine the key perceptual challenges presented by light environments for different animals, it is also necessary to attempt to understand whether evolved structures and physiological mechanisms are optimal solutions for the extraction of information from a particular environment or are trade-offs between competing optimal solutions. This theme has been explored in some detail by Mike Land and Dan Nilsson (Land and Nilsson 2012), particularly with respect to the trade-off between visual sensitivity and visual resolution. It is also possible that trade-offs and complementarities exist with respect to gaining information across different sensory modes, for example between visual and acoustic information in foraging under nocturnal conditions (Chapter 6) or between visual and tactile information with respect to the exploitation of food resources buried within substrates by shorebirds and kiwi (Martin and Piersma 2009; Martin et al. 2007b).

Because of the complexity, and often imperfect knowledge, of both behavioural and environmental parameters that are brought to bear upon evolutionary and functional interpretations of eye structure, it is sometimes possible to regard such explanations as examples of 'just so stories'. These are 'a speculative style of argument that records anatomy and ecology and then tries to construct historical or adaptive explanations' (Gould 1985), rather than testable hypotheses. With respect to a specific explanation regarding a single species, or perhaps two species chosen as a contrasting pair, this is a frequently made and widely discussed criticism (Freckleton et al. 2002; Garland et al. 2005). One solution to such difficulties is to employ a broad comparative approach rather than to rely upon only detailed analysis of single species (Felsenstein 1985; Harvey and Pagel 1991), but this requires a large data set of quantified characteristics.

For any comparative studies of eye structure and function, there is a need to draw study species from a well-established phylogeny, a range of species which occupy different habitats, and species which present a variety of behavioural repertoires. A sample which draws on these criteria is likely to include animals whose visual systems have been shaped by a range of different perceptual challenges, and hence features which have evolved in response to different challenges may become evident. Fortunately, the 10,000 or so species of extant birds (Gill and Donsker 2016) provide a good taxon for such studies since on the whole their ecology and behaviours tend to be known in broad terms, and for many species there have been detailed behavioural and ecological studies (Gill 2007). The taxonomy of birds is reasonably well established at the level of the family and the behaviours and ecology of many species have been well studied and described. However, at higher taxonomic levels there is considerable debate (Hackett et al. 2008; Jetz et al. 2012; Livezey and Zusi 2007), and there are also frequent revisions at the species level as exemplified by the regularly published up-dates to the World Bird List issued by the International Ornithological Congress (Gill and Donsker 2016).

8.2 Which Tasks Drive the Evolution of Sensory Systems in Birds?

Among birds, and perhaps all animals, the tasks which have high informational demands, and are subject to strong natural selection on a daily basis, are locomotion, foraging, and the detection of predators. Less frequent but highly selected behaviours, which are also likely to have important informational demands, involve reproduction and the care of young.

The question is whether any of these are 'key' drivers of sensory capacity. That is, are the informational demands of some tasks primary, while the informational demands of others secondary, such that the information that guides the secondary tasks can, by and large, be met within the parameters set by the primary demands?

The discussion which follows suggests that this is indeed the case and concludes with the proposal that the primary driver of sensory capacity in birds is foraging. Furthermore, because of the unique use of the bill in the foraging of birds, the key driver can be further refined to the quite specific informational demands for the control of bill position and the timing of the bill's arrival at a target. The second most important task that drives vision in birds is the detection of predators, but it seems that accurate bill positioning and predator detection usually make antagonistic informational demands with the result that there is frequently a trade-off between these two demands, but getting the bill in the right place and at the right time takes precedence over predator detection. This is, perhaps, because predator detection can usually be enhanced by behavioural adaptations involving scanning as well as the use of senses additional to vision. But accurate bill position and timing can only be achieved using visual information. The outcome of the trade-offs between the control of bill position and detecting predators depends upon details of the foraging ecology of each species with the result that there are many subtle variations between species. The final conclusion of this chapter is that all other behaviours, including flight and reproduction, are conducted within the constraints set by the sensory information that is necessary to guide foraging and the requirements of predator detection.

8.2.1 Key Tasks and Perceptual Challenges Faced by Birds

To develop this argument attention must first be given to the tasks which are likely to set the primary demands for information in birds: flight, foraging, predator detection, and reproduction.

Flight

Whatever the exact origins of birds (Chiappe 2006; Zheng et al. 2013), it does seem to be clear that the control of flight was a task that the very first birds must have accomplished, and so it has been assumed that even among the earliest birds, flight may have required a high degree of specialization of visual systems to provide information that is both spatially accurate and processed at high speed (Alonso et al. 2004). Both attributes are thought necessary in order to cope with the demands of travelling at relatively high speeds and today's flightless birds almost certainly had ancestors that flew (Bunce et al. 2009; Phillips et al. 2010). It would not seem unreasonable, therefore, to suppose that the gathering of information necessary for the control of flight is likely to have been important throughout the evolution of the sensory systems of birds. However, as described in previous chapters, many birds fly without the benefit of fine spatial information. Furthermore, it seems highly likely that fine spatial resolution may have evolved, as in the case of diurnal raptors, for the detection of objects (prey items) at great distances rather than to perceive fine detail close by. Indeed eagles, which have the highest known spatial resolution of any eye (Chapter 2, Appendix 1), are birds of open habitats and do not frequent spatially complex habitats, and their key informational demand when foraging is probably the detection of distant prey.

Foraging

The task of finding and ingesting food would seem to pose a constant perceptual challenge for most birds. Constrained by the requirements for a combination of high power output and low body weight (King and King 1980), many birds forage almost continuously or at frequent intervals through their waking period each day, and usually for a relatively specific range of food items (Gill 2007). Extant birds exploit a very wide array of food types and the efficient detection and ingestion of each type must pose particular perceptual challenges. Food types exploited by birds range from the minute to the relatively large, and from immobile items to prey that are highly mobile and evasive. Algae and diatoms, plant leaves, many different types of fruits and seeds, animals of all main faunal types from flying and buried invertebrates, to medium sized mammals and birds, and carrion—these are all exploited by different species of birds (Gill 2007).

Each dietary type is associated with particular methods of food acquisition nearly all of which involve using the bill as the sole tool. Pecking, lunging, probing, excavation, aerial pursuit, pursuit under water, grazing, filtering mud and water, trawling water and the air: all of these foraging tasks pose a rich array of perceptual challenges. Furthermore, these tasks must be dealt with frequently, almost continuously, by a bird throughout its life. Retrieving sufficient information from the environment to allow birds to do these tasks must be the subject of exacting natural selection.

Such selection is likely to be equal in its effects to the selection pressures which can lead to rapid changes in the structures employed to actually procure food items, especially the shape and size of the bill. The efficient acquisition and manipulation of food items can require such subtle structural changes to bill shape and size that they can evolve 'in real time' (Grant and Grant 2002; Grant and Grant 2014; Weiner 1994). However, having the right bill shape and size is of little value if it cannot be targeted, or the timing of bill opening controlled with accuracy and precision. With the exception of the small number of species that can feed by filtering substrates (e.g. some ducks, some procellariiforms, flamingos) or by trawling from the air (e.g. nightjars and swifts), the tasks of timing and controlling the bill's position in foraging always needs to be done highly accurately and precisely. Such tasks have to be achieved every time a food item is ingested. Many birds which feed on immobile objects, such as seeds and fruits, and birds which feed upon insects sitting on surfaces, will need to control the timing and positioning of the bill almost continuously throughout their waking life. However, if the foraging task is done less frequently it may be even more exacting since less frequent feeders are likely to be taking larger but mobile and evasive prey. To feed in this way, accuracy of bill position and timing are also paramount as there is often only a single opportunity to take a particular item, otherwise it escapes.

Predator detection

Avoiding being detected and consumed by a predator is a challenge which is probably faced by the majority of bird species almost constantly during their waking hours (Cresswell 2011; Fernandez-Juricic et al. 2011a; Sansom et al. 2009). Only species which evolved in habitats almost free of predators may have been free of this constant source of selection. Such a situation may have never existed or if it did so was only a relatively temporary situation. Even on newly formed islands such as those of New Zealand, where birds were free of mammalian predators for 80 million years, avian predators were present for much of this time (Wilson 2004; Worthy and Holdaway 2002), and on other smaller, more recently, formed oceanic islands, such as the Galápagos and Hawaiian Archipelagos predator–prey relationships among birds were soon established.

Reproduction

Behaviours associated with reproduction can be subject to exacting selective pressures and have long been the focus of research at the heart of understanding evolutionary processes (Davies et al. 2012). Reproduction can occupy a large proportion of a bird's lifetime although the actual amount of time in any one day devoted to behaviours specifically serving reproduction, as opposed to maintenance and provisioning, may be relatively small. Many of the more intriguing aspects of bird behaviour can, however, involve display postures and plumage characteristics which are used as signals associated with reproduction and their detection clearly has an important informational component. The specific investigation of these in the context of sensory ecology has been the subject of detailed and insightful analyses; see for example the work of Endler et al. (2005), Endler and Mielke (2005), and Hagelin (2007).

Although the sensory information which supports them has been studied in detail, these behaviours are, perhaps, not subject to such strong and continuous selective forces that are likely to have applied to the more ubiquitous and almost continuous informational demands of locomotion, foraging, and predator detection. Indeed, reproduction and the informational demands that it brings have to be conducted within the daily context of these more ubiquitous behaviours. Furthermore, in the majority of bird species, behaviours that are associated specifically with reproduction involve the use of the bill as a tool, for example for the gathering of nest material and nest construction. In many species, placement of the bill when feeding young must also be done with accurate positioning and timing. These are the same kinds of demands that apply in foraging and, of course, predator detection will be a constant demand in all phases of reproduction.

8.3 Competing Tasks and Competing Information

The perceptual challenges posed by locomotion, foraging, and predator detection are likely to apply almost constantly in the daily lives of most birds and will have applied throughout their evolution. During every day of a mature bird's life, it is likely to move frequently around a familiar patch, and at less frequent times move over greater distances. Most birds forage and are exposed to predators throughout all of their waking hours. Frequently, the perceptual challenges associated with moving, foraging, and exposure to predation will occur simultaneously, or there may be rapid switching between them and competition for information. For example, a foraging bird may require information to guide the detection and procurement of a food object while at the same time needing information on predatory dangers. Such apparent competition between tasks and the information necessary for their execution have been studied in some detail, for example by studying how foraging birds behave when a predator is introduced under controlled conditions (Devereux et al. 2006; Fernandez-Juricic et al. 2008).

The information requirements of these frequent and various everyday challenges may be quite different from each other. For example, in a particular species the gaining of specific information to optimize the detection and procurement of food may not optimize the way information is gained for the detection of predators. Indeed, it has often been assumed that because these informational demands are so dissimilar, the tasks of foraging and predator detection cannot be conducted simultaneously and require a bird to switch between discrete behaviours. Examples include breaking off foraging with the head down and lifting or reorienting the head to scan for predators (Fernández-Juricic et al. 2004; Fernandez-Juricic et al. 2008; Guillemain et al. 2002; Martin 2012; van den Hout and Martin 2011).

As already shown, trade-offs and compromises can apply within a single sensory modality and apply especially to vision (Chapter 2). The multifaceted nature of vision and its different 'capacities', which are measured independently of each other, means that it is often difficult to understand the trade-offs and compromises that have occurred, although it is possible to understand that different facets of visual performance cannot all be maximized simultaneously (Land and Nilsson 2012). Conflicts in the information required for different tasks may result in compromises and trade-offs between different types or qualities of information and this was first systematically explored by (Lythgoe 1979) with respect to whether the spectral sensitivity of photoreceptors in a fish (when it is trying to detect another fish) should match or be offset from the spectral distribution of the ambient light (Lythgoe and Partridge 1989).

It is argued below (8.5) that differences in the visual fields of birds provide a basis for understanding how the different perceptual challenges presented by foraging and by predator detection have been traded-off within visual systems. Studies of visual fields also exemplify the fine tuning of vision to different perceptual challenges. Such tuning has resulted in functionally significant differences in visual ecology even between species which are in the same genus. This indicates that sensory capacities can be fine-tuned by the competing demands of different tasks.

Similar arguments can be made with reference to other properties of bird eyes. The ways in which the abundance of photoreceptor types differ within a retina were discussed in Chapter 2. A good example is provided by the distinct patterns seen in the retinas of doves (Figure 2.6), but they have also been described more generally in other species (Chapter 2, 2.7.1). Although the functions of these patterns are yet to be well understood in detail, it would seem safe to presume that they are an indication of different perceptual challenges for the conduct of particular tasks that are carried out by different sectors within the field of view of an eye.

The fact that different sectors of the field of view are associated with different tasks is more readily understood by the different distributions of retinal ganglion cells and the positions of foveas, as exemplified in Figures 2.7, 2.8, and 2.9. These patterns are better understood and catalogued than the distribution of photoreceptors, and are usually explained by reference to the detection of particular targets, especially prey items and/or predators (Chapter 2, 2.7).

Differences in optical structures, particularly the relative contributions of the lens and cornea to the formation of the retinal image and differences in the maximum brightness of the retinal image (Chapter 2, 2.8) are less easily interpreted as related to particular tasks. Optical structures may be more generally attributed to either the general patterns of behaviour with respect to ambient light levels (e.g. nocturnal versus diurnal activity, Figure 2.13) and whether the bird is solely terrestrial or whether it is amphibious (Figure 2.11).

Data on visual fields is drawn here from studies in 61 bird species (from 31 families and 20 avian orders) and is briefly summarized in Appendix 2 which also gives the citations for all of the studies. These species differ markedly in their general ecology. Some of the data has been summarized previously (Martin 2014).

8.4 General Characteristics of the Visual Fields of Birds

From a human perspective, we 'know' that the world surrounds us. However, at any one moment that is not how we experience it. Humans experience the visual world as 'in front' and we seem to constantly move forwards, into it. This is a result of the particular configuration of our visual field: it is the 'human eye view' (Figure 2.14). Visual fields define the space around the head of an animal from which information can be extracted at any one instant. Human eyes are placed in the front of the skull; basically our eyes look horizontally, neither up nor down, not sideways or back, just forwards. Furthermore, what the left eye sees is very similar to what the right eye sees, that is, we have a large area of binocular overlap, with each eye looking at the same scene from a slightly different viewpoint. The whole of the human visual field lies well within the hemisphere in front of the face. However, compared with most vertebrates, including all birds, human eye placement and our resultant visual field is unusual (Figure 2.14). Birds are, in effect, surrounded by their visual world and they 'flow through' it, rather than move into it (Martin 2012). As a bird moves through the world an object can be tracked from directly in front to the rear of the head.

In birds, the eyes are on the side of the skull. Each eye looks outwards at a different scene, and the overlap in each eye's visual field is relatively small, typically between 20° and 30° but as narrow as 5°–10° in some birds (Figure 2.14), although the field of an individual eye may be wider than 180°. In no birds do the eyes look directly forwards and for many birds the eyes not only look sideways but they are also positioned more towards the top of the skull and the axes of the eyes project slightly upwards, rather than horizontally. The result is that, for the large majority of birds, the visual world is all around; there is little or no blind area above or to the rear of the head (Figure 8.1). Some bird species, e.g. some ducks (Anatidae) and some shorebirds (Scolopacidae), have totally comprehensive visual coverage of the hemisphere above the head, and also extensive coverage to the sides and front below the horizontal, such that at any one instant they can extract information from the total volume around them except from the space occupied by their own bodies. For some birds, the eyes may point slightly downwards with the result that when the head is held horizontal they are able to examine objects at their feet (as in herons Ardeidae) or comprehensively scan below them when foraging on the wing (as in eagles and vultures Accipitridae).

Despite these very marked departures from the human world view, it has been widely held that frontal eyes are the optimal or preferred position of eyes in the vertebrate skull. Thus, Walls (1942) concluded that, '*Vertebrates have had a powerful incentive to develop binocularity wherever their snouts and their beaks and their requirements for periscopy would permit*' (p. 326), and, '*two eyes are better than one, and that vertebrates in general have seemingly striven to enlarge binocular fields at the expense of uniocular ones. Animals which have clung to strong laterality have done so in obedience to powerful factors, such as defencelessness or total absence of cover in the environment which makes the retention of periscopy vitally important. The various degrees of partial frontality are compromises between the urge for binocularity and the need for periscopy*' (p. 291). It is worth quoting these extracts at some length because, although based very much upon an anthropocentric perspective, such assumptions

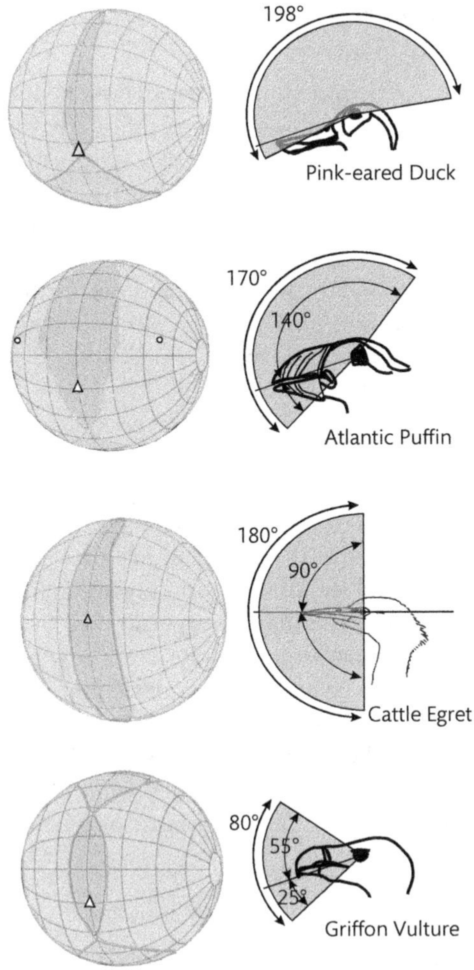

Figure 8.1 Variation in visual fields. Visual fields are complex three-dimensional constructs whose parameters can vary along a number of different dimensions. These diagrams capture some of the diversity between species by focusing upon the size and positions of the binocular fields, the region where the fields of two eyes overlap (Figure 2.15). The conventions in these diagrams follow those used in Figure 2.14. The diagrams in the left hand column show how the binocular regions can vary in width and position relative to the projection of the bill, and also in their vertical extent. The right hand column emphasizes these differences and shows the vertical extent of the binocular region and its position. In the duck, puffin, and egret, the binocular fields are long, extending through approximately 180°. In the duck, this arrangement gives the bird comprehensive visual coverage of the world above it, but it cannot see below the level of its bill. In the egret, the bird can see all of its frontal hemisphere, including its feet, but has a blind area behind the head (Figure 8.5). The puffin has extensive coverage centred obliquely upwards. The vulture has a much smaller binocular region and an extensive blind area above, below, and behind the head. Data from: Pink-eared Duck *Malacorhynchus membranaceus* (Martin et al. 2007a), Atlantic Puffin *Fratercula arctica* (Martin and Wanless 2015), Cattle Egret *Bubulcus ibis* (Katzir and Martin 1994), Griffon Vulture *Gyps fulvus* (Martin et al. 2012).

have achieved wide currency in natural history and even visual science; in effect broad binocularity is seen as desirable, as a good thing to have.

8.5 Functional Interpretations of the Visual Fields of Birds

A framework for explaining both the general and detailed features of the visual field topographies found among birds needs to eschew the more anthropocentric perspective taken in the previous paragraph, and also critically examine the '*wing guided by an eye*' model of birds.

The argument presented here is that both the general and detailed features of bird visual fields, down to the level of individual species, have been driven primarily by the key informational challenge of foraging. This key challenge is the accurate positioning of the bill (or feet) when taking food or prey, which must be simultaneously traded-off against the informational demands of predator detection. There is an additional driver but it applies only to certain species: the need to avoid imaging the sun upon the retina. It is argued below (8.6.2) that the perceptual requirements for the control of locomotion are met within the requirements of selection for efficient foraging and predator detection. Evidence in support of this general argument has been built up from a series of detailed studies of visual ecology and visual fields in the species listed in Appendix 2. Some of these arguments have been described in detail in a series of previously published papers (e.g. Martin 1999b; Martin 2007; Martin 2009; Martin and Katzir 1999; Martin and Osorio 2008).

8.6 The Key Functions of Bird Visual Fields

A number of strands of evidence support the idea that controlling bill position, including the accurate timing of its arrival at a target, is the most demanding task that vision is used for by birds. The second most demanding task is the detection of predators. These tasks make competing demands and the configurations of visual fields are primarily the result of this competition. The strands of evidence in support of this are first summarized briefly and discussed in more detail later.

8.6.1 Control of Bill Position in Foraging

The foraging of most birds requires exact positioning of the bill (or in some species the feet) with respect to a target, regardless of whether the items are taken by pecking or lunging. Control of bill position (both the direction of travel towards a target and time to contact the target) can be achieved from the optic flow-field produced as the head moves towards the target. Optic flow describes the way in which the image of the world moves across the retina as the head moves through space. It is regarded as a foundation of perception in both vertebrates (including humans) and insects (Lee 1980; Srinivasan 1996), and its fundamental role in the control of various aspects of flight behaviour, especially timing of approach to a target in birds has been established (Bhagavatula et al. 2011; Davies and Green 1994; Lee et al. 1991). In birds, a target is usually detected visually in the lateral field of view of a single eye, probably employing a region with the highest quality optics and the highest retinal resolution (Martin 2009) (Chapter 2, 2.7). After detection, visual control is passed to the frontal field, within which the direction of the bill projects, but this may occur at only a relatively close distance from, or short time before, contact with the target. This allows the accurate direction of the bill towards the target and accurate timing of its arrival, so that bill opening can be co-ordinated with arrival at the target and the object grasped in the bill. In some birds, pecking has a ballistic phase and the eyes are closed during the final approach towards the target. In other birds, prey may be taken in the feet which are swung up into the central projection of the binocular field just prior to prey capture.

8.6.2 Panoramic Vision

Comprehensive visual coverage of the celestial hemisphere is found in a number of bird species. It is coupled with a small degree of binocular overlap (< 10°) which extends through 180° from directly in front to directly behind the head. Having complete visual coverage of the world all around the head would seem to be the ultimate adaptation to the demands of predator detection. Given its potentially great utility panoramic vision might be expected to be relatively common among bird species. It is, however, found only in a few key species. While many birds have extensive visual fields, most birds have a blind area behind the head leaving them more vulnerable to predator attack. The presence of these blind areas and their absence in only certain species is evidence that controlling bill position and the detection of predators are tasks which have different informational demands that are in competition.

Total panoramic vision appears to have evolved independently in two quite different bird orders: ducks (Anseriformes) and shorebirds (Charadriiformes). Only a few species in these taxa have totally panoramic vision, but those that do have it share a common feature in that their foraging does not require visual control of bill position; instead, foraging relies upon tactile cues from bill tip organs (Chapter 4, 4.1.2). It appears that, freed from the constraint for the accurate visually guided control of bill position, natural selection has favoured the evolution of comprehensive visual coverage about the head to aid predator detection. Crucially, however, the width of binocular overlap in these species is minimal, between 5° and 10° (Figure 8.1). Yet, these birds are capable of fast flight often in complex habitats, and this suggests that a frontal binocular field of this minimal width is sufficient for the control of flight.

8.6.3 Differences in Visual Fields between Closely Related Species

There is evidence from both ducks and shorebirds that the gaining of comprehensive vision can evolve relatively rapidly at the level of individual species. Thus, significant differences in vigilance behaviour, which are coupled to differences in visual fields, occur between two ducks within the same genus: between the non-visual (tactile and filter) feeding Northern Shovelers *Anas clypeata* and visual feeding Eurasian Wigeons *A. penelope* (Figure 1.3). The foraging of Wigeons involves selective grazing guided by visual cues (Guillemain et al. 2002). Shovelers have comprehensive visual coverage of the celestial hemisphere while Wigeons have a narrow blind sector to the rear of the head but a wider binocular field which embraces the projection of the bill tip. Thus, two species of the same genus, which can be observed exploiting different resources in the same locality, differ in their visual field configurations, foraging technique, and vigilance behaviour. This demonstrates that subtle, but behaviourally significant, differences in visual ecology can occur between closely related species.

Similar differences in visual fields have also been found between more distantly related species of ducks (Blue Ducks, Pink-eared Ducks), which also differ in their use of visual and tactile cues when foraging (Martin et al. 2007a). Among shorebirds within the same family, Scolopacidae, differences exist in visual fields that are correlated with differences in the use of tactile or visual cues in prey detection and capture. Red Knots employ vision for foraging during part of their annual cycle and they have a blind area to the rear of the head, while Eurasian Woodcocks can always rely upon tactile cues and they have totally panoramic vision (Martin and Piersma 2009) (Figure 8.2).

Although not involving comprehensive vision and tactile foraging, there are examples among the Ciconiiformes of significant differences in visual fields between species in the same family which have been correlated with differences in foraging technique. Thus among the ibises (Threskiornithidae) significant differences in visual fields are found between species whose foraging ecology involves surface pecking in dry terrestrial habitats (Northern Bald Ibis *Geronticus eremita*) and species which probe in the soft substrates of marsh habitats (Puna Ibis *Plegadis ridgwayi*) (Martin and Portugal 2011).

8.6.4 The Perceptual Demands of Bill Control versus Predator Detection

The above examples indicate that closely related species, even within the same genus, which employ different perceptual cues for foraging, can differ in their

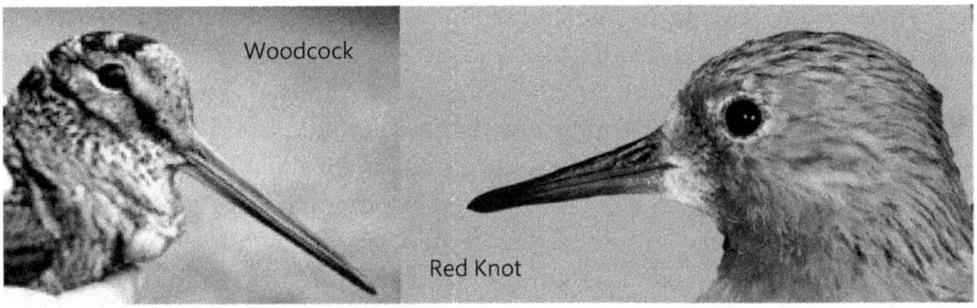

Figure 8.2 Visual fields and dietary change. Both woodcocks and Red Knots are examples of shorebirds which have bill tip organs capable of 'remote touch'. This can be used for detecting prey items buried in soft substrates without visual guidance (Chapter 4). Neither species needs to use their bills for nest construction or feeding young. However, in woodcocks the visual field is such that the bill falls at its very periphery, but, these birds gain comprehensive vision about the head and hence they are able to detect predators from any direction without having to move (Figure 6.10). In Red Knot, the eyes are more forward and the binocular field encompasses the projection of the bill much in the style of the visual fields found in birds which use vision to guide their bill position. The reasons why the probing knots probably needs this arrangement is that for the part of the year when they are breeding they forage for surface and aerial insects. The informational demand for the control of bill position in these tasks is exacting and is achieved by having the bill placed centrally within the field of view. Woodcocks, however, do not show such dietary switching during their annual cycle. Thus it seems that the possibility of gaining comprehensive vision about the head is not possible for knots because of the need for accurate control of bill positions during part of its annual cycle.

visual field characteristics in subtle but functionally important ways. This suggests that visual fields (and the optical and anatomical structures which underpin them) are subject to strong selective forces driven by the informational demands of foraging. These evolutionary outcomes can be considered comparable to the much more widely studied subtle differences in bill structures that are driven by the mechanical demands of procuring different food types in birds (Grant and Grant 2002). These examples reinforce the hypothesis that the configuration of visual fields are driven primarily by the informational challenges of foraging which are traded-off against the requirement for predator detection. Only in those species which do not need to use vision to guide their bill position during foraging, such as some of the ducks and probing shorebirds, is comprehensive visual coverage of the world attained.

Not requiring visual cues to guide foraging is not in itself sufficient to lead to the evolution of comprehensive vision. It is also necessary that the bill does not require fine visual control for any task, not just foraging. Thus comprehensive vision is, in fact, found only among birds which also do not need to position their bills accurately for two other key tasks: nest construction and the provisioning of young. Both the ducks and shorebirds use simple nests which do not require elaborate construction, and their young are precocial, i.e. their young hatch in an advanced stage of development and are able to self-feed. They are never provisioned by their parents; parental care is limited to brooding and protection from predators. Most other birds must use their bills for foraging, for nest building, and for the provisioning of young, all tasks which require accurate positioning and timing of the bill.

A telling example that makes this clear is provided by flamingos (Phoenicopteridae) (Martin et al. 2005). They are filter feeders, having highly specialized structures within their bills to remove minute resources from filtered water and mud, yet unlike the filter-feeding ducks they do not have comprehensive vision. The reasons for this seem to be that they build nests which are based upon a mound of mud, and crucially their young have to be fed very accurately by 'crop milk' (a secretion from the oesophagus) which is dripped into their open mouths. Thus, despite their filter feeding, which also involves inverting the head and holding it at the level of their feet, flamingos require vision that allows accurate bill placement so that young can be provisioned. This results in a relatively broad binocular field into which the bill projection falls, and a blind area behind their head (Figure 8.3).

8.7 What Is the Function of Binocular Vision in Birds?

Stereopsis and the perception of relative depth have become regarded as the prime function of binocular vision in humans and other primates, and it has often been assumed that these same functions apply to all instances of binocular vision. However, it seems unlikely that this is the case among birds.

It is argued below that, with the possible exception of owls, binocularity in birds does not have a higher order visual function which results in the perception of solidity and relative depth. Such perception may be something found uniquely among mammals. It is argued that binocularity in birds is, in fact, a consequence of the requirement for having a portion of the visual field that looks in the direction of travel; hence each eye must have a contralateral projection, that is, each eye must look across the central plane of the head. It is true that this results in binocular vision, but its function is not to do with the perception of relative depth, rather it is to do with the direct control of bill position, and in some birds the control of the position of their feet (Martin 2009). This suggests that binocular vision plays only a minor role in the control of locomotion including flight. In the majority of birds, the function of binocularity would seem to lie in what each eye does independently rather than in what the two eyes might be able to do together.

8.7.1 Binocular Vision in Birds

There are a number of strands of evidence which support the idea that the function of binocular vision lies primarily in the control of bill direction and its time to contact a target.

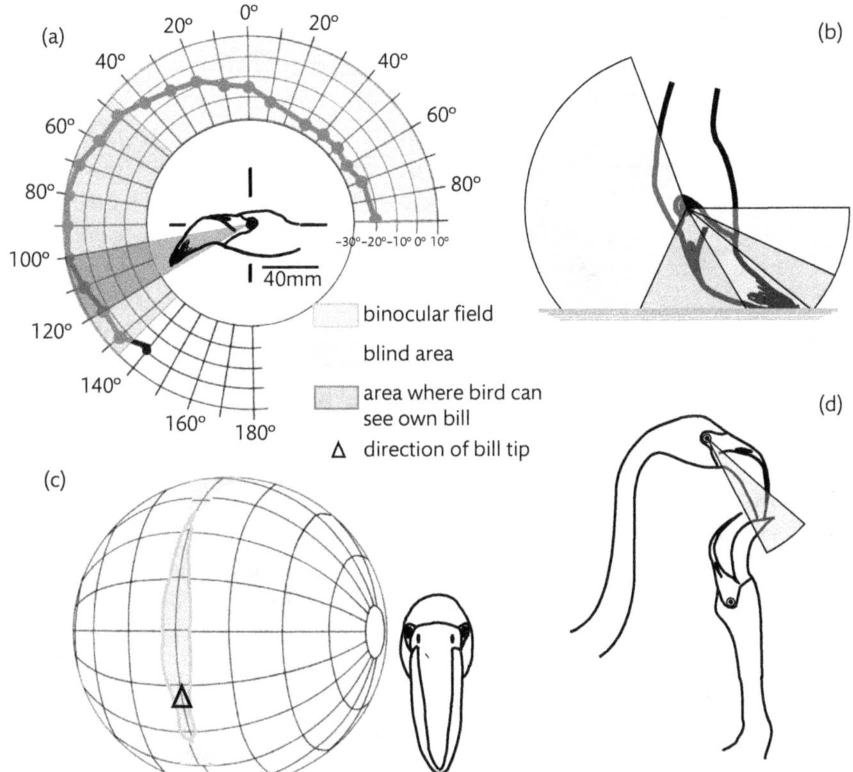

Figure 8.3 The visual demands of filter feeding versus chick feeding. Flamingos feed in a highly specialized way. The head is inverted and held at the water surface close to the feet (b). The tongue acts as a pump to take water into the mouth which is then filtered by forcing the water out through lamellae at the edge of the bill. Precise control of bill position is not required for this. Since their feeding technique would appear to make them very vulnerable to predatory attack, it would be predicted that the birds should have extensive visual coverage above and behind the head. This, however, is not the case. The birds have a narrow and vertically elongated binocular field which encompasses the bill and it is clear that they can see their own bill tip (a and c). In fact, flamingos have a blind area above and behind the head, and as they walk with the head upside down and pointing backways when filter feeding, they are blind in the direction of travel. One way in which this vulnerability is overcome is to feed in large flocks so that there can be collective detection of predator attacks. Also flamingos frequently swing their heads from side to side when feeding. This would gain momentary visual coverage of the direction of head travel. The reasons why these birds have a visual field configuration similar to those found in precision pecking birds is probably due to the visual demands of chick feeding. Chicks are dependent upon adults for up to 12 weeks while their specialized bill feeding apparatus develops. During that time they are fed by the adult dripping liquid nutrient (a product of the adult's oesophagus) directly into the chick's mouth (d); this requires precision placement of the bills. Having the eyes placed sufficiently forward to achieve this seems to have resulted in a trade-off against comprehensive visual coverage above and behind the head. This illustrates the idea that precise bill placement under the control of visual information is a key driver of visual field configuration that is traded-off against the demand for comprehensive vision for the detection of predators.

Binocular field width

In birds which use vision to control bill position when taking food items, the visual fields show a narrow and vertically elongated frontal binocular field in which the bill is placed either centrally or slightly below the centre (Figure 8.1). Such an arrangement is found in a remarkably wide range of bird species (Appendix 2) that differ both in their ecology and in their evolutionary origins. In all of these birds there is a blind area behind the head. The width of this blind area differs between species and the blind area may extend to above the head and even into the frontal hemisphere (e.g. vultures (Martin et al. 2012) and bustards (Martin and Shaw 2010)). However, the topographies of the frontal binocular fields are very similar in all species which employ vision to guide bill (or feet) position, reaching a maximum width in the horizontal plane in most non-passerines species of between 20° and 30° and in passerine species of nearly 60° (Appendix 2).

Such visual field arrangements are found in birds that feed in many different ways: for example, in species which peck at small mainly immobile items (e.g. Ostriches *Struthio camelus*, Eurasian Stone-curlews *Burhinus oedicnemus*, Rock Doves, Eurasian Wigeons, Southern Ground Hornbills *Bucorvus leadbeateri*); those which take prey by lunging at or pursuing evasive prey that are taken directly in the bill (e.g. herons and penguins); those which take evasive prey in the feet (e.g. eagles, vultures), and those which snatch prey from a surface while swimming or flying (e.g. petrels and albatrosses).

This similarity in binocular field configuration across such diverse groups of birds, and across such diverse feeding techniques, suggests a degree of convergence upon a binocular field width that is optimal for a particular purpose. That is, a 20°–30° binocular field is as broad as it needs to be to fulfil a particular function that is common to all of these species. Beyond such a width, there is little advantage to be gained, and in the majority of birds it is the extent of the peripheral fields that are maximized to reduce vulnerability to attack by predators. There would seem to be a trade-off between the demands for accurately controlling bill or feet position when approaching a target (which requires some degree of contralateral vision), and the requirement to gain as comprehensive a view of the world as is possible (Figure 2.15).

In all of the birds mentioned above, although the bill is placed approximately at or just below the centre of the binocular field, there are some subtle variations that reveal further fine-tuning of the visual field in the frontal (binocular) region to particular visual tasks and ecological conditions. For example, in the majority of birds studied so far the bill does not actually intrude into the visual field. The result is that ostriches and herons, for example, cannot actually see their own bill tip, nor can they see what is held in the bill, much in the same way that humans cannot see their nose or mouth. This may explain why true pecking behaviour typically involves a ballistic final phase in which objects are approached with eyes closed, usually from a characteristic distance for the species (Zeigler et al. 1993). However, in birds that employ precision-grasping as opposed to ballistic pecking, it appears that the bill does intrude into the visual field, and this means that these birds can also see what lies between their mandibles when an object is grasped. Thus hornbills, which pick up small items with forceps-like action in the tips of their large decurved bills, can visually inspect objects that lie between their bill tips before ingesting them (Martin and Coetzee 2004). This is also found in Common Starlings which use a specialized technique of 'open-billed probing' to uncover prey in the surface layers of a substrate (Beecher 1978). Another example of where the visual field allows inspection of an item held between the mandibles is found in Great Cormorants. As described in Chapter 7, the poor underwater spatial resolution of these birds may not allow the identification of prey items before they are captured (White et al. 2007). Items are brought above the water surface to be inspected and positioned accurately before they can be swallowed, and this is facilitated by their ability to see between their opened mandibles (Martin et al. 2008).

New Caledonian Crows *Corvus moneduloides* are among the small group of birds which forage using tools, and these birds use sticks for the extractive foraging of grubs from cavities in wood (Hunt 1996). The wide binocular field (60°), which is the widest yet described in birds, is better described as the most extensive contralateral projection (30°) of a visual field in birds. These allow New Caledonian Crows to see along the line of their stick tool which is held in the tip of the bill and usually propped against the cheek below the eye (Figure 8.4). The extensive contralateral projection of the field actually allows these crows to see along the shaft of the tool and hence be able to control the position of its tip (Troscianko et al. 2012). There is no point in having a tool which cannot be accurately controlled, and binocular vision does not help them see down a hole, but the contralateral projection of the visual field allows the birds to see with one eye down into the hole and at the same time look down the shaft of the tool and control its placement in the hole.

Vertical extent of binocular fields

The vertical extent of frontal binocular fields differs between species (Appendix 2). Thus while diverse species have a common maximum binocular field width of between 20° and 30°, the vertical extent of these binocular regions may vary between 60° (e.g. bustards Otididae), 80° (e.g. ostriches, stone-curlews, hornbills, vultures), 120° (e.g. storks Ciconiidae), to 180° (herons Ardeidae). This has consequences for the extent to which the birds can see above and below them. Thus, a heron standing with its bill horizontal and the head slightly forward of the body can see what is at its feet (Figure 8.5). Seeing what lies perpendicularly beneath the bill clearly has the advantage that a foraging heron can remain motionless, monitoring what is going on below it, while it waits for a prey item, such as a frog or fish, to come within striking range. Because these prey species have evolved rapid escape responses, herons may get only a one-strike chance to catch each item. Therefore, monitoring what is going on below, without having to move the head or body,

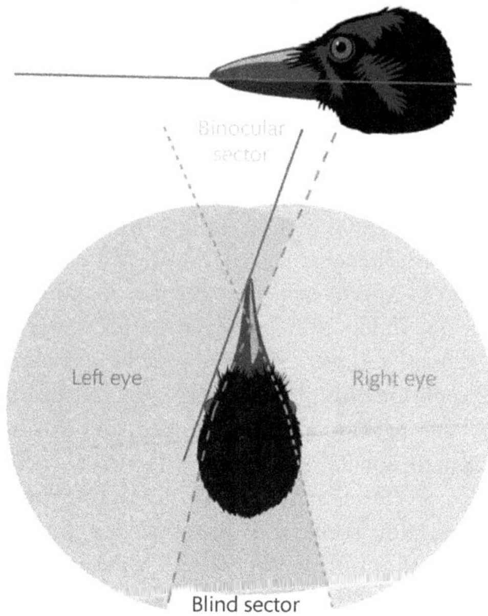

Figure 8.4 Vision and tool control. The use of tools by non-human animals is often regarded as evidence of higher cognitive abilities. However, to use a tool adequately also requires the tool to be properly positioned and controlled. New Caledonian Crows *Corvus moneduloides* face this problem in the control of stick tools which they use to probe for insects larvae in holes in timber. The tool is, in effect, an extension of the bill and has the same demands for precise placement and timing as does pecking. The tool, however, also needs to be held rigidly and this is achieved by holding it in the bill tip and propping it against the cheek below the eye, as show in this diagram. This results in the tool projecting laterally rather than directly forwards as an extension of the bill. It is argued that controlling the position of the tool tip is achieved by the contralateral eye looking down the shaft of the tool. For this to happen, the tool must lie in the binocular sector since binocularity is, in fact, achieved by the contralateral projection of the visual fields of each eye (Figure 2.15). Binocular vision per se does not seem to play a part in the control of the tool. Thus, in this diagram it is the sector of the visual field of the left eye which projects across the sagittal plane of the skull that controls the position of the tool. Once a tool has entered a hole, the right eye cannot play any part in its control. However, the left eye, by looking down the shaft, can control the tool. Drawings by Jolyon Troscianko and Graham Martin and Troscianko et al. (2012).

and waiting for a prey item to come within striking range, is clearly a significant advantage for a heron.

Abolishing binocular vision

All of the discussion above and the data in Appendix 2 give the impression that the configurations of visual fields in birds are fixed for a particular species and

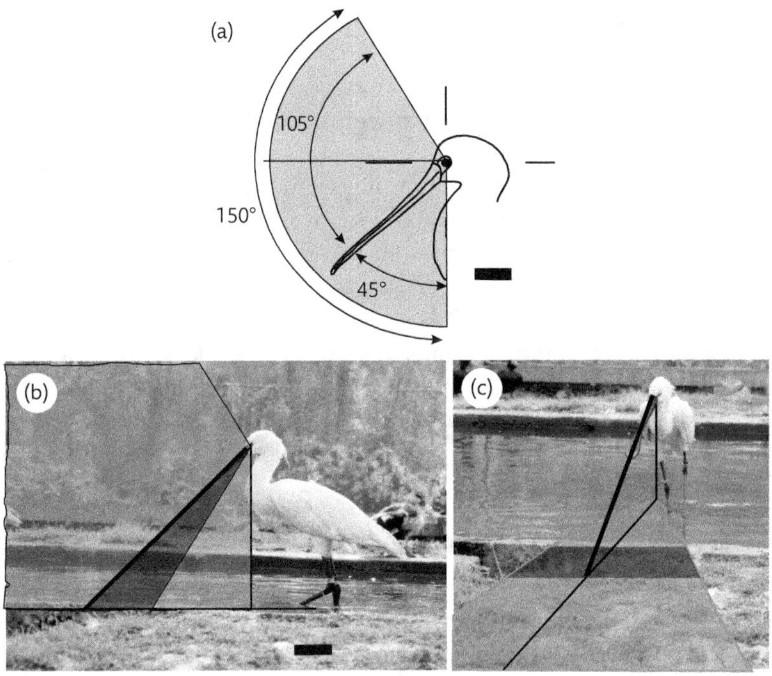

Figure 8.5 Seeing beneath the bill. Herons and spoonbills have vertically long and narrow binocular regions that give them comprehensive visual coverage of the frontal hemisphere. In Western Cattle Egrets *Bubulcus ibis*, the field extends nearly 90° below the bill so that they can see what is at their feet when the bill is held horizontal. This allows the bird to wait motionless for prey to come close enough to be caught in a single bill strike (Fig. 8.1). In Spoonbills the bill is typically held angled downwards but they are still able to see what is directly below them (a). The way in which the binocular region projects onto the surface below and in front of the birds when foraging is shown in (b) and (c).

individual. This may be true for certain species such as owls in which the eyes are indeed more or less fixed because they are not enclosed deep within an eye socket. In the majority of birds, the eyes move by up to about 20° and furthermore eye movements are not conjugate, that is, the eyes do not move together and in the same direction as they do in humans. When we look left to right as we scan a line of text, our eyes move together. In birds, the eyes can move independently and by different degrees and this has been described in many of the species listed in Appendix 2. It was first definitely investigated in detail more than 30 years ago (Wallman and Pettigrew 1985) and has also received detailed investigation more recently when some of the constraints upon independent movements were described (Land 2015; Voss and Bischoff 2009). The result of these movements is that visual fields can in fact take up many configurations and for this reason the visual fields that are summarized in Appendix 2 describe the fields which the eyes form when they are at rest, typically when the head is held in a position characteristically adopted by the species. An indication of the kinds of more extreme positions that the eyes can take up spontaneously are illustrated in Figure 8.6 and the accompanying visual field diagram shows that if the two eyes are simultaneously moved 'backwards', that is the eyes achieve their maximum divergence, binocular vision may be all but abolished. This has important implications for understanding the function of binocular vision and especially the idea that the image of an object will fall on fixed corresponding points in the two eyes, as they do in the case of humans with our two eyes looking forwards and the eyes moving in unison. The fact that there cannot be simple corresponding points that can underlie binocular vision in birds is already clear in Figures 2.3 and 2.15 which showed sections through the visual fields of owls and ostriches. It provides further support for the idea that stereopsis and the perception of relative depth are not functions of binocular vision in the majority of birds (Martin 2009).

Binocular field widths, nocturnality, and predation

The widest binocular fields in birds are not found in owls Strigidae or diurnal raptors Accipitridae as is commonly supposed and as was asserted by both Walls and Rochon-Duvigneaud (Rochon-Duvigneaud 1943; Walls 1942). It is not clear at what date the idea that a link between nocturnality and broad binocularity arose, and it seems to be based upon casual observations of owls, with the assumption that the eyes of owls are frontally placed and have a binocular field of similar width to that of humans. The eyes of owls are, in fact, laterally placed in the skull although they are more frontal than in other birds. However, in Tawny Owls the optic axes diverge by 55° and the binocular field has a maximum width of 48°. It has been argued that the advantage of binocularity with respect to sensitivity is probably marginal, especially in the context of the extreme range of light levels which can occur at night (Figure 6.3). Using two eyes instead of one to view a scene provides an increase in sensitivity of only 0.15 log10 units (× 1.4 fold) (Thorn and Boynton 1974) which is a very small increase in sensitivity in the context of the more than 1 million-fold range over which light levels can change at night (Chapter 6). This suggests that increased sensitivity is unlikely to be a strong driver of binocular overlap and would, in any case, apply to any species with binocular vision.

The maximum binocular field width of Tawny Owls is similar to that reported in a number of bird species (Appendix 2) including the majority of passerines investigated to date. As described in Chapter 6, it has been argued (Martin 2009) that the width of owls' binocular fields may, in fact, be the product of an interaction between enlarged eyes (associated with maximized light gathering), and the elaborate outer ear structures which are unique to owls and are part of the mechanism underlying the enhanced sound localization abilities of owls (Knudsen 1980;

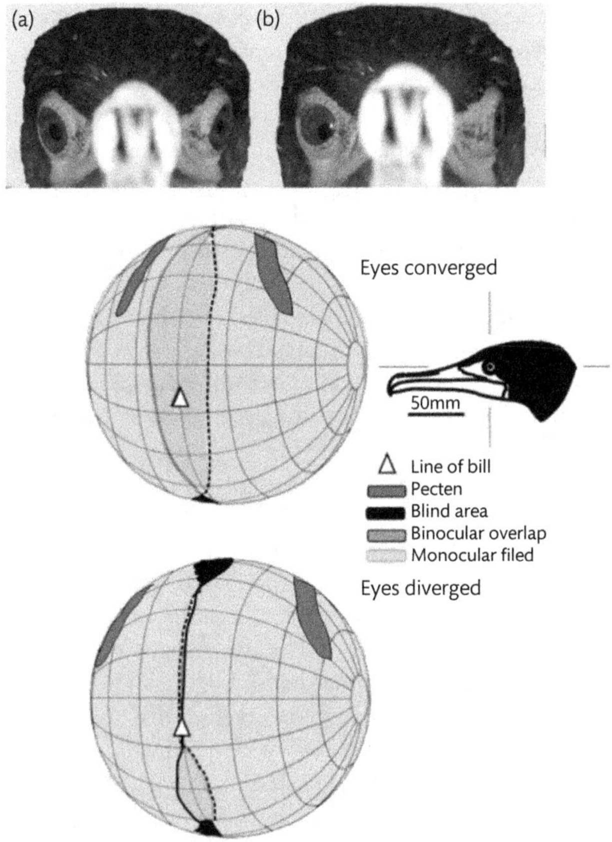

Figure 8.6 Altering binocularity. When eye movements are present in birds, the eyes can move independently of each other, i.e. one eye can move forwards, another backwards, or both eyes can be brought forward or diverged. This means that the configuration of visual fields can change and this is most clearly seen at the edges of the visual fields where the fields of the two eyes overlap to produce the binocular region. These two photographs of a Great Cormorant were taken within a few seconds of each other. In (a) the eyes are in their normal resting position and produce the type of binocular field seen in the upper visual field diagram (the diagram follows the same convention as used in Figure 2.14). The lower diagram shows one possible visual field that can result from the movements of the eyes, in this case when the eyes are fully diverged, i.e. both eyes taking up the positions similar to that of the left eye in photograph (b). This would result in a visual field like that shown in the lower diagram in which binocular overlap is virtually abolished. As the eyes are moved to different intermediate positions, a range of visual fields, between those shown in the two diagrams, would seem to be possible.

Norberg 1978). These large outer ear structures may simply prevent more lateral placement of the eyes. Put simply, more extensive visual coverage laterally would not be possible because the outer ear structures would get in the way.

The wider binocular field of owls may also have an adaptive explanation that is associated with prey capture rather than the nocturnal habit. In owls, the bill lies below the binocular field and in catching prey the feet are swung up before the bird's face, directly into the binocular field, and directly into the flight path of the bird. It seems likely that the binocular field is used in conjunction with accurate sound localization to guide the feet in the final stages of prey capture. In the final approach towards a prey item, the feet are raised and the talons spread wide to fill the visual field in front of the head, which is also the region where sound localization is most accurate (Martin 1986a). Clearly timing the spreading of the talons in order to seize prey is as important as opening the bill to grab a prey item in other birds.

As discussed in Chapter 6, true nocturnality (completing all aspects of the annual cycle between sunset and sunrise) in birds is uncommon, restricted mainly to owls, nightjars and their allies, and kiwi. However, many birds forage at night or at night-time light levels below water (Chapter 7). It is also noteworthy that not all owls are nocturnal hunters. The nocturnal owls are mainly perch and pounce hunters which can employ accurate sound location as well as vision in prey capture in daylight (Payne 1971). On the other hand, nightjars feed mainly by trawling for insects in the open air space, although the frogmouths (Podargidae) and potoos (Nyctibiidae) also perch and pounce, but without the apparent use of hearing (Cleere 1999) (Chapter 6). It is clear (Appendix 2) that the visual fields of bird species which forage at nocturnal light levels, or are nocturnally active during key parts of their annual cycle (e.g. Black-Crowned Night Herons *Nycticorax nycticorax*, Paraques *Nyctidromus albicollis*, Oilbirds, Manx Shearwaters, woodcocks, Golden Plovers *Pluvialis apricaria*, Red Knot, stone curlews, King Penguins, kiwi, Black Skimmers *Rynchops niger*) show no evidence that nocturnality is associated with wide binocularity. In all of these species binocular field widths fall within the usual range found in non-passerine species.

Oilbirds are instructive since they are among the most nocturnal of birds (Chapter 6) and their eyes appear to show extreme adaption of both optics and retina towards increased sensitivity. However, their binocular field width is similar to those of other non-passerine species which are typically active only during day light (Martin et al. 2004a; Martin et al. 2004b).

As described in Chapter 7, King Penguins are in effect, nocturnal foragers and their eyes have a low minimum f-number and a wide cornea which correlate with the maximization of light gathering. Yet their visual fields show the same narrow and vertically long binocular field of other birds whose bill position is guided by visual cues and which forage at daytime light levels.

8.8 Visual Fields, Eye Size, and Imaging the Sun

Systematic data on eye size (eye axial length) are not available for all the species listed in Appendix 2. However, eye size data are available for some species and it seems clear that there is a significant positive relationship between eye size and the width of the blind region above the head (Martin and Coetzee 2004). In these same species, however, there is no simple relationship between eye size and the width of the frontal binocular field. It may be concluded therefore that these two key parameters of visual fields, binocular field width and blind area width above the head, have evolved independently of each other.

The explanation for this is that imaging the sun upon the retina poses a particular set of perceptual problems that have also influenced the visual fields of birds. Extensive visual fields that allow birds to see all, or nearly all, of the world around them at any one instant, may not be without cost. Comprehensive visual coverage of the world about the head must mean that when the sun is in the sky its image will always fall upon the retina. In humans, the problems of imaging the sun are generally recognized to be three-fold: there can be temporary damage to the retina from even brief direct exposure to the sun; the lens and cornea can be damaged due to exposure over a long period; and there can be direct disruptive effects to vision caused by glare and afterimages which leave temporary 'holes' in the field of view (Marshall 1991; Sliney 1991).

Smaller birds, with smaller eyes, must be imaging the sun on their retinas for much of their waking time (Martin and Katzir 2000). However, it could be that birds can position themselves such that the sun is always imaged at the periphery of the visual field so that the effective aperture of the eye for the sun's image is a slit resulting in a high f-number and hence low image brightness (van den Hout and Martin 2011). So the problem may not be as great as it initially appears, although life-time exposure of the cornea and lens is potentially a major problem, but perhaps most birds do not live long enough for the damaging effects to accumulate. However, that some birds have sunshade devices which help reduce the possibility that the sun is imaged on the retina is clear (Martin and Katzir 2000). At one extreme are large-eyed species, including eagles, vultures, ostriches, hornbills, and albatrosses, which have a broad blind area (up to 80° wide) above the head, while smaller-eyed species including herons, pigeons, ducks, and starlings, have no blind area above the head. Furthermore, the large-eyed birds with a blind area above the head also have various optical adnexa which can function as 'sunshade' devices. These include brow ridges and thick and elongated eye lashes (Figure 8.7). Such structures are absent in other birds. In fact, it has been suggested that birds can be divided into two groups: 'large-eyed sun avoiders', which have sunshades and a blind area above the head, and 'smaller-eyed sun viewers', which have no sunshade devices and no blind area above the head (Martin 2007).

Figure 8.7 Sunshades in birds. The size of a blind area above the head is significantly correlated with eye size (Martin and Coetzee 2004). Birds with smaller eyes have near comprehensive visual coverage of the world above their head, while birds with large eyes have sizeable sectors in which they are blind above them. Bird species which have comprehensive visual coverage above and around the head, for example, some ducks and shorebirds (Figures 6.10 and 8.1) cannot avoid imaging the sun upon their retinas. However, large-eyed birds seem to be sun-avoiders (as are humans). In support of this is the presence of sunshade-type structures in birds with larger eyes. These may take the form of ridges above the eyes and extensive thick eye lashes which stop light from above falling directly on the pupil. These are exemplified in these photographs of a Southern Ground Hornbill *Bucorvus leadbeateri* (a and b) which show the shading effect of the prominent eyelids and also show that the shadows of the eye lashes fall across the cornea. The photograph of a Short-toed Snake Eagle *Circaetus gallicus* (c) shows the bird beneath an overhead sun. The bird has adopted its characteristic head posture in which the ridges above the eye ensure that the corneas are almost fully shaded from the sun. The close-up of the ridge above the eye of a Griffon Vulture *Gyps fulvus* (d) shows that the ridge is a substantial structure that protrudes prominently from the skin of the skull and that there are also eye lashes. Photo of vulture courtesy of A. Román Muñoz, Universidad de Málaga.

The reasons for this dichotomy may lie in the reasons for having a large eye in the first place. As argued in Chapter 2, in daytime-active birds large eyes probably evolved primarily to give high spatial resolution and, in general, the resolution of the retina matches the level of detail that can be produced by the larger image (Land and Nilsson 2012). However, a large eye also has a large entrance aperture and will produce a relatively bright retinal image. In humans, the image of the sun on the retina is, in fact, sufficiently bright that it can act as a secondary light source within the eye, scattering light and degrading contrast, and hence resolution across the whole of the retinal image (Dickinson 1991). Clearly, if selective pressure has been to evolve a large eye to maximize resolution, it would be maladaptive to compromise that resolution by degrading the retinal image with light scattered from an image of the sun. Smaller eyes, on the other hand, can only ever have lower acuity and so veiling glare produced by light scattered from a retinal image of the sun may do relatively little to degrade the image generally and lower spatial resolution across the visual field.

It is important to note that there is no relationship between eye axial length and the maximum width of the frontal binocular field in these same species. This provides good evidence that the dimensions of the field above the head are not simply traded-off with the width of the binocular field as a result of geometrical constraints arising from the intersection of the visual fields of the two eyes. Rather, it reinforces the idea that within a species the characteristics of the frontal binocular field and the width of blind area above the head have evolved independently in response to different perceptual challenges.

8.9 Summary: The Key Drivers of Visual Fields in Birds

There are two key tasks that drive the configuration of visual fields in birds. The primary driver is the perceptual challenges of foraging, specifically the control of bill (or feet) position and timing. The second driver is the detection of predators and this requires vision over as wide a sector of space as possible around the head. Because each eye has a limited visual field, the largest visual field of a single eye is about 180°; increased binocularity must result in a smaller total visual field and a blind sector behind the head. Thus, these drivers make competing demands. They can be considered primary and secondary because only under the specific circumstance of bill position not having to be controlled by vision, does predator detection drive comprehensive visual coverage. In addition to these two main drivers, it seems that a third driver also operates, at least in eyes beyond a certain size; the need to avoid imaging the sun upon the retina.

There is a further important difference between the two key tasks that drive vision. The control of bill position requires information extracted from the world that lies in front of and relatively close to the bird. The detection of predators, on the other hand, requires information that lies laterally, or even to the rear of the bird's head, and is concerned with information from locations that are remote from the bird. Thus, the regions within the visual field where there is high spatial resolution, which at high light levels are enhanced by colour vision, are indicated by the high densities of photoreceptors and ganglion cells (Chapter 2, 2.7.1, Figures 2.6 and 2.7). These regions projects outwards, laterally, on both sides of the bird, not directly forwards, and in the majority of birds forward vision is not served by regions of high photoreceptor and ganglion cell density (Fernandez-Juricic et al. 2011b; Tyrell et al. 2013). The lateral fields may be served by one or two foveas and/or a linear band of high density ganglion cells which align roughly with the horizon (Figure 2.7). Such arrangements have been correlated with the regions from which predators are most likely to attack, close to the horizon, as in the 'terrain hypothesis' (Fernandez-Juricic et al. 2011c; Hughes 1977; Lisney et al. 2012b). There is also evidence that predatory birds, such as Peregrine Falcons *Falco peregrinus* detect their prey at a distance using lateral vision, that is using the regions of high photoreceptor density which project laterally and slightly forward, and that when approaching prey Peregrines frequently do so along a curved path which keeps the prey in the central field of view of a single eye and they pass control to the frontal binocular region just prior to prey capture (Tucker 2000; Tucker et al. 2000). That is, the bird does not usually sight the prey into its binocular field until just before prey capture. Thus distant prey is probably initially detected using lateral high resolution vision while the control of the bill and feet close to the time of prey capture probably employs frontal, binocular vision, and this comes into play only at close range. However, there is evidence that other falcon species may use frontal vision quite early on in the pursuit of prey and switch between the different foveas during a pursuit flight by turning the head (Kane and Zamani 2014).

Such use of lateral vision for detecting food items with control passing to forward vision for final prey capture in the bill has been reported in other species. Examples include terns foraging over mud flats for crabs (Land 1999), thrushes searching on the ground for earthworms (Montgomerie and Weatherhead 1997) (Figure 2.10), and domestic chicks when detecting grain from amongst grit (Rogers 2008).

8.10 Binocular Vision, Optic Flow-fields, and Contralateral Vision

For any visual system, the most vital information, more important than recognition of an object, is accurate determination of an object's position. Indeed, it was argued in Chapter 2 that the main driver in the early evolution of vision systems was towards increasing accuracy in spatial resolution which meant increasingly accurate determination of the direction in which objects lay with respect to the viewer.

The next most important piece of information that vision provides is the time it will take to contact that object, that is, when will the object arrive at the observer or when will the observer arrive at the object? The actual identity and distance of an object from an observer may be of little importance compared with knowing its direction and the time it will take to make contact with it. Such information is directly available from optic flow-fields which are regarded as a 'foundation of vision' (Lee 1980; Lee and Lishman 1977). It has been shown convincingly that birds use flow-field information to control apparently exacting tasks. For example, Northern Gannets *Morus bassanus* and hummingbirds (Trochilidae) when carrying out manoeuvres that require accurate visual information regarding location and speed of approach to a target employ flow-field information (Lee et al. 1991; Lee and Reddish 1981), and there is increasing evidence that birds use flow-field information to guide flight (Bhagavatula et al. 2011) and landing (Lee et al. 1993). In mammals, optic flow-field information is analysed in the accessory optic system (Giolli et al. 2006) and pretectum (Gamlin 2006), and a similar accessory system of the visual part of the brain has been identified in birds (McKenna and Wallman 1985; Pakan and Wylie 2006).

Flow-field information is potentially available from any pattern of optical flow across the retina; it is not necessarily associated with binocular portions of a visual field. However, information on direction of travel towards a target and time-to-contact a target can be derived most efficiently when travel is directly towards a target. This will result in an optical flow-field which expands symmetrically about the image of the target. This would seem to be the situation in the key tasks described above as being key drivers of avian vision. For example, when a bird is pecking or lunging at an object, both position and time-to-contact need to be accurately specified. The essential consideration, however, is that a flow-field pattern which expands symmetrically about an object directly ahead of the bird can only be achieved if the visual field of each eye extends contralaterally, i.e. across the median sagittal plane of the bird. Thus the important consideration is contralateral vision, not binocular vision. Binocular vision can, perhaps, be seen simply as a by-product of the need to have eyes that look across the median sagittal plane of the head. Such an arrangement means that movement towards a target by the bill produces a symmetrically expanding optic flow-field. Thus binocular vision is not an adaptation that has evolved specifically to achieve simultaneous views of the same object from slightly different positions but rather to place the bill at the centre of a symmetrical flow-field. Binocular vision is perhaps irrelevant in birds; what is important is contralateral vision.

It can be concluded that binocularity in birds functions to provide information on the direction of travel and time-to-contact a target. However, this information can be provided by each eye independently and for this reason it might be more appropriate to refer to 'contralateral vision' rather than 'binocular vision', since the latter brings with it assumptions concerning the precept of solidity and stereopsis with which binocular vision in birds does not appear to be generally associated. Thus, in the majority of birds the function of binocularity would seem to lie in what each eye does independently rather than in what the two eyes might be able to do together.

8.11 Summary: The Drivers of Visual Fields and their Fine Tuning

The above discussions have argued that the two key drivers of vision in birds are the control of bill position and its timing, and the tasks of detecting food items and predators. The control of bill position is based upon information from the world close to the birds and from within the binocular region. The task of detecting predators and food items is done initially with the lateral fields and these extract information from the world that lies more remote from the birds.

There is much subtle variation in the parameters of visual fields of birds and these variations seem to be functional. The tasks of detecting predators and of placing the bill accurately make contradictory demands upon vision and these have resulted in trade-offs in the form of visual fields and also resulted in localized elevations of photoreceptor densities within areas of the retina which extract information from different parts of the visual field. Thus, the overall driver of frontal visual field characteristics is the demand for the accurate positioning of the bill, and this means that each eye must have a certain portion of its field which projects forwards, with the result that there is typically a blind area behind the head which at any one moment constitutes an area in which predators cannot be detected. It is only in those few bird species which do not have to use vision to achieve precise control of bill position that natural selection has favoured full visual coverage about the head. Visual field characteristics above the head are driven by the need to avoid imaging the sun on the retina in larger-eyed, but not smaller-eyed, species. As comparisons between ducks, shorebirds, and ibises have shown, small differences in foraging techniques can give rise to different perceptual challenges and these have resulted in subtle differences in visual fields between some closely related species, even within the same genus and certainly within the same family.

Given the apparent power of these drivers of vision, it seems likely that other aspects of vision will also show very fine and subtle tuning to the demands of either foraging and/or predator detection. Indeed, many of the patterns of photoreceptor and ganglion cell distribution patterns described in the retinas of birds (Fernandez-Juricic et al. 2011b; Mitkus et al. 2014; Tyrell et al. 2013) (Chapter 2, 2.7.1) are probably best explained by general differences in the importance of these perceptual tasks between species. However, the 'terrain hypothesis' which relates ganglion cell distributions to life in either open or enclosed habitats types (Hughes

1977), does not seem to apply within specific taxa such as galliform birds (Lisney et al. 2012b) (Figure 2.9).

Among birds, there is a strong phylogenetic signal with respect to the maximum width of the binocular field, with passerine species showing broader widths than non-passerines, and within the passerines the broadest fields are found among the Corvidae (Troscianko et al. 2012). However, sample size with respect to the total number of passerines is small and more comprehensive species' sampling of passerines, as well as non-passerines, may reveal some very interesting examples of the fine tuning of vision in birds.

Finally, the informational function of binocular vision in birds seems to lie not in binocularity per se (i.e. two eyes receiving slightly different information simultaneously about the same objects) but in the contralateral projection of the visual field of each eye. This ensures that each eye receives information from a symmetrically expanding optic flow-field about the bill from which the crucial information of direction of travel and time-to-contact can be extracted, almost instantaneously, as elegantly demonstrated, for example, in the control of plunge diving in gannets (Lee and Reddish 1981) or the landing of doves on a perches (Lee et al. 1993).

8.12 What Drives Colour Vision in Birds?

The answer to this question was touched upon in Chapter 2. There it was argued that colour vision in general is a mechanism which has evolved to enhance spatial resolution. Thus, rather than relying just upon the contrast that results from differences in the intensity of light reflected from surfaces, colour vision can provide additional information about the presence and nature of objects by detecting differences in the wavelengths of light. In essence, the finer the differences in wavelength that can be detected by an animal's vision system, the finer the spatial detail that it can potentially detect in its environment. For example, two patches of plumage could reflect equal numbers of photons but these photons may be from different parts of the spectrum. An animal with colour vision can readily detect this difference, while one without colour vision could not.

It was also pointed out that the suite of photoreceptors types and, in particular, the photopigments of the cone receptors, and their associated oil droplets, vary little across bird species. That is, the basic mechanism of colour vision appears to be one that evolved early and has been highly conserved (Chapter 2, 'Photoreceptors and visual pigments'). This suggests that colour vision in birds has general all-purpose properties, which are not tuned to specific tasks performed by different species.

It also seems likely that while the colour vision system of birds is based upon four photoreceptor channels (tetrachromatic colour vision), there may be only two broad types of system found among birds which differ primarily by how far into the short wavelength end of the spectrum the sensitivity of one of the photopigments extends. The SWS1 (short wave sensitive type 1) cone photopigment comes in two types, and these types are found in different species. There are pigments with λ_{max} at 365 nm, referred to as ultraviolet sensitive (UVS) and those in which λ_{max} is at 410 nm, referred to as violet sensitive (VS) (Wilkie et al. 2000). It is possession of the UVS pigment which gives the oscine passerines, gulls, ostriches, and parrots their visual sensitivity into the UV part of the spectrum.

Since the discovery of UV sensitivity in birds, there has been considerable interest in its function and a number of studies have focused on identifying particular functions for UV sensitivity and the enhancement of colour vision and hence spatial discrimination that it can provide. For example, there have been a number of studies on the identification of differential reflection in the UV part of the spectrum in objects used in display behaviours (Endler et al. 2014), in plumage patterns (Andersson and Amundsen 1997; Bennett et al. 1996; Bennett et al. 1997; Cuthill et al. 1999; Cuthill et al. 2000; Vorobyev et al. 1998), in particular types of fruits (Burkhardt 1982), a general role in the foraging of passerines (Church et al. 1998), and even in the detection of scent marks left by small mammals in grass which were reputed to be used by raptors to locate profitable foraging patches (Viitala et al. 1995). It seems that while the detection of UV reflective plumage patterns is well established as a cue in the reproductive behaviour of certain birds, it is not a universal feature of bird plumage, and this is because not all bird species have photoreceptors containing the UVS photopigment (Odeen and Hastad 2013). Demonstration of the use of differential UV reflection of fruits against foliage as a specific aid to foraging (Burkhardt 1982) still needs to be demonstrated in detail but it is a distinct possibility (see Chapter 2, Figure 2.5). The claimed use of UV vision in the foraging of raptors seems highly unlikely. This is because it has now been shown that not only do raptors lack the UVS photopigments but the media of their eyes filter out light at wavelengths below about 380 nm thus preventing UV signals from even reaching the retina (Lind et al. 2013; Lind et al. 2014). Furthermore, the urine trails of voles which were reputed to provide a UV cue to the presence of suitable prey have similar reflection at these wavelengths as water (Lind et al. 2013).

Thus, it has been possible to identify examples of the specific usage of information that can be derived only from the UV end of the spectrum in certain birds. However, this does not mean that UV is a special source of information compared with the other parts of the spectrum that are widely used in plumages, fruits, and foliage reflective patterns, etc. Indeed as explained in Chapter 2 ('Photoreceptors and visual pigments'), UV sensitivity is part of the colour vision mechanism and not a special channel of information. This means that the tasks involving the detection of UV reflectivity cannot be considered to have driven or serve to maintain UV vision in these species, any more than the rest of the spectrum. Rather colour vision in birds would seem to have general utility for enhancing spatial resolution across the spectrum. In some bird taxa that spectrum extends to the shorter UV wavelengths that human eyes cannot detect. Thus colour vision is probably best viewed in birds, even in those which see into the UV, as maintained or driven by unspecified ubiquitous behaviours which require fine spatial resolution. It seems more likely that plumage signals in the UV, and fruit and foliage differential reflectivity in the UV, have evolved in response to UV sensitivity of birds. That is, it is bird colour vision which has driven the spectral qualities of certain targets, rather than the other way around.

8.13 What Drives Bird Senses?

This chapter began by posing a general question, 'What drives bird senses?' Unfortunately while the chapter seemed to promise much, it has not provided a comprehensive answer. Although it has been possible to advance ideas about key tasks that drive vision, it does not seem possible to say more about other senses beyond the many observations of their links with foraging which have been described in previous chapters. It does seem clear that tactile sensitivity of bill tip organs, olfaction, taste, and specific aspects of hearing in some species are clearly linked to informational challenges involved in foraging for a wide range of different foods in different situations. Thus it is reasonable to assert that the task of foraging is the key driver of bird senses. Providing information sufficient for successful foraging must set the general limits of most sensory capabilities in birds. All other behaviours take place within those limits, although there are notable exceptions such as the use of active sonar in swifts and oilbirds which seem to be driven only by the requirement of mobility in the absence of light within caves.

From this analysis of bird senses, it is reasonable to propose that from a sensory ecology perspective a bird should be characterized as '*a bill guided by an eye*'. Such a phrase does seem to capture the essence of the task that a bird essentially does most of its waking life, i.e. guiding their bill towards targets. Furthermore, it seems clear that this task is driven by visual information in the majority of species. In many species or situations, the spatial resolution used for bill control is relatively low and in a few species bill position is guided by tactile information. Of course, as seen in the previous chapters, a very wide range of non-visual sensory information is available to birds, and this is used to various degrees by all birds when faced with specific perceptual challenges. Clearly non-visual information is important to different degrees by species as they exploit the resources of different habitats. However, while this information is crucial for successful completion of a bird's life cycle it is not subject to the same high degree of selection that applies continuously to getting the bill to the right place at the right time. To characterize a bird as 'a bill guided by an eye' is an attractive phrase, which hopefully catches the imagination and stimulates further questions.

Clearly, more detailed comparative analysis of sensory capacities and the structures underlying them will reveal more extraordinary stories of how birds gain and use information extracted from the world about them. This would reinforce the opening observations of Chapter 1 that 'birds' eye views' are extremely diverse and that birds truly live in different worlds to us. It would also reinforce the fact that the world that humans take for granted is just one of many. For a human to experience a real bird's eye view is impossible. However, trying to fully explore what this means is rewarding for we soon find that our experience of the world provides but a small insight into the world that supports our existence.

9
The Sensory Ecology of Collisions and Entrapment

Power lines, fences, communication masts, and buildings have long been recognized as posing major problems for certain bird species and local populations (Figure 9.1). In some instances, birds collide with objects that may be partially obscured by vegetation, for example, fences in woodlands (Catt et al. 1994; Summers and Dugan 2001), but collisions also occur when these static artefacts extend prominently into the open airspace above surrounding vegetation and appear very conspicuous to humans (Avery et al. 1980; Bevanger 1998; Drewitt and Langston 2008; Manville 2005). Some collisions occur at low light levels but others occur in full daylight. Birds from a wide range of species may also collide with panes of glass and diving birds may be entrapped in fishing nets. Gillnet fisheries sometimes set nets that are many kilometres long, and at various depths. It is estimated that entrapment of diving birds claims at least 400,000 bird lives annually (Lewison et al. 2014; Zydelis et al. 2013). Many birds also die as a result of being hit by fast-moving objects such a road vehicles, aircraft, and trains (Kelly et al. 2000; Lima et al. 2015; Sodhi 2002; Thorpe 2003) and some die as a result of collisions with the rotating blades of wind turbines (Drewitt and Langston 2008; Hodos 2003; Rothery 2009).

These instances of bird mortality through collisions and entrapment are not trivial. It has been claimed that mortality caused by collisions with static and moving human artefacts are the largest unintended human cause of avian fatalities worldwide and many hundreds of millions of birds are estimated to be killed annually (Banks 1979; Bishop and Brogan 2013; Erritzoe et al. 2003; Klem et al. 2004; Loss et al. 2014). A very large proportion of fatal collisions with artefacts occur between birds and static structures in open air space, and there is evidence that such collisions, with large and prominent obstacles, may threaten the survival of specific populations or even the survival of certain endangered species (Shaw et al. 2010). For example, in Europe over a 16-year period, it was estimated that approximately 25% of juvenile and 6% of adult White Storks died annually from power line collisions and electrocutions (Schaub and Pradel 2004). In the Overberg region of South Africa, higher rates of power line mortality have been estimated. Twelve

The Sensory Ecology of Birds. Graham R. Martin: Oxford University Press (2017).
 DOI 10.1093/acprof:oso/9780199694532.001.0001

Figure 9.1 Collision hazards. Structures known to be important collision hazards to birds can be large, substantial, and obvious to human observers. Wind turbines and power lines are well-known hazards and in some locations have been recorded as causing significant mortality rates to large and slow-flying species. Small birds may be particularly vulnerable to collisions with glass surfaces and with large, illuminated oil platforms, while simple fences can be hazardous to medium-sized woodland birds, especially grouse. Fishing nets pose significant entrapment hazards to a very wide range of diving birds; in fact, almost any bird species which forages by diving has been recorded as being trapped in nets. Some species, such as the auks, are particularly prone to net entrapment.

percent of Blue Cranes *Anthropoides paradiseus*, a species classified as globally Vulnerable (BirdLife 2009), and 30% of Denham's Bustards *Neotis denhami* are killed annually by power line collisions (Shaw 2009). Ludwig's Bustard *Neotis ludwigii*, White Stork, Grey Crowned crane *Balearica regulorum*, and Kori Bustard *Ardeotis kori* are amongst the other most commonly reported power line collision victims in Southern Africa (Eskom 2008). For Ludwig's Bustard, it is estimated that the rate of mortality from collisions is probably unsustainable, ultimately threatening the survival of this species (Jenkins et al. 2010).

The fact that flying birds are prone to collisions with large static objects may seem surprising because collisions often involve large-eyed diurnally active species in which visual resolution is expected to be high (Chapter 2). Furthermore, collisions frequently occur when birds are apparently in control of their flight, not buffeted by winds, and under conditions of good visibility (Drewitt and Langston 2008). The fact that diving birds get caught in fishing nets is perhaps not so surprising since they are designed to be inconspicuous to fish. Understanding the sensory bases of collisions and entrapment should indicate ways to mitigate these problems: an application of sensory ecology.

Collisions with moving objects would seem to pose different informational and perceptual challenges compared with those that can result in birds flying into static objects. This is in part due to the high speeds with which moving objects, such as motor vehicles, trains, and aircraft, may approach a bird compared with the relatively much slower speed of approach when a bird is propelling itself towards a static object. There may also be a significant cognitive difference between these two situations. As argued below many collisions with static objects may occur because birds are simply not detecting the target. On the other hand, fast-approaching vehicles, even when they are detected, may all be interpreted in the same way, as approaching predators (Bernhardt et al. 2010; Lima et al. 2015). This categorical perception initiates standardized predator escape behaviour, and this occurs at a set distance from the bird, regardless of the speed of the vehicle (DeVault et al. 2015). The result is that there is often not sufficient time to make a successful 'escape'. In essence, it is the speed of the approaching vehicle that kills. This is because the escape responses of birds to oncoming vehicles are too slow and too late in the face of 'apparent predators' travelling at speeds in excess of those of the natural predators with which the birds have co-evolved (DeVault et al. 2015).

9.1 Why do Flying Birds Collide with Static Objects?

Analysis of data on collision incidents has tended to focus primarily upon collision susceptibility that result from flight behaviour, especially flight manoeuvrability with respect to velocity of approach to an obstacle (Bevanger 1998; Drewitt and Langston 2008; Janss 2000). Visual and perceptual aspects of collisions have, until recently, received little investigation beyond the general observation that some collisions may occur when visibility is reduced due to lower light levels or weather conditions, such as rain or mist, which reduce the amount of visual information available for the control of flight. However, recent analyses have attempted to look at the problem from the perspective of sensory ecology, and these have provided alternative insights into the problem (Martin et al. 2012; Martin and Shaw 2010).

Measures to reduce the probability of collisions have usually involved marking obstructions with devices designed to increase the probability of their detection from a greater distance, the assumption being that the obstruction itself is below the limit of visual resolution within the flight avoidance distance of many bird species. For example, power lines have been marked with objects such as reflective balls, flapping flags, and wire coils (Bevanger 1994; Janss and Ferrer 1998); fences have been marked with flags (Summers and Dugan 2001); and there have been laboratory simulations of the effectiveness of marking turbine blades with patterns designed to reduce 'motion smear' (also known as 'motion transparency' and 'motion blur') (Hodos 2003; McIsaac 2001). However, despite more than 30 years of using static markers on power wires, the probability of mortality caused by power line collisions remains high for certain species (Drewitt and Langston 2008; Janss and Ferrer 2000).

Until recently solutions for reducing collisions have been based upon a human perspective of the problem. Put simply, it has been a matter of trying to find solutions to bird collision problems based upon making the perceived hazard more conspicuous to human observers (Martin 2011). Furthermore, work on the development of hazard markers has been constrained by the need to find solutions which have low initial cost, are easy to apply, and are easy to maintain. What is likely to be conspicuous to a bird flying in a particular environment or situation has rarely, if ever, been considered. Solutions have been off-the-shelf, based upon local availability of devices, or common practice.

In light of topics discussed in the previous chapters, it is clear that the human view provides just one way of appreciating the world and that the differences between human and birds' eye views are sufficient to probably render a human view of the problem of bird collisions both inaccurate and misleading.

9.2 Information Available to Flying Birds

It is clear from the preceding chapter that the information that birds extract visually from their environment can be quite different from that extracted by humans in the same circumstance and it is worth rehearsing the four key areas of visual capacity in which the vision of birds can differ from humans.

9.2.1 Colour Vision

The fundamental property of avian colour vision compared to mammals, and particularly primates, lies in the broader extent of the visible spectrum in birds and the subtlety of colour discriminations that can be made within that spectrum. The few detailed psychophysical studies of colour discrimination in birds suggest that birds are capable of subtle discriminations throughout their visible spectrum, including the UV or near-UV (Wright 1979). This has been supported by general models of how the different types of retinal cone photoreceptors in birds mediate colour discrimination (Endler and Mielke 2005; Vorobyev 2003; Vorobyev and Osorio 1998).

9.2.2 Spatial Resolution

Two key findings come from studies of spatial resolution in birds. First, acuity in birds can be high compared to those in other vertebrates with eyes of similar size, suggesting that the eyes of the majority of diurnally active birds can be characterized as adapted to maximize resolution rather than sensitivity (Land and Nilsson 2012). However, despite earlier claims based upon anecdotal observations, the highest known acuity in birds is not exceptionally superior to that of humans (Table 2.1, Appendix 1). For example, earlier claims of exceptional acuity in falcons and eagles have been more recently revised downwards to suggest that the highest acuity of falcons (0.75 minutes of arc) is approximately equal to that of the eye of a young human (0.4 minutes of arc) (Fox et al. 1976; Gaffney and Hodos 2003; Hirsch 1982; Reymond 1987), while that of the largest eagles (0.2 minutes of arc) is perhaps twice that of human eyes (Reymond 1985). For many bird species, acuity is below that of the human fovea at similar light levels, for example Rock Dove (1.7 minutes of arc), Rook *Corvus frugilegus* (1.0 minutes of arc), domestic fowl (3.4 minutes of arc) (Hodos 1993, and Table 2.1).

Second, acuity varies markedly within the visual field of an eye. In humans, there is a single region of high acuity vision which projects directly forwards, typically in the direction of travel. This region is of small angular size (≈ 2° diameter) compared with the total visual field of a single eye (160°) and is mediated by the foveal region of the retina (Westheimer 1972). Acuity decreases rapidly towards the periphery of the eye's visual field. In birds, there may be two areas of high acuity in each eye (Chapter 2, 2.7.1). One typically projects laterally with respect to the axis of the head, approximately along the optical axis of the eye; there may be other frontal or ventrally projecting areas of higher acuity but they do not project directly forwards, and in the eyes of some bird species there is an area of higher acuity that extends in a band across the field of view (Coimbra et al. 2014a; Coimbra et al. 2015; Mitkus et al. 2014; Tyrell et al. 2013) (Figure 2.7). Typically, the regions of highest acuity occur laterally, not frontally, with respect to the head and when behavioural techniques are used in assessments of a bird's acuity, a bird usually chooses to use its lateral field of view. Anatomical evidence also corroborates the use of these laterally projecting regions for tasks involving the determination of the highest visual acuity by freely moving birds (Reymond 1985; Reymond 1987).

9.2.3 Relative Depth, Distance, and Time-to-contact

Determination of the position of an object in relative depth from an animal, as well as its absolute distance, is a complex perceptual process for any visual system (Bruce et al. 2003; Goldstein 1984). However, being able to determine relative depth and distance may not be key to understanding collision avoidance in birds. Rather, what might be crucial is determining time-to-contact an obstacle since this will determine whether avoiding manoeuvres are possible. This is probably also true of humans. The most vital visual information beyond recognition of an object is the object's position and, if there is relative speed between the object and the observer, information on time-to-contact. The actual distance of an object from a bird may be of little importance compared with its direction and the time it may take to make contact with it.

Such information is available from optic flow-fields and it has been shown that birds do make use of such information (Bhagavatula et al. 2011; Davies and Green 1994; Lee et al. 1991). As described in Chapter 8, flow-fields can specify very accurately both the direction of travel and the time to contact an object that is being approached. They may underpin many tasks undertaken by humans, such as driving, cycling, running, jumping, and ball-catching; these are tasks in which the observer has to adjust speed of approach to achieve accurate timing of arrival at a given point (Lee 1980). However, the information extracted from the optic flow-field is contained not in highly detailed spatial information but in information extracted from moving images at relatively low resolution.

9.2.4 Fields of View

Visual fields and the variation of visual capacities within them are likely to have a direct impact on collision susceptibility. This is because visual fields determine what part of an animal's environment can influence its behaviour at any one instant. Especially important are differences in the characteristics of the sections of a bird's visual field that are used to detect and analyse objects of interest, and the characteristics of the section of the visual field that projects forward and hence 'looks' in the direction of travel.

In humans, the section of the visual field that is used for detailed analysis of objects looks directly forward. However, this is not the case in birds. For humans, the detailed world lies ahead, whereas for birds the detailed world lies laterally. As argued in Chapter 8, in birds the function of the forward-facing binocular field appears primarily to be the control of behaviours requiring the accurate positioning and timing of bill-opening towards nearby objects (particularly the control of bill position for food procurement and/or chick provisioning); the control of locomotion with respect to more distant objects is a less important determinant of binocular field characteristics (Martin 2009). Indeed, for birds such as the filter-feeding ducks or probing shorebirds, the binocular field can be very narrow (< 10°) in the direction of travel when flying.

9.3 Comparing Bird and Human Views of their Worlds

Clearly, as described in Chapter 8, the human and bird's eye views of the world differ markedly and for the purpose of analysing collision susceptibility in birds the following differences seem key.

In humans, highest spatial acuity and most acute colour discriminations lie directly ahead and there are extensive blind regions above and behind the head. The best appreciation of relative depth also lies directly ahead. In essence, humans see the world as being 'in front', and we move 'into' it. In birds, the eyes are placed laterally in the skull providing visual coverage of the world ahead, but there is also typically extensive coverage above and behind the head. Binocular and blind areas differ in extent and position, depending upon the ecology of the species. In general, the region of binocular vision is small and even in fast-flying species can be < 10°. There are marked differences in visual capacities within a field of view but typically retinal regions that provide the highest resolution and colour discrimination capacities project laterally, not forwards. Frontal vision is primarily concerned with near tasks such as the control of bill position in foraging (pecking/lunging), chick provisioning, and nest building, not the control of locomotion. Control of locomotion is achieved within constraints imposed by the task of controlling bill position in the conduct of these near tasks.

9.4 The Functions of Lateral Vision in Birds

Birds use their lateral visual fields for many key tasks. Typically, lateral vision has been seen as serving the detection of predators or conspecifics (Fernandez-Juricic et al. 2008). However, it is now clear that lateral vision in birds has a prime role in sophisticated aspects of foraging and predator detection tasks, including the response to novel stimuli and the reliable separation of pertinent from distracting stimuli (Rogers 2008). Moreover, it is also clear that the avian brain is functionally lateralized in the conduct of such tasks and that this is revealed by birds preferentially using their left and right eyes for different tasks (Rogers 2008). The use of lateral vision is seen clearly in tasks in which birds choose to examine different types of objects and scenes preferentially with the left or right eye, rather than binocularly (Dharmaretnam and Andrew 1994; Franklin and Lima 2010; Koboroff et al. 2008; Mench and Andrew 1986; Rogers 1991). Furthermore, use of the left and right eye is secondary to selective activation of the contralateral hemisphere of the brain, and this can change during free viewing depending on the task (Vallortigara 2000). This refers not only to eye use but more generally to allocation of attention in the left and right visual hemispaces (Diekamp et al. 2005). Therefore birds seem to be highly lateralized with respect to both eye and brain function.

Preferential use of lateral vision has been described both in tasks involving close objects, which require approach or a pecking response towards the object, and in tasks that involve flight towards distant objects. Thus a Peregrine Falcon approaching its prey seems to be primarily under the control of lateral rather than frontal/binocular vision, in that the bird typically approaches along a curving path that would allow the prey object to be kept in the vision of the laterally projecting fovea of one eye until the final closure upon the prey object, when transfer is passed to frontal vision at close range (Tucker 2000; Tucker et al. 2000). However, this behaviour may not apply to all falcon species (Kane and Zamani 2014) and it is necessary to consider carefully the distance/time to contact at which frontal vision takes control of the final approach to prey. A similar though less dramatic example of the use of lateral vision for the detection of prey has been investigated in Gull-billed Terns *Gelochelidon nilotica* foraging for crabs above mud-flats (Land 1999). These birds use lateral vision and turn the heads distinctly sideway to look down at the ground, but transfer to frontal vision just before prey capture as they dive downwards. This use of lateral vision to detect an object, and the control of behaviour passing to the frontal field only when the object is in close proximity is similar to that described in thrushes (Montgomerie and Weatherhead 1997) (Figure 2.10), Zebra Finches (Bischof 1988), Rock Doves (Bloch et al. 1988), and Chickens (Dharmaretnam and Andrew 1994), all of which take items by detecting them on the surface on which they are standing.

All these examples suggest that lateral and binocular vision are used for specific tasks and are not interchangeable in their function. Therefore for tasks requiring high spatial resolution, and perhaps separation of pertinent from distracting stimuli, birds seem to fixate initially upon a target with one of their lateral fields of view, and behavioural control typically passes to frontal (binocular) vision for final seizure of object/food only at close range: this may apply to a very wide range of species and tasks.

9.5 When Birds are Flying in Open Airspace, What are they Doing?

When flying in open airspace, are birds looking ahead for obstacles? The complexity of visual fields and the topographical distribution of visual capacities within them suggest that it is wrong to assume that birds are looking forward into the open airspace and attending to what might lie ahead of them. What birds might actually be attending to in flight presents a more complex set of possibilities than the situation in humans.

9.5.1 Looking but Failing to See

Humans seem able to do no other than look at what lies in front of their heads. If this is the case, why do humans ever experience collisions? Humans exemplify an additional problem for the analysis of collisions—the need to differentiate between vision and visual perception. Even when looking ahead in the direction of travel, it has been established that car drivers may at times 'look but fail to see', and that people may frequently drive beyond their 'perceptual limit'. In other words, they drive in a manner that relies on information which is not in fact immediately available via their visual system (Hills 1980); such a framework has been used in the analysis of collisions (Clarke et al. 1995; Clarke et al. 1998). Collisions are generally avoided, however, because the available visual information is supplemented by experience of the nature of roads and traffic and specific knowledge of road layout. The driver simply interprets a meagre or sparse set of information in a useful way. Drivers often, perhaps frequently, assume that the road will continue much as before, unless specific signals suggest otherwise. Indeed, much road engineering is concerned with making roads as predictable as possible within a geographical region, so that only minimal cues are needed to drive safely (Hills 1980). The fact that drivers are often beyond their perceptual limit is indicated by what happens when there is an unpredictable obstruction in the road. At such times, it is more likely that a collision will happen. This is because the rate at which a driver is gaining information about the environment ahead is not sufficient to match the challenge posed. This is one of the main reasons why known driving hazards have to be so well indicated, to warn or prime the driver that the world ahead will be less predictable and that they should reduce speed to adjust their rate of gain of information to match the requirements of the changing or changed circumstances.

Car driving may not seem particularly relevant to birds in flight. However, it does seem likely that the same principles concerning the role of cognition in the correct interpretation of cues and the adjustment of the rate of gain of information to match the challenges of the environment are of wide application (Gibson 1986). The fact that some birds rely habitually on sparse or a paucity of information when foraging at low light levels (Chapter 6) or underwater (Chapter 7) has already been discussed. The correct interpretation of sparse information would seem to underlie the execution of those certain tasks by nocturnal and diving birds. When we consider collision vulnerability in birds, it is worth considering whether a similar phenomenon is in operation. Do birds have available to them a rich array of information to guide their flight or are they sometimes executing tasks at the limits of the information that is available to them? Is there often a paucity of information? Two further key questions arise.

First, when flying can birds adjust their rate of gain of information to meet the perceptual challenge of the environment? For example, under conditions of reduced visibility (which will result in a reduced rate of information gain), can birds slow down and adjust their rate of gain of information to meet the perceptual challenge? Second, in open habitats are flying birds always looking ahead?

9.5.2 Can Flying Birds Adjust their Rate of Gain of Visual Information?

It is well established that the aerobic range of flight speeds for any bird is restricted. The well-established U-shaped function of aerodynamic power requirement as a function of flight speed has wide applicability (Biewener 2003). It shows that for most birds, slow flight, even for short periods, is not possible and this becomes more acute for birds with high wing loading and consequently higher average flight speeds (Biewener 2003; Norberg 1990). In essence, birds cannot readily slow down. Sustained slow flight for a bird which has a high average flight speed is costly or aerodynamically impossible and, hence, being able to reduce speed in order to match the rate of gain of information to increasing perceptual challenges is unlikely to occur. In other words, when the environment restricts the information available (e.g. rain, mist, low light levels), birds cannot easily fly more slowly in order to meet the increased perceptual challenge. Thus if birds are to fly under non-ideal perceptual conditions, or visibility conditions change during a flight, they cannot act in the way that a careful car driver can and reduce their speed in order to gain information at a rate sufficient to match the new perceptual challenge.

9.5.3 Are Flying Birds Always Looking Ahead?

Vision in the direction of travel that is mediated by the forward-projecting binocular region may provide far less spatial information than vision laterally. Extrapolations about what a bird 'ought' to be able to see ahead that are based upon measures of highest acuity (which are almost certainly based upon lateral vision; see Appendix 1) are necessarily over-estimates of visual performance. Furthermore, as argued in Chapter 8, frontal vision may have quite restricted functions that are concerned primarily with nearby objects rather than more distant ones. That is not to say that birds may not 'see' objects that lie ahead at a distance, but it is likely that frontal binocular vision does not match the best visual performance derived from the laterally projecting sections of the visual fields. Furthermore, the open airspace above vegetation is a highly predictable environment, usually clear of hazards, and birds may not be perceptually primed either through learning or evolutionary selection to detect hazards that extend into this airspace from below. Much in the same way that a car driver requires perceptual priming when a predictable road becomes less predictable, birds may also fail to detect objects because they too 'look but fail to see' what lies ahead when flying in open airspace. This is not to imply that birds are unaware of their surroundings when in flight above vegetation, since it is likely that they are perceptually primed to detect aerial predators, but that these are specific targets that are likely to appear above or behind the flying bird, not in the frontal visual hemifield. The fact that birds appear to be primed for the detection of predators is supported by the rapid response shown when the silhouette of a moving predator are presented above them, as in studies of escape or avoidance behaviours (Devereux et al. 2006).

Do birds sometimes actually fail to see the way ahead? Certainly the evidence that some birds can spontaneously abolish their binocular (frontal) field suggests that, at least momentarily, birds of some species could simply not look ahead during flight. In addition, birds can turn their heads sideways (yaw) to bring the laterally projecting visual field to look more forwards with respect to the direction of travel. More important, however, are examples where birds have frontal binocular fields that are of restricted vertical extent, and with extensive blind areas above and below them. In these cases, only relatively small amplitude (25°–35°) downward pitch movements of the head from those typically adopted in level flight will bring these blind areas to project forward in the direction of travel.

This has shown to be the case in a number of collision-prone species including Blue Cranes and Kori Bustards, Griffon Vultures and Short-toed Snake Eagles (Martin et al. 2012; Martin and Katzir 1999; Martin and Shaw 2010). In all of these birds, the binocular field is small in extent, both in width and vertically and is approximately centred about the projection of the bill tip. There is a large blind area above the head, probably associated with avoiding imaging the sun (Chapter 8, 8.8), and this projects forwards into the frontal hemisphere of the visual field (Figure 2.14). In these birds, a relatively small change in the pitch of the head will bring this blind area to project forwards in the direction of travel (Figure 9.2). In the vultures, anecdotal sources, including field observations, video clips, and still photographs of birds in flight, show that birds are often seen with their head pitched downwards or even facing backwards during foraging flight (Figures 9.3 and 9.4).

Why should birds in flight pitch their head down in this manner? In the case of the terns, vultures, and eagles this seems to be directly linked with foraging behaviour, since prey and carrion are principally detected on the ground or in water below. In the case of bustards and cranes, birds may not be looking for individual prey items but for foraging patches, groups of other birds, or searching for roost sites. However, in all cases the birds are more interested in what is below them than what lies ahead in the (presumed) open airspace. It seems clear that when these birds are 'looking downwards' when foraging they are simply not looking where they are going. They may occasionally raise their head to see forwards but even then they may 'look but fail to see'. This is because throughout the whole of their evolution, when flying above natural terrains at a height defined by the tallest trees, the airspace has been empty, and sparse spatial information has been sufficient to ensure safe flight. There would seem to be a parallel here with the reliance on sparse information that is sufficient to guide amphibious birds when underwater (Chapter 7).

9.6 The Sensory Ecology of Collisions

The evidence and arguments presented above suggest that bird collisions may be the result of both visual and perceptual constraints on the information that is available to birds as they fly in the open airspace. Analyses of the reasons why any

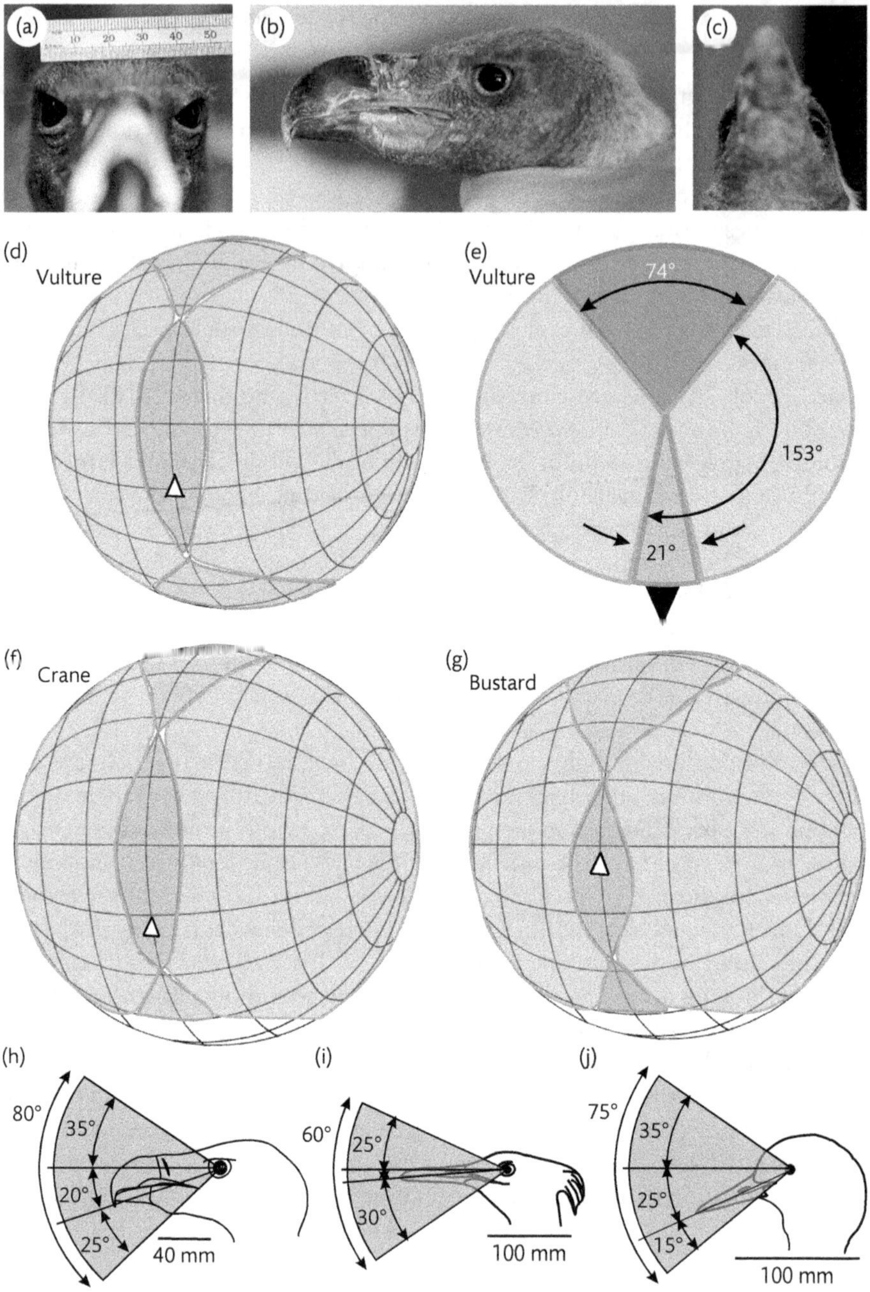

Figure 9.2 Vision and collision vulnerability. Vultures, bustards, and cranes are among the birds which are particularly vulnerable to collisions with large hazards, such as wind turbines and power lines. The photographs of a Griffon Vulture *Gyps fulvus* (a, b, c) show the prominent eyebrow ridges (see also Figure 8.7) which play a role in restricting the size of the visual field and in particular result in a large blind area above and behind the head and a

small binocular field (d, e, h). This in turn means that visual coverage to the front of the bird is much reduced compared with many birds species (see Figures 8.1 and 8.5). Similar visual field configurations are also found in Blue Cranes *Grus paradisea* (f and j) and Kori Bustards *Ardeotis kori* (g and i) which are also vulnerable to collision with power lines. The significance of these restricted frontal visual fields to the collision vulnerability of these birds is made clear in Figure 9.3.

particular species may find particular situations hazardous should acknowledge the following:

1. Birds live in quite different visual worlds to that occupied by humans. It is not possible simply to extrapolate from knowledge of the human perception of a hazard to understand the problem faced by a bird.
2. In flight, some birds may be blind ahead of them. Turning the head in both pitch and yaw to look downwards either with the binocular field or with the central part of an eye's visual field, may not be unusual. This may leave birds blind in the direction of travel.
3. Frontal vision in most birds is not high resolution vision; highest resolution occurs in the lateral fields of view.
4. Frontal vision in birds may be tuned for the detection of movement concerned with the extraction of information from the optic flow-field, rather than high spatial detail.
5. Birds probably employ lateral vision for the detection of conspecifics, foraging opportunities and predators. Attention to these may be more important for a bird than simply looking ahead during flight in the open airspace.
6. Birds in flight may predict that the environment ahead is not cluttered. Even if they are 'looking ahead', they may fail to see an obstacle since they may not predict obstructions. Perceptually they have no 'prior' for human artefacts such as buildings, power wires, or wind turbines.
7. Birds have only a restricted range of flight speeds that can be used to adjust their rate of information gain as the sensory challenges of the environment change due to reduced visibility caused, for example, by rain, mist, or lower light levels.

9.7. A Sensory Ecology Perspective on Collision Mitigation

9.7.1 Collisions with Static Hazards

Armed with this information, are there solutions to the problem of collisions? Some general principles can be suggested.

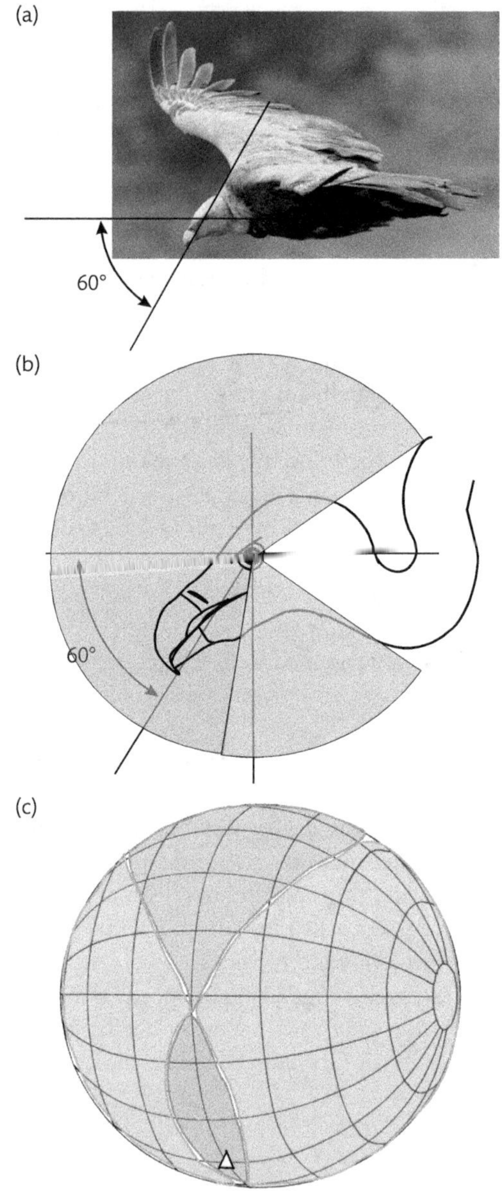

Figure 9.3 Pitching forward and flying blind. Unlike some of the New World vultures (Cathartidae) which have the benefit of olfaction to help them find carcases (Figure 3.6), Old World vultures (Accipitridae) forage exclusively using vision. They fly in the open airspace and look downward scanning for carrion, and they also keep a look out laterally for other vultures foraging in the area which may start descending towards a food item. This means that these birds are primarily concerned with what is below them, or out to their sides, rather than what is in front or above them. The presence of eyebrow ridges may help to keep

the sun out of their eyes while scanning below, but as shown in Figure 9.2 they also restrict the size of the visual field. In this diagram, the blind area is depicted in blue. Frequently, a foraging bird may pitch its head through an angle of more than 60°Compared with the horizontal, as shown in (a). This is sufficient to bring the blind area to face in the direction of travel (b and c). Thus, when these vultures are foraging they are frequently, probably habitually, flying blind as regards their direction of travel. This makes them particularly vulnerable to collisions with large human artefacts, such as turbines and power lines, which protrude into the open airspace above the level of local vegetation.

Visual factors

Although birds cannot be guaranteed to be looking at or attending to an obstacle that extends into open airspace, it is still valuable to employ markers to increase their conspicuousness in key situations where collisions rates are high. Stimuli used to draw attention to the obstacle should, however, be as conspicuous as possible. If possible hazard markers should parallel the apparent overload of conspicuous and repeated information that is used to warn car drivers of change to the road ahead caused by roadworks or other hazards.

The warning stimulus should ideally be physically large, well in excess of the size calculated to be just detectable at a given distance based upon acuity measures

Figure 9.4 An extreme case of flying blind. A Griffon Vulture taking a keen interest in what is below and clearly not looking where it is going! This may seem an extreme case of flying blind, but it is not atypical of this species. Photograph courtesy of Olivier Duriez.

(Appendix 1), and incorporate movement. These features should increase the chances that a stimulus will be detected at a distance sufficient for a change in flight path to be initiated. This recommendation takes account of the idea that forward vision in birds may be tuned primarily for extracting information from optic-flow rather than static stimuli, and that estimates of acuity typically refer to the highest performance of spatial resolution which occurs in the lateral, not the frontal, fields of view.

It is worth recalling that diving birds will forage guided by a paucity of spatial information (Chapter 7). It may be that birds are similarly prepared to fly in what they predict will be open airspace when there is a paucity of spatial information available to them.

Although birds have an extended visible spectrum compared to humans, it seems unlikely that stimuli specifically employing reflectance in the short wavelength (UV) end of the spectrum will have particular salience. In fact, the target that is likely to remain conspicuous under all possible viewing conditions (e.g. reduced light levels, reduced contrast due to mist or rain) should simply be of high black and white contrast so that it reflects highly or absorbs strongly across the full spectrum of ambient light. This is because the degree to which a coloured target is conspicuous depends upon the spectral characteristics of ambient and background illumination as well as the distribution of spectral information within the target itself (Endler and Mielke 2005), and these can vary markedly with situation, time of day, and cloud cover.

Perceptual factors

Collisions with obstacles in the open airspace may be as much a perceptual problem as a visual one. Therefore, if possible, solutions should be found that alert birds well in advance: their attention may need to be primed just as much as the car driver's when approaching a hazard. However, what constitutes a warning or alerting stimulus may be difficult to determine and is likely to vary with species, but auditory as well as visual warning stimuli might be effective since they may generally stimulate renewed or increased attention to the environment and the birds may search generally for a possible hazard.

Diverting and distracting birds

At locations where collision incidents are high, it may be more efficient to divert or distract birds from their flight path with respect to a known hazard rather than attempt to make the hazard more conspicuous. This is because to be effective, a warning on the obstacle itself may have to be very large and moving, and that for some collision-vulnerable species it may be best to assume that birds are more likely to be looking down and laterally rather than forwards. Therefore, any warnings or additional marker may not in any case be seen. To reduce collisions with known hazards, a signal on the ground may be more important than a signal on the obstacle. Foraging patches, conspecific models, or alerting sounds placed a suitable distance from the hazard may be necessary. Birds might be induced to land and then be more attentive to possible obstacles as they take off again into the airspace, rather than simply flying through it with reduced attention to what lies ahead. Alternatively, for low tension power lines, for example, planting barriers of trees on either side of the line should force birds to fly higher and avoid the hazard completely. Such environmental manipulations are likely to have general ecological benefit, but in many instance they may not be possible as the owners of the powerline hazard are likely to have limited control of the land in the vicinity of the obstruction.

Tailored rather than general solutions

There is unlikely to be a single effective way to reduce collisions for multiple species at any one site. Warning or diversion and distraction solutions may need to be tailored for particular target species. Solutions may need to take account of the foraging ecology and social behaviour of the species as well as its visual capacities in order to understand why it flies in the open airspace at particular locations.

9.7.2 Collisions with Moving Hazards: Aircraft and Wheeled Vehicles

Can anything be recommended from a sensory ecology perspective to mitigate bird collisions with moving hazards, especially aircraft and wheeled vehicles? A great deal of work has been done researching the causes of bird–aircraft collisions (bird strikes), primarily because such collisions represent a substantial safety concern and financial burden to aviation worldwide. Furthermore, the rate of damaging bird strikes is increasing, and this is in part thought to be because aircraft are become quieter and the general dispersive effect of loud noise on birds at airports have decreased (Blackwell et al. 2012).

Manipulating the environment

Much work has been done on managing the environment of airports to make them less attractive to birds (DeVault et al. 2013), but there remain questions as to whether aircraft can be made more conspicuous so as to enhance aircraft detection and predator escape responses of birds. Experiments with Canada Geese have shown that model aircraft with landing lights switched on are more readily detected than aircraft with their lights off (Blackwell et al. 2012). Because 99% of bird strikes occur under 3000 m (10,000 feet) (Dolbeer et al. 2015), landing lights are now typically on, day and night, when aircraft are flying below this level and this is now called for by regulation (FAA 2014) (paragraph 4-110 3-23c).

Manipulating the hazard

There is some evidence that aircraft fuselage colour has an effect on the occurrence of bird strikes (Fernandez-Juricic et al. 2011a). However, clear recommendations are difficult to make since these are broad correlational studies and may not identify situations and atmospheric conditions which are particularly hazardous. As argued above with respect to static collision hazards, the effect of ambient light levels and atmospheric conditions can make a marked difference to the salience of particular colours and spatial patterns under particular conditions. Therefore practical solutions, in terms of fuselage colour or the use of patterns similar to those suggested above with regard to making static hazards more conspicuous, may have only marginal effects on the total mortality attributable to bird-vehicle and bird-aircraft collisions.

Are vehicles perceived as predators?

The overwhelming problem for animal–vehicle collisions seems primarily to be behavioural rather than sensory. It is also in part dependent upon particular collision circumstances, for example, whether a bird is moving across the vehicle's path or the vehicle is moving towards the bird. It has been found that birds appear to interpret a fast-approaching vehicle as a predator (Bernhardt et al. 2010; DeVault et al. 2015), and that this typically triggers an escape response when the 'predator' appears to be at a set distance, regardless of its speed. When faced by a potential predator, birds seem to employ some kind of 'distance rule' for initiating an escape response (Lima and Dill 1990), and this also applies to oncoming vehicles/predators. Therefore, vehicle speed is crucial in determining whether a bird can escape.

Experiments with Brown-headed Cowbirds *Molothrus ater* showed that their escape response was usually initiated with less than 0.8 s before collision (the time needed for escape) when vehicles approached at speeds greater than 120 km h^{-1}. This is a 'late response' and results in a high risk of collision when birds are faced with highway traffic or aircraft during take-off and landing (DeVault et al. 2015). Anecdotal observations would suggest that other bird species behave in the same way when confronted by fast-moving vehicles and that they apply distance rules in a similar fashion. However, they may initiate their escape responses at different distances since it is well established that different species have different flight initiation distance when approached by more conventional predators (Blumstein 2003). Clearly evidence for other species would be valuable. Mitigation measures based upon these findings are, however, rather few. As in the case of static collision hazards, the reduction of fatalities primarily requires location-based management of the environment to reduce the encounter rate between birds and the hazard. Increasing the conspicuousness of the moving hazard, beyond the current practices of employing landing lights on aircraft at all times and the growing frequency of employing lights on road vehicles at all times, may not be possible. It is clear that speed does kill; lowering the speed of vehicles would seem to have greater benefit than any efforts to increase the general conspicuousness of vehicles.

9.8 Entrapment: The Problem of Gillnets and Diving Birds

The fact that birds are caught in fishing nets, especially gillnets, is not surprising. After all, nets are set to trap animals. The fact that gillnet bycatch of seabirds is a worldwide problem which results in the deaths of at least 400,000 birds annually is also not surprising (Lewison et al. 2014; Zydelis et al. 2013). Nets are deployed at a huge scale all around the globe. The annual rate of bycatch is thought to be unsustainable for some bird species, and there is evidence that in some localities gillnet bycatch has resulted in severe reductions in the numbers of breeding birds (Osterblom et al. 2002; Regular et al. 2013). Gillnet bycatch is not unique to birds but is a well-established issue for other animals groups including sea turtles (Chelonioidea), pinnipeds, cetaceans, and blue water fish (Tunas *Thunnus* spp. and billfish Istiophoridae and Xiphiidae) (Fritsches and Warrant 2006; Lewison et al. 2014; Myers and Worm 2003; Reeves et al. 2013; Wallace et al. 2013). It is recognized that there is an urgent need to reduce this bycatch, but at the same time, in order to obtain support and adoption by the fishing industry, it is desirable not to reduce the efficiency of gillnet fishing. This may be an impossible task. Although spatial and temporal closures of fisheries have been shown to be effective in managing the impact of gillnets (Regular et al. 2013), these may be difficult to establish and enforce, and they do not actually solve the problem of bycatch per se; rather they just limit when and how frequently it occurs. Can a sensory ecology perspective throw any light on possible technical solutions to the bird gillnet bycatch problem? Can nets be made less efficient at catching birds while maintaining their ability to catch target fish?

9.9 Gillnet Bycatch Bird Species

As with aerial collisions, bird species that are prone to entrapment in gillnets come from a range of orders and families (Zydelis et al. 2013). They include species which differ markedly in their foraging ecology, suggesting that the problem of gillnet bycatch may derive from rather broad and general sensory ecology factors rather than specific factors which predispose only certain species.

Thus bycatch-prone bird species that make up the majority of annual bycatch mortality primarily exhibit the following combinations of diet and foraging behaviours: surface diving to take sessile prey items (molluscs) which are removed from underwater substrates at depths down to 50 m (various ducks species;

Anatidae); surface diving to take fish from the water column at depths down to 150 m and mainly in open waters (various penguin and auk species; Spheniscidae and Alcidae); surface diving for fish, usually only to shallow depths (< 10 m) and mainly in coastal waters (various species of divers; Gaviidae).

The fuller list of species which are less commonly trapped in nets encompasses a much wider range of diets and foraging behaviours (Zydelis et al. 2013). There are species which take evasive prey (fish) and slow-moving prey (crustaceans and molluscs, e.g. squid) following plunge dives to shallow depth (Pelecanidae, Sulidae, Diomedeidae, Procellariidae, Hydrobatidae, Phaethontidae, Pelecanoididae). Species which take evasive prey at shallow to mid-depths from surface dives with the prey often disturbed from substrates or hiding places (Phalacrocoracidae, Podicipedidae), and species which take small generally non-evasive prey from or close below the water surface (Laridae). These food items may be detected by sight from above the water surface (terns and gulls), but the skimmers take prey by surface trawling guided by tactile, rather than visual cues (Martin et al. 2007). The important point to note is that practically any bird which forages where gillnets are deployed can be caught in them. As far as birds are concerned, gillnets are a catch-all device, and no one diet or foraging strategy seems to make the birds particularly prone to entrapment.

9.10 The Role of Vision in Seabird Gillnet Bycatch Species

It is highly probable that all of the species that take prey from the water column, or close to the seabed, use vision to detect prey and to direct their prey catching behaviour, or as suggested in the case of some auk species, they may rely upon random encounters with prey (Chapter 7, 7.7.3). Only among the ducks is there evidence that tactile cues may be used for prey detection. However, there is no direct evidence for tactile prey detection in either Great Scaup or Long-Tailed Ducks, which are particularly prone to gillnet by catch (Zydelis et al. 2013), but comparison with some other duck species (Chapter 4) makes this seems likely. Captive long-tailed ducks have, in fact, been observed 'ploughing' underwater substrates with their bills, suggesting that they are using tactile cues to detect hidden prey (S. Portugal, personal communication).

In this context, it is important to recall some of the key sensory conclusions concerning under water foraging by birds (Chapter 7).

1. Even those species which are visually guided in their prey capture cannot be assumed to have high spatial resolution.
2. While many of these species do forage during daytime, there is much evidence that many, including all those in the high impact category, especially auk species, may also forage regularly during twilight or night-time (Camphuysen

1998; Cramp and Simmons 1977; Gremillet et al. 2005; Nilsson 1969; Piersma et al. 1988; Regular et al. 2010; Regular et al. 2011; Systad and Bustnes 2001; Wanless et al. 1999).

3. Even when foraging in day time, some taxa (especially penguins and auks) may typically forage at depths at which ambient light levels are attenuated to equal those commonly experienced at the water surface under twilight and moon-lit night-time conditions (Hedd et al. 2009; Martin 1990a; Martin 1999a; Regular et al. 2010; Regular et al. 2011; Wilson et al. 1993).
4. Some species may even rely upon random encounters with prey rather than detect prey at a distance (Regular et al. 2011).

Thus diving birds are characterized as having available to them a paucity of information that is devoid of fine spatial detail in underwater environments and that prey is detected and caught using minimal sensory information. In light of this, it is not surprising that gillnets are a hazard to diving birds; unless near the surface in clear water and in bright daytime light conditions, gill nets will not be visible until the birds are very close to them, and under many circumstances, the nets are undetectable.

9.11 Distracting Birds from Nets

The ultimate solution, as in the case of aerial collisions, would be to distract birds away from the vicinity of obstructions so that there is no net–bird interaction. In terrestrial habitats, this may be possible because obstacles are fixed structures and their surrounding habitat features can be manipulated (Martin and Shaw 2010). However, this is not an option with gillnets which are deployed in new situations on an almost daily basis.

9.12 A Sensory Ecology Solution to Gillnet Bycatch

Mitigating gillnet entrapment is perhaps even more difficult than mitigating the collisions of flying birds with apparently 'obvious' artefacts that intrude into the open airspace. In the open airspace birds may not be keeping a look out directly ahead and their attention can be focused upon what lies laterally or even behind them. Underwater, spatial resolution is considerably lower than in air and, at best, only gross details can be resolved (Figures 7.5–7.8).

9.12.1 Making Nets Conspicuous

The simple, and perhaps not surprising, conclusion for gillnet bycatch reduction is that the actual nets need to be made more visible, i.e. the hazard has in some way to be made more conspicuous. However, there is an obvious tension in that nets need to be invisible to target fish species but at the same time nets need to be more detectable to bycatch species. A further problem is the possibility that if the nets are detectable to birds they could, in fact, attract the birds to them. If birds can see the net, they may investigate it and render themselves more vulnerable to being caught. This cannot be answered without knowledge of what birds do when they encounter a net or a warning stimulus; do they avoid or are they attracted, do they show curiosity or aversion?

Lighting nets

It may seem attractive to use light sources placed upon the nets to draw attention to them and/or to make them more visible. Indeed, such devices have been employed with some success aimed at reducing gillnet bycatch of turtles at night (Wang et al. 2013; Wang et al. 2010). However, this might not be advisable with birds because of the effect of bright lights on the adaptation of the retina to the ambient light. When foraging either at night or at depth vertebrate retinas will have a high degree of dark adaptation (Chapter 7, 7.7.2). Exposure to a light source within the twilight–daylight luminance range will produce a rapid reversal in adaptation of the retina. This will result in impairment in the retina's ability to resolve detail, at least within a portion of the visual field, and this impairment will last for a period considerably longer than the brief exposure to the light. This is because of the relatively long time necessary for the eye to readapt to the ambient light level (Warrant 2008). An eye which is not properly adapted to the ambient light regime has lower resolving power than one that is well adapted. Therefore exposure to a light is likely to increase the probability that parts of the net not immediately illuminated will be less visible. This is the effect that bright security lights can have at night with humans; they create regions around the light source in which details are less likely to be detected (see Chapter 7, 7.4).

Warning birds of the presence of nets

To avoid the problems of disrupting the dark-adapted state of the bird's retina, it has been argued that a measure is required which can warn birds of the presence of a net and which is effective over a wide range of the natural light levels that occur at depth. The most important requirement is that any warning stimulus should be conspicuous over a wide range of naturally occurring low light levels and also be effective under a range of contrast conditions. To achieve this, the warning stimuli need to be both relatively large and/or have high internal contrast (Kleyheeg-Hartman et al. 2014; Martin 2011; Martin and Shaw 2010; Summers and Dugan 2001). To achieve these aims, it has been proposed that it would be sufficient to introduce 'warning panels' to the surface of gillnets (Martin and Crawford 2015).

9.13 Warning Panels

9.13.1 Patterns on Warning Panels

It is instructive to briefly work through the argument used to justify the characteristic of what might be suitable warning panels. The visual acuity of cormorants is the only data available on birds underwater and provides the only bench mark from which to argue. Cormorant acuity is ≈ 10 min of arc at light levels within the higher end of the twilight range (Chapter 7). Since acuity decreases with light level and with reduced contrast (Figures 7.5 and 7.6), the best resolution of cormorants provides only the starting point for the design of a stimulus that will be detectable under the wide range of natural light conditions that birds experience when foraging under water. Thus a good guide as to the visual size of an object that would be detectable under a range of naturally occurring conditions would be to assume that the stimulus would need to be at least 10 times above threshold, i.e. it should subtend 100 min of arc. It is noteworthy that the European Union eye sight test for car driving requires vision capable of detecting a stimulus that is at least six time larger than the acuity of the human normative threshold (British Standard BS AU 145d; DVLA INF188/1); this means that it is considered that safe driving during daylight depends upon stimuli that are at least six time larger than the spatial threshold; driving at night is guided by lights, and night driving under natural light regimes without the benefit of lights is impossible under modern traffic conditions.

9.13.2 The Size of Warning Panels

A stimulus that can be detected at a particular level of acuity can be translated into the size of an object (of high contrast) that can be just detected at a given viewing distance (this is the basis of the car driving eyesight tests). Thus, in order that a high contrast grating stimulus panel, which meets the 100 min of arc threshold, is visible at a viewing distance of 2 m, the stripes should have a width of 60 mm. Two metres is chosen as the minimum detection distance as it would seem to be a possible minimum limit at which a foraging bird might detect a prey item in its lateral field and then be able to change swim direction to take the prey in its bill. Clearly data on the actual prey catching techniques of vulnerable bycatch species could refine this distance. To be effective as a grating, the whole stimulus should ideally contain at least 10 stripes (or 100 checkerboard squares), and so would need to be 60 cm × 60 cm in overall size.

Such panels would appear to the human eye in daylight as highly conspicuous. However, it should be borne in mind that the 60 mm stripe width is similar to the size fish prey that penguins, cormorants, auks, and divers regularly take. Therefore, this pattern provides a stimulus which contains the basic spatial element of the kinds of prey that these birds take. However, it would be predicted that the elements of the panel could be detected at a distance of only 2 m over a wide range of lower naturally occurring light levels (twilight to starlight) and low stimulus contrast conditions.

9.13.3 The Colour of Warning Panels

It should also be noted that there is no advantage in including any element of spectral information (colour) into such a stimulus panel. At the low light levels at which birds predominantly forage underwater their visual systems are not capable of colour vision. Vision would be mediated only by the rod receptors which have a sensitivity peak at about 500 nm. Furthermore, although the rod receptor photopigment has a detectable spectrum of approximately 400–600 nm (Chapter 2), at depth there would be very little light at the longer wavelengths (Figure 7.3).

9.13.4 The Detection Distances of Warning Panels

Clearly such grating (or chequerboard) warning panels of these dimensions and contrast would be detectable by the birds at a distance greater than 2 m under higher light level conditions. For example, at high daylight light levels and in clear water, it would be predicted that cormorants would be able to detect the elements of the pattern at a distance of 20 m if the water was clear enough. Thus this stimulus should be visible to bycatch-prone birds from between 20 m (under conditions of high light levels and low turbidity) and 2 m (under low natural light levels and high turbidity).

There is, however, no guarantee that the birds will detect or attend to warning panels from these longer distances since at least some diving species appear to be prepared to forage more or less in the absence of any detailed visual information (Chapter 7). All that can be done therefore is to provide a stimulus that is likely to be detectable under a range of light levels and from a minimum distance at which a diving bird is likely to be able to respond to its presence. However, given the range of light levels over which the stimulus needs to be detected, 2 m would provide a reasoned starting point from which to begin experimentation. Greater or smaller ideal detection distances would imply a change to the dimensions of the pattern and stimulus panel size, but would not change the nature of the stimulus.

9.13.5 How Many Panels?

Such a warning panel is designed to be detected by bycatch-prone species which encounter it at random while diving under natural foraging conditions and therefore it should be effective when placed at the actual net surface. Once the stimulus panel is detected it might be assumed that the bird is likely to move away from it to avoid collision. However, it is conceivable that the bird could approach to investigate the pattern. The question also arises as to how frequently animals will need to encounter such stimulus panels in order to avoid a large area of net. The assumption has been made that such panels should be visible under a range of viewing conditions at a distance of 2 m, and this will be achieved as a bird swims past a panel (using lateral visual field viewing) or approaches it at an acute angle. It would seem reasonable to suggest therefore that the panels should be placed in the plane of the net in a regular grid with centres at 4 m.

9.13.6 Would Warning Panels be Effective?

The above analysis and possible solution may not be workable because they may interfere too much with the efficiency of the nets to catch fish: fishermen would have to compromise their yields in order to reduce bycatch. However, this analysis does indicate some of the parameters that need to be considered if nets are to be made conspicuous to bycatch prone bird species. The main problem arises because diving birds have low spatial resolution and it is exacerbated by the ability of these birds to forage effectively guided only by a paucity of information (Chapter 7).

9.14 Collisions with Glass

Collisions of birds with glass panes has long been recognized as a problem of significant scale and was first quantified as a hazard to migrating birds, particularly passerines 40 years ago (Banks 1976; Banks 1979). It has more recently been estimated that bird mortality due to collisions with glass could be millions, maybe billions, of birds annually worldwide (Drewitt and Langston 2008; Klem 2009). As such, it is a bird mortality problem whose total magnitude eclipses the combined effect of collisions and entrapment, although each hazard type has its impact primarily on different suites of species.

The problem of collisions with glass arises from two different properties of a glass pane: transparency and reflection. As a transparent structure, birds may simply see what is beyond the glass and attempt to fly to it. As a reflective structure, birds may perceive the reflection as a space into which they can fly: in effect, they attempt to fly into what is behind them. In either case, the birds are unaware of the hard glass surface.

9.14.1 Mitigation Measures

In response to this collision problem mitigation measures are frequently employed usually involving devices to disrupt either the transparency or reflective properties of the glass by placing structures before or on the glass surface. Self-adhesive patterns to apply to glass are widely available but it is not clear whether their use has reduced the incidence of collisions.

There has been surprisingly little experimental investigation of the efficacy of such measures and this arises because of the difficulty of replicating under controlled conditions situations that are analogous to what occurs in the field. As in the case of gillnet bycatch reduction, the aim of any mitigation is to reduce collisions while still maintaining the essential properties of the hazard that are of benefit to humans. People are reluctant to forego their large windows, much in the same way that fishermen are reluctant to forego any of their fish catch.

Patterns on glass surfaces

The experimental work that has been has been conducted suggests that patterns applied to glass have limited efficacy unless they occur at such a density that they disrupt the essential benefits of glass to humans (Klem et al. 2004; Klem and Saenger 2013; Rossler et al. 2015). Rossler et al (2015) were able to show that patterns must be repeated right across the width of a glass panel. They found that randomly placed designs were of little effect and that vertical stripes 10 cm apart were more effective than the same pattern in horizontal orientation. They also showed that adding colour to the pattern made them no more effective than black or white patterns. However, one particular limitation of these experimental studies is that to get a behavioural response, the birds had to be encouraged to make an escape flight from a dark interior towards a brightly lit exterior and this does not really replicate the normal circumstances under which birds collide with glass panes. A more limited approach used by (Klem et al. 2004) replicated the situation in more domestic circumstances where garden bird feeders were placed within different distances of glass panes. This was in effect a manipulation of the configuration of the environment around windows rather than treatment of the window pane itself.

UV patterns on glass

One intriguing solution that had been proposed to this impasse is the use of UV reflective or absorbing markers on glass. UV marker pens and self-adhesive strips have been readily available and have been marketed as a way of reducing bird strikes. The idea behind this is that since humans are not sensitive to UV, the application of UV marks would not be detected by humans while they would be detected by birds, at least by passerine species which have relatively high sensitivity to UV light (Odeen et al. 2011; Odeen and Hastad 2013). Careful modelling of the sensitivity of bird eyes to UV light, the reflectivity and absorption of UV strips, and the amount of UV light available in the natural environment show that such markings might indeed be visible to passerines but only under high UV lighting (Hastad and Odeen 2014). They would not be visible to non-passerines even under high UV. Furthermore, the amount of UV reflected from the images of foliage and other structures in the area immediately surrounding panes of glass will frequently render the UV patterns undetectable. Thus a simple UV-based marker solution does not seem to be viable other than under tightly circumscribed natural high light conditions. Flooding UV reflective or absorbing patterns with UV light to make them more conspicuous to birds would, of course, not be viable because of the potential hazard to human vision.

9.15 The Sensory Ecology of Collisions and Entrapment: Conclusions

This examination of the problems of collisions and entrapment has been discussed from the perspective of a proper 'birds' eye view'. It is a view based upon what is known about the information that birds may have available to them when faced with these man-made perceptual challenges. To do this, it has been possible to draw together a number of strands from the preceding chapters.

While much is known in general terms about the wide array of information that is potentially available to birds collectively, little is known about the information which guides even quite specific tasks in any one species. Sensory ecology give us general clues as to what guides birds in their daily behaviour, while an evolutionary perspective gives clues as to what tasks have ultimately driven the gathering of particular sensory information in birds.

Although incomplete and partial, this knowledge does at least allow analysis of behaviour from the perspective of the birds themselves rather than the perspective of a human looking on. Clearly, comprehensive solutions to the collision and entrapment problems are not yet available and considerably more work is required. It would be hoped, however, that humans will cease to view possible solutions of these problems from the perspective of their own sensory information and will attempt to approach it from the perspective of birds. It cannot be overemphasized that the human perspective is but one of many, a truth recognized two millennia ago but so readily forgotten when attempting to understand bird behaviour or attempting to mitigate problems generated for birds by human activity.

10

Postscript: Conclusions, Implications, and Comment

The ranges of information and ideas that have been discussed under the heading of avian sensory ecology are both broad and complex. The information is challenging to our assumptions about the natural world, and to the way we understand the biology of birds. The ideas are stimulating; they liberate our assumptions about birds; and they also challenge the ways we understand the world that we personally inhabit. A sensory ecology perspective questions the nature of the 'reality' that we share with other animals. The idea of a bird's eye view with which the book opened is found to be superficial: just a metaphor used in a lazy way.

Listed below are nine main conclusion or key points; the reader will have others to add. These points both capture some of the main findings from the study of the sensory ecology of birds, and highlight the very many gaps in our knowledge.

1. Echoing the conclusion of Casey Albert Wood, writing 100 years ago, it is clear that no report of a bird species can be considered complete without description of its sensory capacities, and descriptions of how the birds may use the information that their sensory systems provide. In short, knowledge of sensory systems and the information that they provide are fundamental to a species' description.
2. The natural world contains a huge amount of constantly changing information. However, limitations on, and specializations within, sensory systems mean that each species receives only a small part of that information. In essence, information is filtered by sensory systems. The task of sensory ecology is to understand the nature and functions of those filters for each species and for each sensory system. This has important implications, in that we cannot assume what a bird can detect. It is important to measure sensory capacities and to quantify the sensory challenges posed for the conduct of different tasks in different environments.
3. The fluxes of information, and the different perceptual challenges posed by different natural environments, are so large that sensory specializations have been inevitable. No sensory system can function adequately throughout the full ranges of stimuli that are found in the natural world. There have been many

The Sensory Ecology of Birds. Graham R. Martin: Oxford University Press (2017).
 DOI 10.1093/acprof:oso/9780199694532.001.0001

trade-offs in the evolution of particular sensory capacities, and trade-offs and complementarity between different sensory capacities within a species.

4. There are very marked differences in all sensory capacities across all bird species. Generalizations about sensory capacities must be framed with great caution. Even species placed in the same genus can differ significantly in their sensory capacities and hence differ in the information available to control their behaviours.
5. Differences between species' sensory capacities can arise from many and diverse aspects of the structure and physiology of sensory systems. This is especially true in the case of vision because eyes can vary, one from another, in each of their three main functional components: the image-producing system, the image-analysing mechanism, and the placement of the eyes in the skull. The eyes of two different species may appear superficially similar but they can vary markedly in the information that they retrieve from the environment.
6. Many behavioural tasks may have influenced the evolution of sensory capacities in birds but the principal drivers have probably been associated with two key tasks: foraging and predator detection. The key task is probably the control of the position and the timing of the approach of the bill towards a target. The information that is used to control other tasks, such as locomotion and reproduction, may be gained within the informational requirements of foraging and predatory detection.
7. Foraging and predator detection usually make conflicting simultaneous demands for information, and this has resulted in significant trade-offs in the information retrieved within a single sense and between information gained from different senses.
8. Birds may often be guided by information at the limits of their sensory capacities. Information that guides behaviours may often be sparse and partial, and key behaviours may only be possible because of cognitive abilities which allow adequate interpretation of partial information. To achieve this, individual birds may require experience in, and learning about, particular and restricted environments.
9. Human modifications of natural environments, including the introduction of human artefacts, have presented perceptual challenges that cannot always be met by the information available to particular birds. Effective mitigations of the negative effects of human disruptions and intrusions into natural environments must take account of the sensory ecology of the affected species. The effects of environmental changes cannot be understood sufficiently by viewing them through the filters of human sensory systems.

The above conclusions may seem straightforward and, in the light of the information contained in this book, may no longer seem surprising. However, this has not always been the case. These insights, truths even, have been hard won by very many people over a long period of time. They have applied increasing ingenuity to find out how avian sensory systems work and what their capacities are. Great effort and ingenuity has also been applied to the task of quantifying the perceptual challenges of natural environments.

Much of this work has built upon comparisons and it is essential that this work continues. There is a clear implication: don't stop gathering information on the senses of birds! Our knowledge and understanding only scratches the surface. The number of species discussed in this book is but a very small portion of the total diversity of birds. Also for some avian sensory systems, we know very little indeed. It is essential that we carry on building comparative data sets across all sensory systems, compare and contrast species, environments, and behaviours.

The data presented in this book are both partial and sparse. Finding out about the sensory system of birds and the information that they provide is far more than an academic exercise. Humans have changed the world's environments and presented new sensory challenges by way of structures, artificial lights, and sounds. These clearly exceed the kinds of perceptual challenges that sensory systems of many birds have evolved to meet. We need to mitigate these challenges for bird sensory systems. To do so, we must give up sole reliance upon the human view of the world, stop viewing the world through the human sensory filters, and try to truly appreciate the world through the sensory systems of other species. As Epicurus first made clear, reality is defined by our senses, and our real world is not the same as the worlds in which other animals live.

Appendix 1

Visual spatial resolution in birds

The table gives the acuity, i.e. the smallest detail that the eye can detect, in 46 bird species (drawn from 12 Orders and 23 Families). The table gives the highest resolution determined in each species expressed in two ways: cycles/degree and minutes of arc.

Three main methods have been used in these determinations of spatial resolution:

1. Behaviour. This method is based upon using a training technique in which a bird is repeatedly tested for their ability to detect the difference between pairs of simultaneously presented stimulus patterns (viewed from a fixed distance). Usually, high contrast gratings are employed (see inset in Figure 2.16), hence the idea of cycles/degree in which one cycle is defined by a pair of adjacent black and white stripes. The stripe widths are systemically manipulated between individual trials. In some studies, birds were trained to discriminate between identical stripe patterns presented in different orientations (vertical versus horizontal), and in other studies birds were required to discriminate between a striped pattern and a uniform stimulus panel of the same overall brightness as the striped pattern. The ability of a bird to make these discriminations is tested many times in order to find the stripe widths at which the ability to make consistently correct discriminations starts to deteriorate. Such tests give a good indication of the smallest spatial detail which can be shown to have direct control of a bird's behaviour. These behavioural techniques require birds to be tested over many hundreds of trials and the experiments can take a number of months to complete. These behavioural techniques have the additional advantage that it is possible to manipulate the brightness (luminance) of the panels. By conducting tests over a range of light levels, it is possible to produce the kinds of acuity–luminance functions that are shown in Figure 2.16.
2. Ganglion cell spacing. This method is based upon anatomical measurements taken from the retina. It uses measurements of the average smallest spacing between retinal ganglion cells, or the highest density of ganglion cells, from sample areas of prepared retinas (see e.g. Figures 2.6 and 2.7). Sample areas typically include the fovea (see Figure 2.8). Account has to be taken of the shrinkage of tissue during the preparation of the retinal material. The method also requires knowledge of the focal length of the eye and this is usually estimated from the axial length of the eye. However, as shown in Figure 2.11, eyes of the same length can have significantly different focal lengths depending upon their optical structure, and this can introduce some error into the acuity estimates. Also, this method of estimating maximum resolution cannot take account of light levels and their effect on resolution and so cannot produce the kind of data used to construct Figure 2.16.

3. Photoreceptor cell spacing. This method is similar to the ganglion method described above but is based upon the smallest spacing between photoreceptor cells, usually single cone photoreceptors in the fovea.

For each species listed in the table resolution has been rounded to the nearest decimal point. The values given are the best estimate of the highest resolving power in the eyes of each species. There may be considerable differences between individual birds within a species sample. These single threshold values do not give an indication of how resolution is influenced by light level or how resolution varies within the visual field.

There are a few patterns in this data with respect to taxonomy. Thus, higher acuity is found in raptorial birds but even among the Accipitridae (Kites, Hawks, Eagles, and Old World Vultures) there is considerable interspecific variation in maximum resolution, and the New World Vultures (Cathartidae) have relatively modest acuity. It is also clear that there is considerable variation in acuity among the passerine species.

The maximum spatial resolution recorded in the eyes of young humans (0.4 minutes of arc) is given at the foot of the table. It is clear that maximum human resolution exceeds that of the majority of bird species. It is noteworthy that an acuity of 1 minute of arc (30 cyc/deg) is regarded as 'normal' or 'adequate' for humans to complete everyday visually guided tasks, including reading this text. The resolution of many bird species cluster around this value. In humans, spectacle corrections are usually prescribed to achieve such a level of acuity. Acuities lower than 1 minute of arc are for clinical purposes bracketed into broad categories; acuity between 1 and 3 minutes of arc is regarded as 'mild vision loss', between 3 and 8 minutes of arc is termed 'moderate visual impairment', and acuity lower than 10 minutes of arc is labelled 'severe visual impairment'. Thus, some of the birds listed here (notable the Anseriformes, Galliformes, Columbiformes, and some of the Passeriformes) would be regarded by human standards to have mild vision loss, while the owls have between mild loss and moderate impairment, and the Great Cormorant when underwater would be regarded as having 'severe visual impairment'.

Species are arranged in the table in the currently accepted taxonomic sequence published in the International Ornithological Congress World Bird List (Gill and Donsker 2016). The names used are also those recommended by the IOC. Species English Names used may not be the same as those used in the sources cited and in one species (House Finch) the scientific name has changed as a result of taxonomic revision.

Table A1. Visual spatial resolution in birds.

Order	Family	Species	Acuity (cyc/deg)	Acuity (minutes of arc)	Method	Sources
Struthioniformes	Struthionidae (Ostriches)	Common Ostrich *Struthio camelus*	19.8	1.5	ganglion	Boire et al. (2001)
Anseriformes	Anatidae (Ducks, Geese and Swans)	Mallard *Anas platyrhynchos*	11.9	2.52	ganglion	Lisney et al. (2013a)
		Canada Goose *Branta canadensis*	9.6	3.1	ganglion	Fernandez-Juricic et al. (2011a)
Galliformes	Phasianidae (Pheasants and allies)	Grey Partridge *Perdix perdix*	10.2	2.9	ganglion	Lisney et al. (2012b)
		Japanese Quail *Coturnix coturnix*	9.7	3.1	ganglion	"
		Red Jungle Fowl *Gallus gallus*	8.3	3.6	ganglion	Ehrlich (1981)
		Common Pheasant *Phasianus colchicus*	12.9	2.3	ganglion	Lisney et al. (2012)
		Indian Peafowl *Pavo cristatus*	20.6	1.5	ganglion	Hart (2002)
Procellariiformes	Hydrobatidae (Northern Storm Petrels)	Leach's Storm Petrel *Oceanodroma leucorhoa*	7.1	4.2	ganglion	Mitkus (2015)
	Procellariidae (Petrels, Shearwaters)	Northern Fulmar *Fulmarus glacialis*	44.7	0.7	ganglion	"
Suliformes	Phalacrocoracidae (Cormorants, Shags)	Great Cormorant *Phalacrocorax carbo*	3.3 (underwater)	9.1	behaviour	White et al. (2007)
Accipitriformes	Cathartidae (New World Vultures)	Turkey Vulture *Cathartes atratus*	15.4	1.9	ganglion	Lisney et al. (2013b)
		Black Vulture *Coragyps aura*	15.8	1.9	ganglion	Lisney et al. (2013b)

(*Continued*)

Table A1. (Continued)

Order	Family	Species	Acuity (cyc/deg)	Acuity (minutes of arc)	Method	Sources
	Accipitridae (Kites, Hawks and Eagles)	Indian Vulture *Gyps indicus*	135	0.2	behaviour	Fischer (1969)
		Griffon Vulture *Gyps fulvus*	104	0.3	behaviour	"
		Egyptian vulture *Neophron percnopterus*	135	0.2	behaviour	"
		Wedge-tailed Eagle *Aquila audax*	142	0.2	behaviour	Reymond (1985)
		Black Kite *Milvus migrans*	37.3	0.8	behaviour	Potier et al. (2016)
		Harris's Hawk *Parabuteo unicinctus*	29.3	1.0	behaviour	"
Columbiformes	Columbidae (Pigeons, Doves)	American Mourning Dove *Zenaida macroura*	7.5	3.9	ganglion	Dolan and Fernandez-Juricic (2010)
		Rock Dove *Columba livia*	18	1.7	behaviour	Hodos et al. (1976)
Strigiformes	Tytonidae (Barn Owls)	Western Barn Owl *Tyto alba*	4.0	7.5	behaviour	Harmening et al. (2009)
	Strigidae (Owls)	Great Horned Owl *Bubo virginianus*	7.5	4.0	behaviour	Fite (1973)
		Tawny Owl *Strix aluco*	11.1	2.7	behaviour	Martin and Gordon (1974)
Coraciiformes	Alcedinidae (Kingfishers)	Laughing Kookaburra *Dacelo novaeguineae*	41	0.7	ganglion	Moroney and Pettigrew (1987)
		Sacred Kingfisher *Todiramphus sanctus*	26	1.2	ganglion	"
Falconiformes	Falconidae (Caracaras, Falcons)	American Kestrel *Falco sparverius*	40	0.75	behaviour	Hirsch (1982)

		Brown Falcon *Falco berigora*	73	0.4	behaviour	Reymond (1987)
Psittaciformes	Psittaculidae (Old World Parrots)	Bourke's Parrot *Neopsephotus bourkii*	9.4	3.2	behaviour	Lind et al. (2012)
		Budgerigar *Melopsittacus undulatus*	11.7	2.6	behaviour	"
Passeriformes	Meliphagidae (Honeyeaters)	Brown Honeyeater *Lichmera indistincta*	20.2	1.5	photoreceptor spacing	Coimbra et al. (2015)
		Red Wattlebird *Anthochaera carunculata*	40.8	0.7	photoreceptor spacing	"
	Acanthizidae (Australasian Warblers)	Yellow-rumped Thornbill *Acanthiza chrysorrhoa*	25.6	1.2	photoreceptor spacing	"
	Corvidae (Crows, Jays)	Blue Jay *Cyanocitta cristata*	19	1.6	ganglion	Fite and Rosenfield-Wessels (1975)
		Eurasian Magpie *Pica pica*	33.3	0.9	behaviour	Dabrowska (1975)
		Rook *Corvus frugilegus*	29.5	1.0	behaviour	"
	Paridae (Tits, Chickadees)	Tufted Titmouse *Baeolophus bicolor*	6.6	4.5	ganglion	Moore et al. (2013)
		Carolina Chickadee *Poecile carolinensis*	5	0.6	ganglion	"
	Zosteropidae (White-eyes)	Silvereye *Zosterops lateralis*	18.5	1.6	photoreceptor spacing	Coimbra et al. (2015)
	Turdidae (Thrushes)	Common Blackbird *Turdus merula*	22.5	1.3	behaviour	Donner (1951)
		European Robin *Erithacus rubecula*	6	5.0	behaviour	"
	Passeridae (Old World Sparrows)	House Sparrow *Passer domesticus*	4.8	6.3	ganglion	Dolan and Fernandez-Juricic (2010)

(Continued)

Table A1. (Continued)

Order	Family	Species	Acuity (cyc/deg)	Acuity (minutes of arc)	Method	Sources
	Fringillidae (Finches)	Common Chaffinch *Fringilla coelebs*	22.5	1.3	behaviour	Donner (1951)
		House Finch *Haemorhous mexicanus*	4.7	6.4	ganglion	Dolan and Fernandez-Juricic (2010)
	Emberizidae (Buntings, New World Sparrows)	Yellowhammer *Emberiza citrinella*	9.7	3.1	behaviour	Donner (1951)
		Common Reed Bunting *Emberiza schoeniclus*	7.8	3.8	behaviour	"
Human			72	0.4	behaviour	Land and Nilsson (2012)

Appendix 2

Visual field data available for birds

The table lists species and the sources of visual field data. For ease of comparison just the maximum width of the binocular fields in each species are given. Visual fields differ in many parameters. The maximum binocular width is just one parameter, although an important one for understanding the form and functions of visual fields. Each of the papers cited should be consulted for comprehensive diagrams of the visual fields.

See also the book's companion website www.oup.co.uk/companion/martin for more comprehensive data which includes the vertical height and position of the binocular region, the width of blind areas above and behind the head, and the widths of the field of a single eye.

Species are arranged in the conventional taxonomic sequence and follow the names and classifications given in the International Ornithological Congress World Bird List (Gill and Donsker 2016). Note: the species English names used here may not be the same as those used in the sources cited.

While there is much variation in maximum binocular field widths, it is worth noting that the broadest fields do not occur in either owls or diurnal raptors as is commonly supposed. Also in no birds does binocular field width come close to that of humans (120°; Figure 2.14). Statistical analysis of this visual field data has shown that the binocular fields of passerines birds are significantly broader than those of non-passerine species (Troscianko et al. 2012).

Table A2. Visual field data available for birds.

Order	Family	Species	Maximum binocular overlap (°)	Sources
Struthioniformes	Struthionidae (Ostriches)	Common Ostrich *Struthio camelus*	20	Martin and Katzir (1995)
Apterygiformes	Apterygidae (Kiwi)	North Island Brown Kiwi *Apteryx mantelli*	11	Martin et al. (2007c)
		Great Spotted Kiwi *Apteryx haastii*	11	"
Anseriformes	Anatidae (Ducks, Geese, and Swans)	Northern Shoveler *Anas clypeata*	20	Guillemain et al. (2002)
		Eurasian Wigeon *Anas penelope*	20	"
		Mallard *Anas platyrhynchos*	20	Martin (1986b)
		Blue Duck *Hymenolaimus malacorhynchos*	34	Martin et al. (2007a)
		Pink-eared Duck *Malacorhynchus membranaceus*	17	"
		Canada Goose Branta canadensis	22	Fernandez-Juricic et al. (2011a)
Sphenisciformes	Spheniscidae (Penguins)	King Penguin *Aptenodytes patagonicus*	29	Martin (1999a)
		Humboldt penguin *Spheniscus humboldti*	45	Martin and Young (1984)
Procellariiformes	Diomedeidae (Albatrosses)	Grey headed albatross *Thalassarche chrysostoma*	27	Martin (1998)
		Black-browed albatross *Thalassarche melanophris*	32	"
	Procellariidae (Petrels, Shearwaters)	Antarctic Prion *Pachyptila desolata*	20	Martin and Prince (2001)

		White-chinned Petrel *Procellaria aequinoctialis*	40	"
		Manx Shearwater *Puffinus puffinus*	18	Martin and Brooke (1991)
Phoenicopteriformes	Phoenicopteridae (Flamingos)	Lesser Flamingo *Phoeniconaias minor*	10	Martin et al. (2005)
Ciconiiformes	Ciconiidae (Storks)	White Stork *Ciconia ciconia*	28	Martin and Shaw (2010)
Pelecaniformes	Ardeidae (Herons, Bitterns)	Squacco Heron *Ardeola ralloides*	20	Martin and Katzir (1994a)
		Western Cattle Egret *Bubulcus ibis*	22	"
		Western Reef Heron *Egretta gularis*	20	"
		Black-crowned Night Heron *Nycticorax nycticorax*	22	Katzir and Martin (1998)
		African Spoonbill *Platalea alba*	26	Martin and Portugal (2011)
		Eurasian Spoonbill *Platalea leucorodia*	26	"
		Puna Ibis *Plegadis ridgwayi*	16.5	"
		Northern Bald Ibis *Geronticus eremita*	27	"
Suliformes	Phalacrocoracidae (Cormorants, Shags)	Great Cormorant *Phalacrocorax carbo*	28	Martin et al. (2008)
Accipitriformes	Accipitridae (Kites, Hawks, and Eagles)	Short-toed Snake Eagle *Circaetus gallicus*	20	Martin and Katzir (1999)
		White-backed Vulture *Gyps africanus*	21	Martin et al. (2012)
		Griffon Vulture *Gyps fulvus*	21	"

(*Continued*)

Table A2. (Continued)

Order	Family	Species	Maximum binocular overlap (°)	Sources
		White-headed Vulture *Trigonoceps occipitalis*	30	Portugal, Martin, and Murn (2017)
		Harris's Hawk *Parabuteo unicinctus*	45	Potier et al. (2016)
		Red-tailed Hawk *Buteo jamaicensis*	33	O'Rourke et al. (2010)
		Cooper's Hawk *Accipiter cooperii*	36	"
		Black Kite *Milvus migrans*	39	Potier et al. (2016)
Otidiformes	Otididae (Bustards)	Kori Bustard *Ardeotis kori*	21	Martin and Shaw (2010)
Gruiformes	Gruidae (Cranes)	Blue Crane *Grus paradisea*	23	"
Charadriiformes	Burhinidae (Stone-curlews, Thick-knees)	Eurasian Stone-curlew *Burhinus oedicnemus*	18	Martin and Katzir (1994b)
	Charadriidae (Plovers)	European Golden Plover *Pluvialis apricaria*	15	Martin and Piersma (2009)
	Laridae (Gulls, Terns, and Skimmers)	Black Skimmer *Rynchops niger*	14	Martin et al. (2007b)
	Scolopacidae (Sandpipers, Snipes)	Red Knot *Calidris canutus*	22	Martin and Piersma (2009)
		Eurasian Woodcock *Scolopax rusticola*	12	Martin (1994)
	Alcidae (Auks)	Atlantic Puffin *Fratercula arctica*	49	Martin and Wanless (2015)
		Common Murre *Uria aalge*	30	"

Columbiformes	Columbidae (Pigeons, Doves)	Rock Dove *Columba livia*	27	Martin and Young (1983)
		Mourning Dove *Zenaida macroura*	11	Blackwell et al. (2009)
Psittaciformes	Psittacidae (African & New World Parrots)	Senegal Parrot *Poicephalus senegalus*	27	Demery et al. (2011)
Falconiformes	Falconidae (Caracaras, Falcons)	American Kestrel *Falco sparverius*	33	O'Rourke et al. (2010)
Strigiformes	Strigidae (Owls)	Tawny Owl *Strix aluco*	48	Martin (1984)
Caprimulgiformes	Steatornithidae (Oilbird)	Oilbird *Steatornis caripensis*	38	Martin et al. (2004a)
	Caprimulgidae (Nightjars)	Pauraque *Nyctidromus albicollis*	25	"
Bucerotiformes	Bucorvidae (Ground Hornbills)	Southern Ground Hornbill *Bucorvus leadbeateri*	30	Martin and Coetzee (2004)
	Bucerotidae (Hornbills)	Southern Yellow-billed Hornbill *Tockus leucomelas*	30	"
Passeriformes	Corvidae (Crows, Jays)	California Scrub Jay *Aphelocoma californica*	35	Fernandez-Juricic et al. (2010)
		American Crow *Corvus brachyrhynchos*	42	"
		New Caledonia Crow *Corvus moneduloides*	61	Troscianko et al. (2012)
		Carrion Crow *Corvus corone*	38	"
		Jackdaw *Corvus monedula*	46	"
		Pied Crow *Corvus alba*	46	"

(Continued)

Table A2. (Continued)

Order	Family	Species	Maximum binocular overlap (°)	Sources
		Rook *Corvus frugilegus*	47	"
		Raven *Corvus corax*	43	"
	Fringillidae (Finches)	House Finch *Carpodacus mexicanus*	46	Fernandez-Juricic et al. (2008)
	Icteridae (Oropendolas, Orioles, and Blackbirds)	Brown-headed Cowbird *Molothrus ater*	51	Blackwell et al. (2009)
	Passeridae (Old World Sparrows)	House Sparrow *Passer domesticus*	46	Fernandez-Juricic et al. (2008)
	Sturnidae (Starlings, Rhabdornis)	Common Starling *Sturnus vulgaris*	43	Martin (1986a)
	Tyrannidae (Tyrant Flycatchers)	Black Phoebe *Sayornis nigricans*	40	Gall and Fernández-Juricic (2010)

References

Alonso, P.D., Milner, A.C., Ketcham, R.A., Cookson, M.J., & Rowe, T.B. 2004. The avian nature of the brain and inner ear of *Archaeopteryx*. *Nature*, 430, 666–669.

Amo, L., Aviles, J.M., Parejo, D., Pena, A., Rodriguez, J., & Tomas, G. 2012a. Sex recognition by odour and variation in the uropygial gland secretion in starlings. *Journal of Animal Ecology*, 81, 605–613.

Amo, L., Caro, S.P., & Visser, M.E. 2011. Sleeping birds do not respond to predator odour. *PLoS One*, 6, doi: 10.1371/journal.pone.0027576.

Amo, L., Galvan, I., Tomas, G., & Sanz, J.J. 2008. Predator odour recognition and avoidance in a songbird. *Functional Ecology*, 22, 289–293.

Amo, L., Jansen, J.J., van Dam, N.M., Dicke, M., & Visser, M.E. 2013. Birds exploit herbivore-induced plant volatiles to locate herbivorous prey. *Ecology Letters*, 16, 1348–1355.

Amo, L., Lopez-Rull, I., Pagan, I., & Garcia, C.M. 2012b. Male quality and conspecific scent preferences in the house finch, *Carpodacus mexicanus*. *Animal Behaviour*, 84, 1483–1489.

Andersson, S. & Amundsen, T. 1997. Ultraviolet colour vision and ornamentation in bluethroats. *Proceedings of the Royal Society B: Biological Sciences*, 264, 1587–1591.

Anon 1952. *Natural illumination charts*. US Navy Research and Development project NS 714–100, Report No 374-1 (September).

Anon 1988. *The Astronomical Almanac*. London, HMSO.

Archer, S.N., Djamgoz, M.B.A., Loew, E., Partridge, J.C., & Vallerga, S. 1999. *Adaptive Mechanisms in the Ecology of Vision*. Dordrecht, Kluwer.

Arthus-Bertrand, Y. 1999. *The Earth from the Air*. London, Thames & Hudson.

Au, W.W.L. & Nachtigall, P.E. 1997. Acoustics of echolocating dolphins and small whales. *Marine and Freshwater Behaviour and Physiology*, 29, 127–162.

Avery, M.L., Springer, P.F., Dailey, & N.S. 1980. *Avian mortality at man-made structures: an annotated bibliography*. US Fish and Wildlife Service, Biological Services Program, National Power Plant Team.

Bailey, A. 2002. *Sextus Empiricus and Pyrrhonean Scepticism*. Oxford, Oxford University Press.

Baker, A.J., Daugherty, C.H., Colbourne, R., & McLennan, J.L. 1995. Flightless brown kiwis of New Zealand possess extremely sub divided population structure and cryptic species like small mammals. *Proceedings of the National Academy of Sciences of the United States of America*, 92, 8254–8258.

Bala, A.D.S., Spitzer, M.W., & Takahashi, T.T. 2003. Prediction of auditory spatial acuity from neural images of the owl's auditory space map. *Nature*, 424, 771–774.

Baldwin, M.W., et al. 2014. Evolution of sweet taste perception in hummingbirds by transformation of the ancestral umami receptor. *Science*, 345, 929–933.

Bang, B.G. & Wenzel, B.M. 1985, 'Nasal cavity and olfactory system.' In: *Form and Function in Birds*. A.S. King & J. McLelland, eds, London, Academic Press, pp. 195–225.

Banks, R.C. 1976. Reflective plate glass: a hazard to migrating birds. *Bioscience*, 26, 414.

Banks, R.C. 1979. *Human-related mortality of birds in the United States*. US Fish and Wildlife Service Special Scientific Report.

Barker, R. 2002. *The Royal Flying Corps in World War I*. London, Robinson.

Basbaum, A.I., Kaneko, A., Shimizu, T., & Westheimer, G. 2008. *The Senses: A Comprehensive Reference, Vols. 1 & 2.* Amsterdam, Elsevier.

Bath, W. 1906. Die Geschmacksorgane der Vogel und Krokodile. *Archiv fur Biontologie*, 1, 5–47.

Baumel, J.J. 1993. *Handbook of Avian Anatomy: Nomina Anatomica Avium.* 2nd ed. Cambridge, MA, Nuttall Ornithological Club.

Beason, R.C. & Nichols, J.E. 1984. Magnetic orientation and magnetically sensitive material in a transequatorial migratory bird. *Nature*, 309, 151–153.

Beason, R.C. & Wiltschko, W. 2015. Cues indicating location in pigeon navigation. *Journal of Comparative Physiology A*, 201, 961–967.

Beecher, W.J. 1978. Feeding adaptations and evolution in the starlings. *Bulletin of the Chicago Academy of Science*, 11, 269–298.

Bennett, A.T.D., Cuthill, I.C., Partridge, J.C., & Lunau, K. 1997. Ultraviolet plumage colors predict mate preferences in starlings. *Proceedings of the National Academy of Sciences of the United States of America*, 94, 8618–8621.

Bennett, A.T.D., Cuthill, I.C., Partridge, J.C., & Maier, E.J. 1996. Ultraviolet vision and mate choice in zebra finches. *Nature*, 380, 433–435.

Benton, M.J. 2005. *Vertebrate Paleontology, 3rd Edition.* Oxford, Blackwell.

Benvenuti, S., Ioale, P., & Massa, B. 1993. Olfactory experiments on Cory Shearwater (Calonectris-Diomedea)—the effect of intranasal zinc-sulfate treatment on short-range homing behavior. *Bollettino di zoologia*, 60, 207–210.

Berkhoudt, H. 1980. The morphology and distribution of cutaneous mechanoreceptors (Herbst and Grandry corpuscles) in bill and tongue of mallard (*Anas platyrhynchos L.*). *Netherlands Journal of Zoology*, 30, 1–34.

Berkhoudt, H. 1985, 'Structure and function of avian taste receptors.' In: *Form and Function in Birds.* A.S. King & J. McLelland, eds, London, Academic Press, pp. 463–496.

Berkley, M. 1976, 'Cat visual psychophysics: neural correlates and comparison with man.' In: *Progress in Psychobiology.* J. Sprague & A. Epstein, eds, New York, Academic Press, pp. 63–119.

Bernhardt, G.E., Blackwell, B.F., DeVault, T.L., & Kutschbach-Brohl, L. 2010. Fatal injuries to birds from collisions with aircraft reveal anti-predator behaviours. *Ibis*, 152, 830–834.

Berthold, P. 2001. *Bird Migration. A General Survey. 2nd Ed.* Oxford, Oxford University Press.

Berthold, P., Gwinner, E., & Sonnenschein, E. 2003. *Avian Migration.* Berlin, Springer.

Bertin, A., et al. 2010. In ovo olfactory experience influences post-hatch feeding behaviour in young chickens. *Ethology*, 116, 1027–1037.

Bevanger, K. 1994. Bird interactions with utility structures—collision and electrocution, causes and mitigating measures. *Ibis*, 136, 412–425.

Bevanger, K. 1998. Biological and conservation aspects of bird mortality caused by electricity power lines: a review. *Biological Conservation*, 86, 67–76.

Bhagavatula, P., Claudianos, C., Ibbotson, M., & Srinivasan, M. 2011. Optic flow cues guide flight in birds. *Current Biology*, 21, 1794–1799.

Biewener, A.A. 2003. *Animal Locomotion.* Oxford, Oxford University Press.

Bingelli, R.L. & Paule, W.J. 1969. The pigeon retina: quantitative aspects of the optic nerve and ganglion cell layer. *Journal of Comparative Neurology*, 137, 1–18.

BirdLife International 2015. *Species factsheet: Pezoporus occidentalis.* Downloaded from http://www.birdlife.org on 07/12/2015.

BirdLife. Species factsheet: *Grus paradisea.* http://www.birdlife.org. 4-12-2009.

Bischof, H.-J. 1988. The visual field and visually guided behavior in the zebra finch (*Taeniopygia guttata*). *Journal of Comparative Physiology A*, 163, 329–337.

Bishop, C.A. & Brogan, J.M. 2013. Estimates of avian mortality attributed to vehicles collisions in Canada. *Avian Conservation and Ecology*, 8, 2.

Bisti, S. & Maffei, L. 1974. Behavioural contrast sensitivity of the cat in various visual meridians. *Journal of Physiology*, 241, 201–210.

Blackwell, B.F., DeVault, T.L., Seamans, T.W., Lima, S.L., Baumhardt, P., & Fernandez-Juricic, E. 2012. Exploiting avian vision with aircraft lighting to reduce bird strikes. *Journal of Applied Ecology*, 49, 758–766.

Blackwell, B.F., Fernandez-Juricic, E., Seamans, T.W., & Dolan, T. 2009. Avian visual system configuration and behavioural response to object approach. *Animal Behaviour*, 77, 673–684.

Blakemore, R. 1975. Magnetotactic bacteria. *Science*, 190, 377–379.

Bloch, S., Jäger, R., Lemeignant, M., & Martinoya, C. 1988. Correlations between ocular saccades and head movements in walking pigeons. *Journal of Physiology (London)*, 406, 173.

Blough, D.S. 1956. Dark adaptation in the pigeon. *Journal of Comparative and Physiological Psychology*, 49, 425–430.

Blough, D.S. 1971. The visual acuity of the pigeon for distant targets. *Journal of the Experimental Analysis of Behavior*, 15, 57–68.

Blumstein, D.T. 2003. Flight initiation distance in birds is dependent on intruder starting distance. *Journal of wildlife Management*, 67, 852–857.

Boire, D., Dufour, J.S., Theoret, H., & Ptito, M. 2001. Quantitative analysis of the retinal ganglion cell layer in the ostrich, Struthio camelus. *Brain Behavior and Evolution*, 58, 343–355.

Bonadonna, F. & Bretagnolle, V. 2002. Smelling home: a good solution for burrow-finding in nocturnal petrels? *Journal of Experimental Biology*, 205, 2519–2523.

Bonadonna, F., Cunningham, G.B., Jouventin, P., Hesters, F., & Nevitt, G.A. 2003. Evidence for nest-odour recognition in two species of diving petrel. *Journal of Experimental Biology*, 206, 3719–3722.

Bonadonna, F. & Mardon, J. 2010. One house two families: petrel squatters get a sniff of low-cost breeding opportunities. *Ethology*, 116, 176–182.

Bonadonna, F. & Nevitt, G.A. 2004. Partner-specific odor recognition in an Antarctic seabird. *Science*, 306, 835.

Bowmaker, J.K. 1977. The visual pigments, oil droplets and spectral sensitivity of the pigeon. *Vision Research*, 17, 1129–1138.

Bowmaker, J.K. & Loew, E. 2008, 'Vision in fish.' In: *The Senses: A Comprehensive Reference, Vol. 1 Vision I*. A.I. Basbaum et al., eds, Amsterdam, Elsevier, pp. 54–76.

Bowmaker, J.K. & Martin, G.R. 1978. Visual pigments and colour vision in a nocturnal bird, *Strix aluco* (tawny owl). *Vision Research*, 18, 1125–1130.

Bowmaker, J.K. & Martin, G.R. 1985. Visual pigments and oil droplets in the penguin, *Spheniscus humboldti*. *Journal of Comparative Physiology A*, 156, 71–77.

Bradley, T.J. 2009. *Animal Osmoregulation*. Oxford, Oxford University Press.

Brinklov, S., Fenton, M.B., & Ratcliffe, J.M. 2013. Echolocation in oilbirds and swiftlets. *Frontiers in Physiology*, 4, doi: 10.3389/fphys.2013.00123.

Brooke, M. de L. 1978. A test for visual location of the burrow by Manx Shearwater *Puffinus puffinus*. *Ibis*, 120, 347–349.

Brooke, M. de L. 2004. *Albratrosses and Petrels Across the World*. Oxford, Oxford University Press.

Brooke, M. de L. 2010. *The Manx Shearwater*. London, Poyser.

Brooke, M. de L., Hanley, S., & Laughlin, S.B. 1999. The scaling of eye size with body mass in birds. *Proceedings of the Royal Society B: Biological Sciences*, 266, 405–412.

Brophy, K. & Cowley, D. 2005. *From the Air: Understanding Aerial Archaeology*. London, The History Press.

Bruce, V., Green, P.R., & Georgeson, M.A. 2003. *Visual Perception: Physiology, Psychology, & Ecology*. 4th ed. Hove, Psychology Press.

Bruderer, B. & Boldt, A. 2001. Flight characteristics of birds: I. radar measurements of speeds. *Ibis*, 143, 178–204.

Buhler, P. 1970. Schademorphologie und Kiefermechanik der Caprimulgidae (Aves). *Zoologie Morphologie Tiere*, 66, 337–399.

Buhler, P. 1981, 'The functional anatomy of the avian jaw apparatus.' In: *Form and Function in Birds, Vol. 2*. A.S. King & J. McLelland, eds, London, Academic Press.

Bunce, M., et al. 2009. The evolutionary history of the extinct ratite moa and New Zealand Neogene paleogeography. *Proceedings of the National Academy of Sciences of the United States of America*, 106, 20646–20651.

Bunn, D.S., Warburton, R.S., & Wilson, R.D.S. 1982. *The Barn Owl*. Calton, Poyser.

Burkhardt, D. 1982. Birds, berries and UV–a note on some consequences of uv vision in birds. *Naturwissenschaften*, 69, 153–157.

Butler, D. 1989. *Quest for the Kakapo*. Auckland, Heinemann Reed.

Caley, K.J. 2007, 'Fossil birds.' In: *Handbook of the Birds of the World, Vol. 12. Picathartes to Tits and Chickadees*. J. del Hoyo, A. Elliott, & D.A. Christie, eds, Barcelona, Lynx Edicions, pp. 11–55.

Camphuysen, K. 1998. Diurnal activity patterns and nocturnal group formation of wintering Common Murres in the central North Sea. *Colonial Waterbirds*, 21, 406–413.

Carboneras, C. 1992, 'Family Gaviidae (Divers).' In: *Handbook of the Birds of the World, Vol. 1. Ostrich to Ducks*. J. del Hoyo, A. Elliot, & J. Sargatal, eds, Barcelona, Lynx Edicions, pp. 162–172.

Carey, J.R. & Judge, D.S. 2000. *Longevity Records: Life Spans of Mammals, Birds, Amphibians, Reptiles, and Fish*. Odense, Odense University Press.

Caro, S.P., Balthazart, J., & Bonadonna, F. 2015. The perfume of reproduction in birds: chemosignaling in avian social life. *Hormones and Behavior*, 68, 25–42.

Cartoni, C., et al. 2010. Taste preference for fatty acids is mediated by GPR40 and GPR120. *Journal of Neuroscience*, 30, 8376–8382.

Catania, K.C. & Remple, F.E. 2004. Tactile foveation in the star-nosed mole. *Brain Behavior and Evolution*, 63, 1–12.

Catchpole, C.K. & Slater, P.J.B. 2008. *Bird Song: Biological Themes and Variations, 2nd Edition*. Cambridge, Cambridge University Press.

Catt, D.C., et al. 1994. Collisions against fences by woodland grouse in Scotland. *Forestry*, 67, 105–118.

Chantler, P., Wells, D.R., & Schuchmann, K.L. 1999, 'Family Apodidae(Swifts).' In: *Handbook of the Birds of the World, Vol. 5. Barn-owls to Hummingbirds*. J. del Hoyo, A. Elliot, & J. Sargatal, eds, Barcelona, Lynx Edicions, pp. 388–457.

Chiappe, L.M. 2006. *The Glorified Dinosaurs: Origins and Early Evolution of Birds*. Hoboken, NJ, John Wiley & Sons.

Church, S.C., Bennett, A.T.D., Cuthill, I.C., & Partridge, J.C. 1998. Ultraviolet cues affect the foraging behaviour of blue tits. *Proceedings of the Royal Society B: Biological Sciences*, 265, 1509–1514.

Clarke, D.D., Forsyth, R.S., & Wright, R.L. 1995. *The Analysis of Pre-accident Sequences*. Crowthorne, Transport Research Laboratory.

Clarke, D.D., Ward, P.J., & Jones, J. 1998. Overtaking road-accidents: differences in manoeuvre as a function of driver age. *Accident Analysis and Prevention*, 30, 455–467.

Cleere, N. 1999, 'Family Caprimulgidae (Nightjars).' In: *Handbook of the Birds of the World, Vol. 5. Barn-owls to Hummingbirds*. J. del Hoyo, A. Elliott, & J. Sargatal, eds, Barcelona, Lynx Edicions, pp. 302–386.

Coimbra, J.P., Collin, S.P., & Hart, N.S. 2014a. Topographic specializations in the retinal ganglion cell layer correlate with lateralized visual behavior, ecology, and evolution in cockatoos. *Journal of Comparative Neurology*, 522, 3363–3385.

Coimbra, J.P., Collin, S.P., & Hart, N.S. 2014b. Topographic specializations in the retinal ganglion cell layer of Australian passerines. *Journal of Comparative Neurology*, 522, 3609–3628.

Coimbra, J.P., Collin, S.P., & Hart, N.S. 2015. Variations in retinal photoreceptor topography and the organization of the rod-free zone reflect behavioral diversity in Australian passerines. *Journal of Comparative Neurology*, 523, 1073–1094.

Coimbra, J.P., Nolan, P.M., Collin, S.P., & Hart, N.S. 2012. Retinal ganglion cell topography and spatial resolving power in penguins. *Brain Behavior and Evolution*, 80, 254–268.

Collier, K.J. 1991. Invertebrate food supplies and diet of blue duck on rivers in two regions of the North Island, New Zealand. *New Zealand Journal of Ecology*, 15, 131–138.

Collier, K.J. 2004. Perturbation pathways affecting the avian lotic predator, blue duck, following a volcanic eruption. *Freshwater Biology*, 49, 818–834.

Collier, K.J., Moralee, S.J., & Wakelin, M. 1993. Factors affecting the distribution of blue duck *Hymenolaimus malacorhynchos* on New Zealand rivers. *Biological Conservation*, 63, 119–126.

Conway Morris, S. 1998. *The Crucible of Creation: The Burgess Shale and the Rise of Animals*. Oxford, Oxford University Press.

Conway Morris, S. 2006. Darwin's dilema: the realities of the Cambrian 'explosion'. *Philosophical Transactions of the Royal Society B: Biological Sciences*, 361, 1069–1083.

Corfield, J., Kubke, M.F., Parsons, S., Wild, J.M., & Koppl, C. 2011a. Evidence for an auditory fovea in the New Zealand kiwi (*Apteryx mantelli*). *PLoS One*, 6, doi: 10.1371/journal.pone.0023771.

Corfield, J., Price, K., Iwaniuk, A.N., Gutiérrez-Ibáñez, C., Birkhead, T., & Wylie, D.R. 2015a. Diversity in olfactory bulb size in birds reflects allometry, ecology, and phylogeny. *Frontiers in Neuroanatomy*, doi: 10.3389/fnana.2015.00102.

Corfield, J.R., Eisthen, H.L., Iwaniuk, A.N., & Parsons, S. 2014. Anatomical specializations for enhanced olfactory sensitivity in kiwi, *Apteryx mantelli*. *Brain Behavior and Evolution*, 84, 214–216.

Corfield, J.R., et al. 2011b. Anatomical specializations for nocturnality in a critically endangered parrot, the Kakapo *Strigops habroptilus*. *PLoS One*, 6(8), e22945.

Corfield, J.R., Kubke, M.F., Parsons, S., & Koppl, C. 2012. Inner-ear morphology of the New Zealand Kiwi (*Apteryx mantelli*) suggests high-frequency specialization. *JARO*, 13, 629–639.

Corfield, J.R., Parsons, S., Harimoto, Y., & Acosta, M.L. 2015b. Retinal anatomy of the New Zealand kiwi: structural traits consistent with their nocturnal behavior. *Anatomical Record*, 298, 771–779.

Cowles, G.S. 1967. On the palate of the red-necked nightjar *Caprimulgus ruficollis* with a description of a new feature. *Ibis*, 109, 260–265.

Cramp, S. 1985. *Handbook of the Birds of Europe, the Middle East and North Africa. The Birds of the Western Palearctic, Vol. 4. Terns to Woodpeckers*. Oxford, Oxford University Press.

Cramp, S. & Perrins, C.M. 1994. *Handbook of the Birds of Europe, the Middle East and North Africa. The Birds of the Western Palearctic, Vol. 8. Crows to Finches*. Oxford, Oxford University Press.

Cramp, S. & Simmons, K.E.L. 1977. *Handbook of the Birds of Europe, the Middle East and North Africa. The Birds of the Western Palearctic, Vol. 1. Ostrich to Ducks*. Oxford, Oxford University Press.

Cramp, S. & Simmons, K.E.L. 1983. *Handbook of the Birds of Europe, the Middle East and North Africa. The Birds of the Western Palearctic, Vol. 3. Waders to Gulls*. Oxford, Oxford University Press.

Cresswell, W. 2011. Predation in bird populations. *Journal of Ornithology*, 152 (Suppl. 1), 251–263.

Cronin, T.W. 2008, 'Visual ecology.' In: *The Senses: A Comprehensive Reference, Vol. 1. Vision I.* A.I. Basbaum et al., eds, Amsterdam, Elsevier, pp. 211–245.

Cronin, T.W. & Douglas, R.H. 2014. Seeing and doing: how vision shapes animal behaviour. *Philosophical Transactions of the Royal Society B: Biological Sciences*, 369: 20130030. http://dx.doi.org/10.1098/rstb.2013.0030

Cronin, T.W., Johnsen, S., Marshall, N.J., & Warrant, E.J. 2014. *Visual Ecology*. Princeton and Oxford, Princeton University Press.

Cserhati, P., Szel, A., & Rohlich, P. 1989. Four cone types characterized by anti-visual pigment antibodies in the pigeon retina. *Investigative Ophthalmology and Visual Science*, 30, 74–81.

Cunningham, G.B., Van Buskirk, R.W., Bonadonna, F., Weimerskirch, H., & Nevitt, G.A. 2003. A comparison of the olfactory abilities of three species of procellariiform chicks. *Journal of Experimental Biology*, 206, 1615–1620.

Cunningham, S.J. 2010. Remote touch prey-detection by Madagascar crested ibises *Lophotibis cristat urschi*. *Journal of Avian Biology*, 41, 350–353.

Cunningham, S.J., Alley, M.R., Castro, I., Potter, M.A., & Pyne, M.J. 2010. Bill morphology of ibises suggests a remote-tactile sensory system for prey detection. *Auk*, 127, 308–316.

Cunningham, S.J. & Castro, I. 2011. The secret life of wild brown kiwi: studying behaviour of a cryptic species by direct observation. *New Zealand Journal of Ecology*, 35, 209–219.

Cunningham, S.J., Castro, I., & Alley, M. 2007. A new prey-detection mechanism for kiwi (*Apteryx* spp.) suggests convergent evolution between paleognathous and neognathous birds. *Journal of Anatomy*, 211, 493–502.

Cunningham, S.J., Castro, I., & Potter, M.A. 2009. The relative importance of olfaction and remote touch in prey detection by North Island brown kiwis. *Animal Behaviour*, 78, 899–905.

Cunningham, S.J., et al. 2013. The anatomy of the bill tip of Kiwis and associated somatosensory regions of the brain: comparisons with shorebirds. *PLoS One*, 8(11), e80036.

Cuthill, I.C., Bennett, A.T.D., Partridge, J.C., & Maier, E.J. 1999. Plumage reflectance and the objective assessment of avian sexual dichromatism. *American Naturalist*, 153, 183–200.

Cuthill, I.C., Partridge, J.C., Bennett, A.T.D., Church, S.C., Hart, N.S., & Hunt, S. 2000. Ultraviolet vision in birds. *Advances in the Study of Behavior*, 29, 159–214.

Dabrowska, B. 1975. Investigations on visual acuity of some corvine species. *Folia Biologica (Krakow)*, 23, 311–332.

Davies, M.N.O. & Green, P.R. 1994. *Perception and Motor Control in Birds: An Ecological Approach*. Berlin, Springer-Verlag.

Davies, N.B., Krebs, J.R., & West, S.A. 2012. *An Introduction to Behavioural Ecology, 4th Edition*. Oxford, Wiley-Blackwell.

Deeming, D.C. & Reynolds, S.J. 2015. *Nests, Eggs and Incubation: New Ideas about Avian Reproduction*. Oxford, UK, Oxford University Press.

Dehnhardt, G., Mauck, B., Hanke, W., & Bleckmann, H. 2001. Hydrodynamic trail following in harbor seals *Phoca vitulina*. *Science*, 293, 102–104.

Dehnhardt, G., Mauck, B., & Hyvarinen, H. 2003, 'The functional significance of the vibrissal system of marine mammals.' In: *Merkel Cell: Structure-Development-Function-Cancerogenesis.* K.I. Baumann, Z. Halata, I. Moll, eds, Berlin, Springer-Verlag, pp. 127–135.

Dehnhardt, G., Sinder, M., & Sachser, N. 1997. Tactual discrimination of size by means of mystacial vibrissae in harbour seals: in air versus underwater. *Zeitschrift fur Saugetierkunde-International Journal of Mammalian Biology*, 62, 40–43.

De Leon, A., Minguez, E., & Belliure, B. 2003. Self-odour recognition in European storm-petrel chicks. *Behaviour*, 140, 925–933.

Demery, Z.P., Chappell, J., & Martin, G.R. 2011. Vision, touch and object manipulation in Senegal parrots *Poicephalus senegalus*. *Proceedings of the Royal Society B: Biological Sciences*, 278, 3687–3693.

Deutschlander, M.E. & Beason, R.C. 2014. Avian navigation and geographic positioning. *Journal of Field Ornithology*, 85, 111–133.

DeVault, T.L., Blackwell, B.F., & Belant, J.L. 2013. *Wildlife in Airport Environments: Preventing Animal-Aircraft Collisions through Science-based Management*. Baltimore, MD, Johns Hopkins University Press.

DeVault, T.L., Blackwell, B.F., Seamans, T.W., Lima, S.L., & Fernandez-Juricic, E. 2015. Speed kills: ineffective avian escape responses to oncoming vehicles. *Proceedings of the Royal Society B: Biological Sciences*, 282: 20142188.

Devereux, C.L., Whittingham, M.J., Fernandez-Juricic, E., Vickery, J.A., & Krebs, J.R. 2006. Predator detection and avoidance by starlings under differing scenarios of predation risk. *Behavioral Ecology*, 17, 303–309.

Dharmaretnam, M. & Andrew, R.J. 1994. Age- and stimulus-specific use of right and left eyes by domestic chick. *Animal Behaviour*, 48, 1395–1406.

Dickinson, C.M. 1991, 'Optical aids for low vision.' In: *Vision and Visual Dysfunction*. W.N. Charman, ed., London, Macmillan, pp. 183–228.

Diekamp, B., Regolin, L., & Gunturkun, O. 2005. A left-sided visuospatial bias in birds. *Current Biology*, 15, R372–R373.

Dolan, T. & Fernandez-Juricic, E. 2010. Retinal ganglion cell topography of five species of ground-foraging birds. *Brain Behavior and Evolution*, 75, 111–121.

Dolbeer, R.A., Wright, S.E., Weller, J.R., Anderson, A.L., & Beiger, M.J. 2015. *Wildlife strikes to civil aircraft in the United States, 1990–2014*. Washington, DC, USA, US Department of Transportation, 152 Federal Aviation Administration, Office of Airport Safety and Standards, Serial Report No. 153 21.

Donner, K.O. 1951. The visual acuity of some passerine birds. *Acta Zoologica Fennica*, 66, 3–40.

Douglas, H.D. 2006. Measurement of chemical emissions in crested auklets (*Aethia cristatella*). *Journal of Chemical Ecology*, 32, 2559–2567.

Douglas, H.D. 2013. Colonial seabird's paralytic perfume slows lice down: an opportunity for parasite-mediated selection? *International Journal for Parasitology*, 43, 399–407.

Drewitt, A.L. & Langston, R.H.W. 2008. Collision effects of wind-power generators and other obstacles on birds. *Annals of the New York Academy of Science*, 1134, 233–266.

Dusenbery, D. 1992. *Sensory Ecology: How Organisms Acquire and Respond to Information*. New York, Freeman.

Eastlake, C.L. 1982. *Goethe: Theory of Colours*. Cambridge, MA, MIT Press.

Eastwood, E. 1967. *Radar Ornithology*. London, Methuen.

Ehrlich, D. 1981. Regional specialization of the chick retina as revealed by the size and density of neurons in the ganglion cell layer. *Journal of Comparative Neurology*, 195, 643–657.

Eimer, T. 1873. Die schnauze des maulwurfs als tastwerkzeug. *Archiv Für Mikroskopische Anatomie*, 7, 181–201.

Emery, N.J. 2006. Cognitive ornithology: the evolution of avian intelligence. *Philosophical Transactions of the Royal Society B: Biological Sciences*, 361, 23–43.

Emmerton, J. & Delius, J. 1980. Wavelength discrimination in the 'visible' and ultraviolet spectrum by pigeons. *Journal of Comparative Physiology A*, 141, 47–52.

Endler, J.A., Gaburro, J., & Kelley, L.A. 2014. Visual effects in great bowerbird sexual displays and their implications for signal design. *Proceedings of the Royal Society B: Biological Sciences*, 281, 10.1098/rspb.2014.0864.

Endler, J.A. & Mielke, P.W. 2005. Comparing entire colour patterns as birds see them. *Biological Journal of the Linnean Society*, 86, 405–431.

Endler, J.A., Westcott, D.A., Madden, J.R., & Robson, T. 2005. Animal visual systems and the evolution of color patterns: sensory processing illuminates signal evolution. *Evolution*, 59, 1795–1818.

Erritzoe, J., Mazgajski, T.D., & Reijt, L. 2003. Bird casualties on European roads. *Acta Ornithologica*, 38, 77–93.

Eskom, E.W.T. 2008. *Eskom Endangered Wildlife Trust Strategic Partnership Central Incident Register. Held and curated by the Wildlife & Energy Interaction Group*. Johannesburg, Endangered Wildlife Trust.

Espaillat, J.E. & Mason, J.R. 1990. Differences in taste preference between Red-Winged Blackbirds and European Starlings. *Wilson Bulletin*, 102, 292–299.

FAA 2014. *Aeronautical Information Manual*. Washington, DC, Federal Aviation Administration.

Fallet, M. 1962. Uber Bodenvogel und ihre terrocolen Beutetiere: Technick der Nahrungssuche-Populationsdynamik. *Zoological Anzeiger*, 168, 187–212.

Fay, R.R. 1992, 'The evolutionary biology of hearing.' In: *Structure and Function in Sound Discrimination among Vertebrates*. D.B. Webster, R.R. Fay, & A.N. Popper, eds, Berlin, New York, Springer-Verlag, pp. 229–264.

Federer, C.A. & Tanner, C.B. 1966. Spectral distribution of light in a forest. *Ecology*, 47, 555–560.

Feinkohl, A. & Klump, G.M. 2013. Azimuthal sound localization in the European starling (*Sturnus vulgaris*): II. Psychophysical results. *Journal of Comparative Physiology A*, 199, 127–138.

Felsenstein, J. 1985. Phylogenies and the comparative method. *American Naturalist*, 125, 1–15.

Fenton, M.B. 1975. Acuity of echolocation in *Collocalia hirundinacea* (Aves: Apodidae), with comments on the distribution of echlocating swiflets and molossid bats. *Biotropica*, 7, 1–7.

Fernández-Juricic, E., Erichsen, J.T., & Kacelnik, A. 2004. Visual perception and social foraging in birds. *Trends in Ecology and Evolution*, 19, 25–31.

Fernandez-Juricic, E., Gaffnet, J., Blackwell, B.F., & Baumhardt, P. 2011a. Bird strikes and aircraft fuselage color: a correlational study. *Human and Wildlife Interactions*, 5, 224–234.

Fernandez-Juricic, E., Gall, M.D., Dolan, T., O'Rourke, C., Thomas, S., & Lynch, J.R. 2011b. Visual systems and vigilance behaviour of two ground-foraging avian prey species: white-crowned sparrows and California towhees. *Animal Behaviour*, 81, 705–713.

Fernandez-Juricic, E., et al. 2011. Testing the terrain hypothesis: Canada Geese see the world laterally and obliquely. *Brain Behavior and Evolution*, 77, 147–158.

Fernandez-Juricic, E., Gall, M., Dolan, T., Tisdale, V., & Martin, G.R. 2008. The visual fields of two ground foraging birds, House Finches and House Sparrows, allow for simultaneous foraging and antipredator vigilance. *Ibis*, 150, 779–787.

Fernandez-Juricic, E., O'Rourke, C., & Pitlik, T. 2010. Visual coverage and scanning behavior in two corvid species: American crow and Western scrub jay. *Journal of Comparative Physiology A*, 196, 879–888.

Fischer, V.A.B. 1969. Laboruntersuchungen und Freiland beobachtungen zum Sehvermogen und Verhalten von Altweltgeiern. *Zoologische Jahrbuecher Abteilung fuer Systematik Biologie*, 96, S81–S132.

Fite, K.V. 1973. Anatomical and behavioral correlates of visual acuity in the Great Horned Owl. *Vision Research*, 13, 219–230.

Fite, K.V. & Rosenfield-Wessels, J. 1975. A comparative study of deep avian foveas. *Brain, Behavior and Evolution*, 12, 97–115.

Fleissner, G., et al. 2003. Ultrastructural analysis of a putative magnetoreceptor in the beak of homing pigeons. *Journal of Comparative Neurology*, 458, 350–360.

Fox, R., Lehmkuhle, S.W., & Westendorf, D.H. 1976. Falcon visual acuity. *Science*, 192, 263–265.

Franklin, W.E. & Lima, S.L. 2010. Laterality in avian vigilance: do sparrows have a favourite eye? *Animal Behaviour*, 62, 879–885.

Freckleton, R.P., Harvey, P.H., & Pagel, M. 2002. Phylogenetic analysis and comparative data: a test and review of evidence. *American Naturalist*, 160, 712–726.

Freire, R., Munro, U., Rogers, L.J., Sagasser, S., Wiltschko, R., & Wiltschko, W. 2008. Different responses in two strains of chickens (*Gallus gallus*) in a magnetic orientation test. *Animal Cognition*, 11, 547–552.

Fritsches, K.A. & Warrant, E. 2006, 'Differences in the visual capabilities of seaturtles and blue water fishes—implications for bycatch reduction.' In: *Sea Turtle and Pelagic Fish Sensory Biology: Developing Techniques to Reduce Sea Turtle Bycatch in Longline Fisheries: NOAA Technical Memorandum NMFS-PIFSC-7*. Y. Swimmer & R. Brill, eds, Honolulu, Hawaii, NOAA, pp. 1–21.

Fritzsch, B. 1992, 'The water-to-land transition evolution of the tetrapod basilar papilla, middle ear and auditory nuclei.' In: *The Evolutionary Biology of Hearing*. D.B. Webster, R.R. Fay, & A.N. Popper, eds, New York, Springer, pp. 351–376.

Fuerst, F.F. & Kare, M.R. 1962. The influence of pH on fluid tolerance and preferences. *Poultry Science*, 41, 71–77.

Fullard, J.J., Barclay, R.M.R., & Thomas, D.W. 1993. Echolocation in free-flying Atiu Swiftlets (*Aerodramus sawtelli*). *Biotropica*, 25, 334–339.

Gaffney, M.F. & Hodos, W. 2003. The visual acuity and refractive state of the American kestrel (*Falco sparverius*). *Vision Research*, 43, 2053–2059.

Gagliardo, A. 2013. Forty years of olfactory navigation in birds. *Journal of Experimental Biology*, 216, 2165–2171.

Galifret, Y. 1968. Les diverse aires fonctionelles de la retine du pigeon. *Zeitschrift fur Zellforschung und Mikroskopische Anatomie*, 86, 535–545.

Gall, M.D. & Fernández-Juricic, E. 2010. Visual fields, eye movements, and scanning behavior of a sit-and-wait predator, the black phoebe (*Sayornis nigricans*). *Journal of Comparative Physiology B*, 196, 15–22.

Gamlin, P.D. 2006. The pretectum: connections and oculomotor-related roles. *Progress in Brain Research*, 151, 379–405.

Garland, T., Jr., Bennet, A.F., & Rezende, E.L. 2005. Phylogenetic approaches in comparative physiology. *Journal of Experimental Biology*, 208, 3015–3035.

Gauthreaux, S.A. 1980. *Animal Migration, Orientation, and Navigation*. New York, Academic Press.

Gerritsen, A.F.C. & Sevenster, J.G. 1985. Foraging behaviour and bill anatomy in sandpipers. *Fortschritte der Zoologie*, 30, 237–240.

Gerritsen, A.F.C., van Heezik, Y.M., & Swennen, C. 1983. Chemoreception in two further *Calidris* species (*C. maritima* and *C. canutus*) with a comparison of the relative importance of chemoreception during foraging in *Calidris* species. *Netherlands Journal of Zoology*, 33, 485–496.

Ghim, M.M. & Hodos, W. 2006. Spatial contrast sensitivity of birds. *Journal of Comparative Physiology A*, 192, 523–534.

Gibson, J.J. 1986. *The Ecological Approach to Visual Perception*. Hove, Erlbaum.

Gill, F. & Donsker, D. 2016. *IOC World Bird List (v 6.2)*. Available at http://www.worldbirdnames.org/, dx.doi.org/10.14344/IOC.ML.6.2.

Gill, F.B. 2007. *Ornithology*. 3rd ed. New York, W.H. Freeman.

Giolli, R.A., Blanks, R.H., & Lui, F. 2006. The accessory optic system: basic organization with an update on connectivity, neurochemistry, and function. *Progress in Brain Research*, 151, 407–440.

Gislen, A., Dacke, M., Kroger, R.H.H., Abrahamsson, M., Nilsson, D.-E., & Warrant, E. 2003. Superior underwater vision in a human population of sea gypsies. *Current Biology*, 13, 833–836.

Goldsmith, T.H. 1980. Hummingbirds see near ultraviolet light. *Science*, 207, 786–788.

Goldsmith, T.H. 1990. Optimization, constraint, and history in the evolution of eyes. *Quarterly Review of Biology*, 65, 281–322.

Goldsmith, T.H. & Butler, B.K. 2003. The roles of receptor noise and cone oil droplets in the photopic spectral sensitivity of the budgerigar, *Melopsittacus undulatus*. *Journal of Comparative Physiology A*, 189, 135–142.

Goldsmith, T.H. & Butler, B.K. 2005. Color vision of the budgerigar (*Melopsittacus undulatus*): hue matches, tetrachromacy, and intensity discrimination. *Journal of Comparative Physiology A*, 191, 933–951.

Goldstein, B. 1984. *Sensation and Perception*. Belmont, CA, Wadsworth.

Goss-Custard, J.D. 1969. The winter feeding ecology of Redshank *Tringa totanus*. *Ibis*, 111, 338–356.

Goss-Custard, J.D. & dit Durell, S.E.A.l.V. 1987. Age-related effects in Oystercatchers *Haematopus ostralegus* feeding on mussels *Mytilus edulis*. III. The effect of interference on overall intake rate. *Journal of Animal Ecology*, 56, 549–558.

Gottschaldt, K.M. 1985, 'Structure and function of avian somatosensory receptors.' In: *Form and Function in Birds, Vol. 3*. A.S. King & J. McLelland, eds, London, Academic Press, pp. 375–461.

Gould, S.J. 1985. *The Flamingo's Smile: Reflections in Natural History*. New York, Norton.

Gould, S.J. 1989. *Wonderful Life: The Burgess Shale and the Nature of History*. New York, W.W. Norton & Company.

Gover, N., Jarvis, J.R., Abeyesinghe, S.M., & Wathes, C.M. 2009. Stimulus luminance and the spatial acuity of domestic fowl (*Gallus. g. domesticus*). *Vision Research*, 49, 2747–2753.

Grant, P.R. & Grant, B.R. 2002. Unpredictable evolution in a 30-year study of Darwin's finches. *Science*, 296, 707–711.

Grant, P.R. & Grant, B.R. 2014. *Forty Years of Evolution: Darwin's Finches on Daphne Major Island*. Princeton, Princeton University Press.

Graves, G.R. 1992. Greater Yellow-Headed Vulture (*Cathartes melambrotus*) locates good by olfaction. *Raptor Research*, 26, 38–39.

Grémillet, D., Kuntz, G., Delbart, F., Mellet, M., & Kato, A. 2004. Linking the foraging performance of a marine predator to local prey abundance. *Functional Ecology*, 18, 793–801.

Grémillet, D., Kuntz, G., Gilbert, C., Woakes, A.J., Butler, P.J., & Le Maho, Y. 2005. Cormorants dive through the Polar night. *Biology Letters*, 1, 469–471.

Grémillet, D., Nazirides, T., Nikolaou, H., & Crivelli, A.J. 2012. Fish are not safe from great cormorants in turbid water. *Aquatic Biology*, 15, 187–194.

Gridi-Papp, M. & Narins, P.M. 2008, 'Sensory ecology of hearing.' In: *The Senses: A Comprehensive Reference, Vol. 3. Audition*. A.I. Basbaum et al., eds, Amsterdam, Elsevier, pp. 62–74.

Griffin, D.R. 1958. *Listening in the Dark*. New Haven, Yale University Press.

Griffin, D.R. & Suthers, R.A. 1970. Sensitivity of echolocation in cave swiflets. *Biological Bulletin*, 139, 495–501.

Guilford, T. & Biro, D. 2014. Route following and the pigeon's familiar area map. *Journal of Experimental Biology*, 217, 169–179.

Guilford, T. & Taylor, G.K. 2014. The sun compass revisited. *Animal Behaviour*, 97, 135–143.

Guillemain, M., Martin, G.R., & Fritz, H. 2002. Feeding methods, visual fields and vigilance in dabbling ducks (Anatidae). *Functional Ecology*, 16, 522–529.

Gwinner, E. 1975, 'Circadian and circannual rhythms in birds.' In: *Avian Biology, Vol. 5*. D.S. Farmer & J. King, eds, London and New York, Academic Press.

Gwinner, H. 2013, 'Male European Starlings use odorous herbs as nest material to attract females and benefit nestlings.' In: *Chemical Signals in Vertebrates*. M.L. East & M. Dehnhard, eds, New York, Springer, pp. 353–362.

Hackett, S.J., et al. 2008. A phylogenomic study of birds reveals their evolutionary history. *Science*, 320, 1763–1768.

Hagelin, J.C. 2004. Observations on the olfactory ability of the Kakapo *Strigops habroptilus*, the critically endangered parrot of New Zealand. *Ibis*, 146, 161–164.

Hagelin, J.C. 2007. The citrus-like scent of crested auklets: reviewing the evidence for an avian olfactory ornament. *Journal of Ornithology*, 148, S195–S201.

Hagstrum, J.T. 2000. Infrasound and the avian navigational map. *Journal of Experimental Biology*, 203, 1103–1111.

Hale, W.D. 1980. *Waders*. London, Collins.

Harmening, W.M., Nikolay, P., Orlowski, J., & Wagner, H. 2009. Spatial contrast sensitivity and grating acuity of barn owls. *Journal of Vision*, 9, doi: 10.1167/9.7.13.

Harmening, W.M. & Wagner, H. 2011. From optics to attention: visual perception in barn owls. *Journal of Comparative Physiology A*, 197, 1031–1042.

Harshman, J., et al. 2008. Phylogenomic evidence for multiple losses of flight in ratite birds. *Proceedings of the National Academy of Sciences of the United States of America*, 105, 13462–13467.

Hart, N.S. 2001. Variations in cone photoreceptor abundance and the visual ecology of birds. *Journal of Comparative Physiology A*, 187, 685–697.

Hart, N.S. 2002. Vision in the peafowl (Aves: *Pavo cristatus*). *Journal of Experimental Biology*, 205, 3925–3935.

Hart, N.S. 2004. Microspectrophotometry of visual pigments and oil droplets in a marine bird, the wedge-tailed shearwater *Puffinus pacificus*: topographic variations in photoreceptor spectral characteristics. *Journal of Experimental Biology*, 207, 1229–1240.

Hart, N.S. & Hunt, D. 2007. Avian visual pigments: characteristics, spectral tuning, and evolution. *American Naturalist*, 169, S7–S26.

Hart, N.S., Partridge, J.C., & Cuthill, I.C. 1998. Visual pigments, oil droplets and cone photoreceptor distribution in the European Starling (*Sturnus vulgaris*). *Journal of Experimental Biology*, 201, 1433–1446.

Hart, N.S., Partridge, J.C., & Cuthill, I.C. 2000a. Retinal asymmetry in birds. *Current Biology*, 10, 115–117.

Hart, N.S., Partridge, J.C., Cuthill, I.C., & Bennett, A.T.D. 2000b. Visual pigments, oil droplets, ocular media and cone photoreceptor distribution in two species of Passerine bird: the Blue Tit (*Parus caeruleus* L.) and the Blackbird (*Turdus merula* L.). *Journal of Comparative Physiology A*, 186, 375–387.

Hart, N.S. & Vorobyev, M. 2005. Modelling oil droplet absorption spectra and spectral sensitivities of bird cone photoreceptors. *Journal of Comparative Physiology A*, 191, 381–392.

Harvey, P.H. & Pagel, M.D. 1991. *The Comparative Method in Evolutionary Biology*. Oxford, Oxford University Press.

Hastad, O. & Odeen, A. 2014. A vision physiological estimation of ultraviolet window marking visibility to birds. *Peerj*, doi: 10.7717/peerj.621.

Hastad, O., Victorsson, J., & Odeen, A. 2005. Differences in color vision make passerines less conspicuous in the eyes of their predators. *Proceedings of the National Academy of Sciences of the United States of America*, 102, 6391–6394.

Hayes, B.P. & Brooke, M. d.L. 1990. Retinal ganglion cell distribution and behaviour in procellariiform seabirds. *Vision Research*, 30, 1277–1289.

Hayes, B.P., Martin, G.R., & Brooke, M. d.L. 1991. Novel area serving binocular vision in the retinae of procellariiform seabirds. *Brain, Behavior and Evolution*, 37, 79–84.

Hecht, S., Shlaer, S., & Pirenne, M.H. 1942. Energy, quanta, and vision. *Journal of General Physiology*, 25, 819–840.

Hedd, A., Regular, P.M., Montevecchi, W.A., Buren, A.D., Burke, C.M., & Fifield, D.A. 2009. Going deep: common murres dive into frigid water for aggregated, persistent and slow-moving capelin. *Marine Biology*, 156, 741–751.

Heffner, H.E. & Heffner, R.S. 1998, 'Hearing.' In: *Comparative Psychology. A Handbook.* G. Greenberg & H.M. Haraway, eds, New York, Garland, pp. 290–303.

Heffner, R.S. & Heffner, H.E. 1982. Hearing in the elephant (*Elephas maximus*)—absolute sensitivity, frequency discrimination, and sound localization. *Journal of Comparative and Physiological Psychology*, 96, 926–944.

Heffner, R.S. & Heffner, H.E. 1992. Hearing and sound localization in blind mole rats (*Spalax ehrenbergi*). *Hearing Research*, 62, 206–216.

Hess, C. 1909. Die Akkommodation bei Tauchervogeln. *Arch vergleich Opthal*, 1, 153–164.

Heyers, D., Manns, M., Güntürkün, O., & Mouritsen, H. 2007. A visual pathway links brain structures active during magnetic compass orientation in migratory birds. *PLoS One*, 2, e937.

Hill, E.M., Koay, G., Heffner, R.S., & Heffner, H.E. 2014. Audiogram of the chicken (*Gallus gallus domesticus*) from 2 Hz to 9 kHz. *Journal of Comparative Physiology A*, 200, 863–870.

Hills, B.L. 1980. Vision, visibility and perception in driving. *Perception*, 9, 183–216.

Hirao, A., Aoyama, M., & Sugita, S. 2009. The role of uropygial gland on sexual behavior in domestic chicken *Gallus gallus domesticus*. *Behavioural Processes*, 80, 115–120.

Hirons, G.J.M. 1985. The effects of territorial behaviour on the stability and dispersion of Tawny Owl (*Strix aluco*) populations. *Journal of Zoology Series B*, 1, 21–48.

Hirsch, J. 1982. Falcon visual sensitivity to grating contrast. *Nature*, 300, 57–58.

Hodos, W. 1993, 'The visual capabilities of birds.' In: *Avian Vision, Brain and Behavior.* H.P. Ziegler & H.J. Bischof, eds, Cambridge, MA, MIT Press, pp. 63–76.

Hodos, W. 2003. *Minimization of motion smear: reducing avian collisions with wind turbines.* Report NREL/SR-500–33249. Washington, DC, National Renewable Energy Laboratory.

Hodos, W., Ghim, M., Potocki, A., Fields, J., & Storm, T. 2002. Contrast sensitivity in pigeons: a comparison of behavioral and pattern ERG methods. *Documenta Ophthalmologica*, 104, 107–118.

Hodos, W., Leibowitz, R.W., & Bonbright, J.C. 1976. Near-field visual acuity of pigeons: effects of head location and stimulus luminance. *Journal of the Experimental Analysis of Behavior*, 25, 129–141.

Holland, J., Dabelsteen, T., Bjørn, C.P., & Pedersen, S.B. 2001. The location of ranging cues in wren song: evidence from calibrated interactive playback experiments. *Behaviour*, 138, 189–206.

Holland, R.A., Wikelski, M., Kummeth, F., & Bosque, C. 2009. The secret life of Oilbirds: new insights into the movement ecology of a unique avian frugivore. *PLoS One*, 4, e8264.

Holyoak, D.T. 2001. *Nightjars and their Allies: The Caprimulgiformes.* Oxford, Oxford University Press.

Hoodless, A.N. & Hirons, G.J.M. 2007. Habitat selection and foraging behaviour of breeding Eurasian Woodcock *Scolopax rusticola*: a comparison between contrasting landscapes. *Ibis*, 149, 234–249.

Hoon, M.A., Adler, E., Lindemeier, J., Battey, J.F., Ryba, N.J.P., & Zuker, C.S. 1999. Putative mammalian taste receptors: a class of taste-specific GPCRs with distinct topographic selectivity. *Cell*, 96, 541–551.

Houston, D.C. 1984. Does the King Vulture *Sarcorhamphus papa* use a sense of smell to locate food? *Ibis*, 126, 67–69.

Houston, D.C. 1986. Scavenging efficiency of Turkey Vultures in tropical forest. *Condor*, 88, 318–323.

Howland, H.C., Howland, M., & Schmid, K.L. 1992. Focusing and accommodation in the Brown Kiwi (*Apteryx australis*). *Journal of Comparative Physiology A*, 170, 687–689.

Hughes, A. 1977, 'The topography of vision in mammals of contrasting life style: comparative optics and retinal organization.' In: *Handbook of Sensory Physiology, Vol. VII/5*. F. Crescitelli, ed., Berlin, Springer-Verlag, pp. 613–756.

Hunt, D.M., Carvalho, L.S., Cowing, J.A., & Davies, W.L. 2009. Evolution and spectral tuning of visual pigments in birds and mammals. *Philosophical Transactions of the Royal Society B: Biological Sciences*, 364, 2941–2955.

Hunt, G.R. 1996. Manufacture and use of hook-tools by New Caledonian crows. *Nature*, 379, 249–251.

Hunt, S., Cuthill, I.C., Bennett, A.T.D., Church, S.C., & Partridge, J.C. 2001. Is the ultraviolet waveband a special communication channel in avian mate choice? *Journal of Experimental Biology*, 204, 2499–2507.

Huppop, O., Huppop, K., Dierscke, J., & Hill, R. 2016. Bird collisions at an offshore platform in the North Sea. *Bird Study*, 63, 73–82.

James, P.C. 1986. How do Manx Shearwaters *Puffinus puffinus* find their burrows? *Ethology*, 71, 287–294.

Janss, G.F.E. 2000. Avian mortality from power lines: a morphologic approach of a species-specific mortality. *Biological Conservation*, 95, 353–359.

Janss, G.F.E. & Ferrer, M. 1998. Rate of bird collision with power lines: effects of conductor-marking and static wire-marking. *Journal of Field Ornithology*, 69, 8–17.

Janss, G.F.E. & Ferrer, M. 2000. Common crane and great bustard collision with power lines: collision rate and risk exposure. *Wildlife Society Bulletin*, 28, 675–680.

Jaworski, J.W. & Peake, N. 2013. Aerodynamic noise from a poroelastic edge with implications for the silent flight of owls. *Journal of Fluid Mechanics*, 723, 456–479.

Jeffery, W.R. 2005. Adaptive evolution of eye degeneration in the Mexican blind cavefish. *Journal of Heredity*, 96, 185–196.

Jenkins, A.R., Smallie, J.J., & Diamond, M. 2010. South African perspectives on a global search for ways to prevent avian collisions with overhead lines. *Bird Conservation International*, 20, 263–278.

Jerlov, N.G. 1976. *Marine Optics*. Amsterdam, Elsevier.

Jetz, W., Thomas, G.H., Joy, J.B., Hartmann, K., & Mooers, A.O. 2012. The global diversity of birds in space and time. *Nature*, 491, 444–448.

Jiang, P.H., et al. 2012. Reply to Zhao and Zhang: loss of taste receptor function in mammals is directly related to feeding specializations. *Proceedings of the National Academy of Sciences of the United States of America*, 109, E1465.

Johnson, C.S. 1986, 'Dolphin audition and echolocation capacities.' In: *Dolphin Cognition and Behvaiour: A Comparative Approach*. R. Schusterman, J. Thomas, & F. Wopod, eds, Lawrance Erlbaum, pp. 43–56.

Jones, C.D. & Osorio, D. 2004. Discrimination of oriented visual textures by poultry chicks. *Vision Research*, 44, 83–89.

Jones, R.B. & Roper, T.J. 1997. Olfaction in the domestic fowl: a critical review. *Physiology & Behavior*, 62, 1009–1018.

Jouventin, P., Mouret, V., & Bonadonna, F. 2007. Wilson's storm petrels *Oceonites oceonicus* recognise the olfactory signature of their mate. *Ethology*, 113, 1228–1232.

Kane, S.A. & Zamani, M. 2014. Falcons pursue prey using visual motion cues: new perspectives from animal-borne cameras. *Journal of Evolutionary Biology*, 217, 225–234.

Kardong, K.V. 2014. *Vertebrates: Comparative Anatomy, Function, Evolution, 6th Edition*. New York, McGraw-Hill.

Kare, M.R. & Mason, J.R. 1986, 'The chemical senses in birds.' In: *Avian Physiology, 4th Edition*. P.D. Sturkie, ed., New York, Springer-Verlag, pp. 59–67.

Karlsson, A.C., et al. 2010. Red junglefowl have individual body odors. *Journal of Experimental Biology*, 213, 1619–1624.

Karten, H.J. & Hodos, W. 1967. *A Stereotaxic Atlas of the Brain of the Pigeon (Columba livia)*. Baltimore, MD, Johns Hopkins Press.

Katzir, G. & Howland, H.C. 2003. Corneal power and underwater accommodation in great cormorants (*Phalacrocorax carbo sinensis*). *Journal of Experimental Biology*, 206, 833–841.

Katzir, G. & Martin, G.R. 1994. Visual fields in herons (Ardeidae)—panoramic vision beneath the bill. *Naturwissenschaften*, 81, 182–184.

Katzir, G. & Martin, G.R. 1998. Visual fields in the Black-crowned Night Heron *Nycticorax nycticorax*: nocturnality does not result in owl-like features. *Ibis*, 140, 157–162.

Kear, J. & Burton, P.J.K. 1971. The food and feeding apparatus of the Blue Duck Hymenolaimus. *Ibis*, 113, 483–493.

Kelber, A., Vorobyev, M., & Osorio, D. 2003. Animal colour vision: behavioural tests and physiological concepts. *Biological Reviews*, 78, 81–118.

Kellogg, W.N. 1962. Sonar system of the blind. *Science*, 137, 399–404.

Kelly, T.C., Buurma, L., O'Callaghan, M.J.A., & Bolger, R. 2000. Why do birds collide with aircraft? A behavioural approach. *International Bird Strike Committee IBSC*, 25, 47–48.

Kight, C.R. & Swaddle, J.P. 2015. Eastern Bluebirds alter their song in response to anthropogenic changes in the acoustic environment. *Integrative and Comparative Biology*, 55, 418–431.

King, A.S. & King, D.Z. 1980, 'Avian morphology: general principles.' In: *Form and Function in Birds*. A.S. King & J. McLelland, eds, London, Academic Press, pp. 10–89.

King, A.S. & McLelland, J. 1985. *Form and Function in Birds. Volume 1*. London, Academic Press.

Klasing, K.C. 1998, 'Anatomy and physiology of the digestive system.' In: *Comparative Avian Nutrition*, K.C. Klasing, ed., Oxford, CAB International, pp. 16–19.

Klem, D. 2009, 'Avian mortality at windows: the second largest human source of bird mortality on earth.' In: *Tundra to Tropics: Connecting Birds, Habitats and People*. Proceedings of the fourth international partners in flight conference 2008. T.D. Rich et al., eds, McAllen, Texas, Partners in Flight, pp. 244–254.

Klem, D., Keck, D.C., Marty, K.L., Ball, A.J.M., Niciu, E.E., & Platt, C.T. 2004. Effects of window angling, feeder placement and scavengers on avian mortality at plate glass. *Wilson Bulletin*, 116, 69–73.

Klem, D. & Saenger, P.G. 2013. Evaluating the effectiveness of select visual signals to prevent bird-window collisions. *Wilson Journal of Ornithology*, 125, 406–411.

Kleyheeg-Hartman, J.C. 2014. *Overzicht en korte beschrijving van beschikbare collision rate models*. Culemborg, Netherlands, Bureau Waardenburg.

Klump, G. 2000, 'Sound localization in birds.' In: *Comparative Hearing: Birds and Reptiles*. R.J. Dooling, R.R. Fay, & A.N. Popper, eds, New York, Springer, pp. 249–307.

Klump, G., Windt, W., & Curio, E. 1986. The great tit's (*Parus major*) auditory resolution in azimuth. *Journal of Comparative Physiology A*, 158, 383–390.

Klump, G.M. & Larsen, O.N. 1992. Azimuthal sound localization in the European starling (*Sturnus vulgaris*): I. Physical binaural cues. *Journal of Comparative Physiology A*, 170, 243–251.

Knudsen, E.I. 1980, 'Sound localisation in birds.' In: *Comparative Studies of Hearing in Vertebrates*. A.N. Popper & R.R. Fay, eds, Berlin, Springer, pp. 289–322.

Knudsen, E.I. & Konishi, M. 1979. Mechanisms of sound localization in the barn owl (*Tyto alba*). *Journal of Comparative Physiology A*, 133, 13–21.

Koboroff, A., Kaplan, G., & Rogers, L.J. 2008. Lateralization and antipredtor behaviour in Australian magpies (*Gymnorhina tibicen*). *Brain Research Bulletin*, 76, 304–306.

Konishi, M. & Knudsen, E.I. 1979. The Oilbird: hearing and echolocation. *Science*, 204, 425–427.

Kooyman, G.L., et al. 1992. Diving behavior and energetics during foraging cycles in King Penguins. *Ecological Monographs*, 62, 143–163.

Krebs, C. 2009. *Ecology: The Experimental Analysis of Distribution and Abundance. 6th ed.* San Francisco, Benjamin Cummings.

Kreithen, M.L. & Quine, D.B. 1979. Infrasound detection by the homing pigeon—behavioral audiogram. *Journal of Comparative Physiology*, 129, 1–4.

Kroodsma, D.E. & Miller, E.H. 1996. *Ecology and Evolution of Acoustic Communication in Birds.* Ithaca, Cornell University Press.

Land, M.F. 1981, 'Optics and vision in invertebrates.' In: *Handbook of Sensory Physiology, Vol. VII/6B.* H. Autrum, ed., Berlin, Springer-Verlag, pp. 471–592.

Land, M.F. 1999. The roles of head movements in the search and capture strategy of a tern (Aves, Laridae). *Journal of Comparative Physiology A*, 184, 265–272.

Land, M.F. 2015. Eye movements of vertebrates and their relation to eye form and function. *Journal of Comparative Physiology A*, 201, 195–214.

Land, M.F. & Nilsson, D.-E. 2012. *Animal Eyes, 2nd Edition.* Oxford, Oxford University Press.

Lange, G. 1968. Uber Nahrung, Nahrungsaufnahme und Verdaugstraky mitteleuropaischer Limikolen. *Beitraege zur Vogelkunde*, 13, 225–234.

Laughlin, S.B. 2001a. Energy as a constraint on the coding and processing of sensory information. *Current Opinion in Neurobiology*, 11, 475–480.

Laughlin, S.B. 2001b, 'The metabolic cost of information—a fundamental factor in visual ecology.' In: *Ecology of Sensing.* F.G. Barth & A. Schmid, eds, Berlin, Springer, pp. 170–185.

Leclaire, S., Mulard, H., Wagner, R.H., Hatch, S.A., & Danchin, E. 2009. Can Kittiwakes smell? Experimental evidence in a Larid species. *Ibis*, 151, 584–587.

Le Duc, D., Renaud, G., Krishnan, A., & Almen, M.S. 2015. Kiwi genome provides insights into evolution of a nocturnal lifestyle. *Genome Biology*, 16, 147.

Lee, D.N. 1980. The optic flow field: the foundation of vision. *Philosophical Transactions of the Royal Society B: Biological Sciences*, 290, 169–179.

Lee, D.N., Davies, M.N.O., Green, P.R., & van der Weel, F.R. 1993. Visual control of velocity of approach by pigeons when landing. *Journal of Experimental Biology*, 180, 85–104.

Lee, D.N. & Lishman, R. 1977. Visual control of locomotion. *Scandianvian Journal of Psychology*, 18, 224–230.

Lee, D.N. & Reddish, P.E. 1981. Plummeting gannets: a paradigm of ecological optics. *Nature*, 293, 293–294.

Lee, D.N., Reddish, P.E., & Rand, D.T. 1991. Aerial docking by Hummingbirds. *Naturwissenschaften*, 78, 526–527.

Levy, B. & Sivak, J.G. 1980. Mechanisms of accommodation in the bird eye. *Journal of Comparative Physiology A*, 137, 267–272.

Lewis, E.R., Narins, P.M., Jarvis, J.U.M., Bronner, G., & Mason, M. 2006. Preliminary evidence for the use of microseismic cues for navigation by the Namib golden mole. *Journal of the Acoustical Society of America*, 119, 1260–1268.

Lewison, R.L., et al. 2014. Global patterns of marine mammal, seabird, and sea turtle bycatch reveal taxa-specific and cumulative megafauna hotspots. *Proceedings of the National Academy of Sciences of the United States of America*, 111, 5271–5276.

Leys, R., Cooper, S.J.B., Strecker, U., & Wilkens, H. 2005. Regressive evolution of eye pigment gene in independently evolved eyeless subterranean diving beetles. *Biology Letters*, 1, 496–499.

Lima, S.L., Blackwell, B.F., DeVault, T.L., & Fernandez-Juricic, E. 2015. Animal reactions to oncoming vehicles: a conceptual review. *Biological Reviews*, 90, 60–76.

Lima, S.L. & Dill, L.M. 1990. Behavioral decisions made under the risk of predation: a review and prospectus. *Canadian Journal of Zoology*, 68, 619–640.

Lind, O. & Kelber, A. 2011. The spatial tuning of achromatic and chromatic vision in budgerigars. *Journal of Vision*, 11, doi: 10.1167/11.7.2.

Lind, O., Mitkus, M., Olsson, O., & Kelber, A. 2013. Ultraviolet sensitivity and colour vision in raptor foraging. *Journal of Experimental Biology*, 216, 1819–1826.

Lind, O., Mitkus, M., Olsson, P., & Kelber, A. 2014. Ultraviolet vision in birds: the importance of transparent eye media. *Proceedings of the Royal Society B: Biological Sciences*, 281, 20132209.

Lind, O., Sunesson, T., Mitkus, M., & Kelber, A. 2012. Luminance-dependence of spatial vision in budgerigars (*Melopsittacus undulatus*) and Bourke's parrots (*Neopsephotus bourkii*). *Journal of Comparative Physiology A*, 198, 69–77.

Lisney, T.J., Iwaniuk, A.N., Bandet, M.V., & Wylie, D.R. 2012a. Eye shape and retinal topography in owls (Aves: Strigiformes). *Brain, Behavior and Evolution*, 79, 218–236.

Lisney, T.J., Iwaniuk, A.N., Kolominsky, J., Bandet, M.V., & Corfield, J. 2012b. Interspecific variation in eye shape and retinal topography in seven species of galliform bird (Aves: Galliformes: Phasianidae). *Journal of Comparative Physiology A*, 198, 717–731.

Lisney, T.J., et al. 2013a. Ecomorphology of eye shape and retinal topography in waterfowl (Aves: Anseriformes: Anatidae) with different foraging modes. *Journal of Comparative Physiology A*, 199, 385–402.

Lisney, T.J., Stecyk, K., Kolominsky, J., Graves, G.R., Wylie, D.R., & Iwaniuk, A.N. 2013b. Comparison of eye morphology and retinal topography in two species of New World Vultures (Aves: Cathartidae). *Anatomical Record*, 296, 1954–1970.

Livezey, B.C. & Zusi, R.L. 2007. Higher-order phylogeny of modern birds (Theropoda, Aves: Neornithes) based on comparative anatomy. II. Analysis and discussion. *Zoological Journal of the Linnean Society*, 149, 1–95.

Locket, N.A. 1977, 'Adaptations to the deep-sea environment.' In: *Handbook of Sensory Physiology*. F. Crescitelli, ed., Berlin, Springer-Verlag, pp. 67–192.

Loss, S.R., Will, T., & Marra, P.P. 2014. Estimation of bird-vehicles collision mortality on U.S. roads. *Journal of wildlife Management*, 78, 763–771.

Luo, D.-G., Kefalov, K.-W., & Yau Luo, D.-G. 2008, 'Phototransduction in rods and cones.' In: *The Senses: A Comprehensive Reference, Vol. 1. Vision Part 1*. A.I. Basbaum et al., eds, Amsterdam, Elsevier, pp. 269–301.

Luther, D. & Gentry, K. 2013. Sources of background noise and their influence on vertebrate acoustic communication. *Behaviour*, 150, 1045–1068.

Lythgoe, J.N. 1979. *The Ecology of Vision*. Oxford, Clarendon Press.

Lythgoe, J.N. & Partridge, J.C. 1989. Visual pigments and the acquisition of visual information. *Journal of Experimental Biology*, 146, 1–20.

Macdonald, D.W. 1976. Nocturnal observations of Tawny Owls *Strix aluco* preying upon earthworms. *Ibis*, 118, 579–580.

Manley, G.A. & Clack, J.A. 2004, 'An outline of the evolution of vertebrate hearing organs.' In: *Evolution of the Vertebrate Auditory System*. G.A. Manley, A.N. Popper, & R.R. Fay, eds, New York, Springer, pp. 1–26.

Manville, A.M. 2005. *Bird strikes and electrocutions at power lines, communication towers, and wind turbines: state of the art and state of the science—next steps toward mitigation*. USDA Forest Service General Technical Report PSW-GTR-191.

Marchant, S. & Higgins, P.J. 1990. *Handbook of Australian, New Zealand and Antarctic Birds, Vol. 1 Part B*. Melbourne, Oxford University Press.

Mardon, J. & Bonadonna, F. 2009. Atypical homing or self-odour avoidance? Blue petrels (*Halobaena caerulea*) are attracted to their mate's odour but avoid their own. *Behavioral Ecology and Sociobiology*, 63, 537–542.

Marshall, J. 1991, 'The effect of ultraviolet radiation and blue light on the eye.' In: *Vision and Visual Dysfunction*. J. Marshall, ed., London, Macmillan, pp. 54–66.

Marshall, J., Mellerio, J., & Palmer, D.A. 1973. A schematic eye for the pigeon. *Vision Research*, 13, 2449–2453.

Marten, K. & Marler, P. 1977. Sound transmission and its significance for animal vocalization.1. Temperate habitats. *Behavioral Ecology and Sociobiology*, 2, 271–290.

Martin, G.R. 1974. Colour vision in the Tawny Owl *Strix aluco*. *Journal of Comparative and Physiological Psychology*, 86, 133–141.

Martin, G.R. 1977. Absolute visual threshold and scotopic spectral sensitivity in the tawny owl *Strix aluco*. *Nature*, 268, 636–638.

Martin, G.R. 1982. An owl's eye: schematic optics and visual performance in *Strix aluco* L. *Journal of Comparative Physiology A*, 145, 341–349.

Martin, G.R. 1983, 'Schematic eye models in vertebrates.' In: *Progress in Sensory Physiology, Vol. 4*. D. Ottoson, ed., Berlin, Springer-Verlag, pp. 43–81.

Martin, G.R. 1984. The visual fields of the tawny owl, *Strix aluco L*. *Vision Research*, 24, 1739–1751.

Martin, G.R. 1985, 'Eye.' In: *Form and Function in Birds, Vol. 3*. A.S. King & J. McLelland, eds, London, Academic Press, pp. 311–373.

Martin, G.R. 1986a. Sensory capacities and the nocturnal habit of owls (Strigiformes). *Ibis*, 128, 266–277.

Martin, G.R. 1986b. The eye of a passeriform bird, the European starling (*Sturnus vulgaris*): eye movement amplitude, visual fields and schematic optics. *Journal of Comparative Physiology A*, 159, 545–557.

Martin, G.R. 1986c. Total panoramic vision in the mallard duck, Anas platyrhynchos. *Vision Research*, 26, 1303–1306.

Martin, G.R. 1990a. *Birds by Night*. London, T & A D Poyser.

Martin, G.R. 1990b, 'The visual problems of nocturnal migration.' In: *Bird Migration: Physiology and Ecophysiology*. E. Gwinner, ed., Berlin, Springer-Verlag, pp. 185–197.

Martin, G.R. 1994. Visual fields in woodcocks *Scolopax rusticola* (Scolopacidae; Charadriiformes). *Journal of Comparative Physiology A*, 174, 787–793.

Martin, G.R. 1998. Eye structure and amphibious foraging in albatrosses. *Proceedings of the Royal Society B: Biological Sciences*, 265, 1–7.

Martin, G.R. 1999a. Eye structure and foraging in King Penguins *Aptenodytes patagonicus*. *Ibis*, 141, 444–450.

Martin, G.R. 1999b, 'Optical structure and visual field in birds: their relationship with foraging behaviour and ecology.' In: *Adaptive Mechanisms in the Ecology of Vision*. S.N. Archer et al., eds, Dordrecht, Kluwer, pp. 485–507.

Martin, G.R. 2007. Visual fields and their functions in birds. *Journal of Ornithology*, 148 (Suppl. 2), 547–562.

Martin, G.R. 2009. What is binocular vision for? A birds' eye view. *Journal of Vision*, 9, 1–19.

Martin, G.R. 2011. Understanding bird collisions with man-made objects: a sensory ecology approach. *Ibis*, 153, 239–254.

Martin, G.R. 2012. Through birds' eyes: insights into avian sensory ecology. *Journal of Ornithology*, 153 (Suppl. 1), S23–S48.

Martin, G.R. 2014. The subtlety of simple eyes: the tuning of visual fields to perceptual challenges in birds. *Philosophical Transactions of the Royal Society B: Biological Sciences*, 369, doi.org/10.1098/rstb.2013.0040.

Martin, G.R., Ashash, U., & Katzir, G. 2001. Ostrich ocular optics. *Brain, Behavior and Evolution*, 58, 115–120.

Martin, G.R. & Brooke, M.D.L. 1991. The eye of a procellariiform seabird, the Manx shearwater, *Puffinus puffinus*: visual fields and optical structure. *Brain, Behavior and Evolution*, 37, 65–78.

Martin, G.R. & Coetzee, H.C. 2004. Visual fields in Hornbills: precision-grasping and sunshades. *Ibis*, 146, 18–26.

Martin, G.R. & Crawford, R. 2015. Reducing bycatch in gillnets: a sensory ecology perspective. *Global Ecology and Conservation*, 3, 28–50.

Martin, G.R. & Gordon, I.E. 1974. Visual acuity in the tawny owl (*Strix aluco*). *Vision Research*, 14, 1393–1397.

Martin, G.R., Jarrett, N., Tovey, P., & White, C.R. 2005. Visual fields in Flamingos: chick-feeding versus filter-feeding. *Naturwissenschaften*, 92, 351–354.

Martin, G.R., Jarrett, N., & Williams, M. 2007a. Visual fields in Blue Ducks and Pink-eared Ducks: visual and tactile foraging. *Ibis*, 149, 112–120.

Martin, G.R. & Katzir, G. 1994a. Visual fields and eye movements in herons (Ardeidae). *Brain, Behavior and Evolution*, 44, 74–85.

Martin, G.R. & Katzir, G. 1994b. Visual fields in the stone curlew *Burhinus oedicnemus*. *Ibis*, 136, 448–453.

Martin, G.R. & Katzir, G. 1995. Visual fields in ostriches. *Nature*, 374, 19–20.

Martin, G.R. & Katzir, G. 1999. Visual field in Short-toed eagles *Circaetus gallicus* and the function of binocularity in birds. *Brain, Behavior and Evolution*, 53, 55–66.

Martin, G.R. & Katzir, G. 2000. Sun shades and eye size in birds. *Brain, Behavior and Evolution*, 56, 340–344.

Martin, G.R., McNeil, R., & Rojas, L.M. 2007b. Vision and the foraging technique of skimmers (Rynchopidae). *Ibis*, 149, 750–759.

Martin, G.R. & Osorio, D. 2008, 'Vision in birds.' In: *The Senses: A Comprehensive Reference, Vol. 1. Vision Part I.* A.I. Basbaum et al., eds, Amsterdam, Elsevier, pp. 25–52.

Martin, G.R. & Piersma, T. 2009. Vision and touch in relation to foraging and predator detection: insightful contrasts between a plover and a sandpiper. *Proceedings of the Royal Society B: Biological Sciences*, 276, 437–445.

Martin, G.R. & Portugal, S.J. 2011. Differences in foraging ecology determine variation in visual field in ibises and spoonbills (Threskiornithidae). *Ibis*, 153, 662–671.

Martin, G.R., Portugal, S.J., & Murn, C.P. 2012. Visual fields, foraging and collision vulnerability in Gyps vultures Ibis. *Ibis*, 154, 626–631.

Martin, G.R. & Prince, P.A. 2001. Visual fields and foraging in Procellariiform seabirds: sensory aspects of dietary segregation. *Brain, Behavior and Evolution*, 57, 33–38.

Martin, G.R., Rojas, L.M., Ramirez Figueroa, Y.M., & McNeil, R. 2004a. Binocular vision and nocturnal activity in Oilbirds (*Steatornis caripensis*) and Pauraques (*Nyctidromus albicollis*): Caprimulgiformes. *Ornithologia Neotropical*, 15 (Suppl.), 233–242.

Martin, G.R., Rojas, L.M., Ramirez, Y., & McNeil, R. 2004b. The eyes of oilbirds (*Steatornis caripensis*): pushing at the limits of sensitivity. *Naturwissenschaften*, 91, 26–29.

Martin, G.R. & Shaw, J.M. 2010. Bird collisions with power lines: failing to see the way ahead? *Biological Conservation*, 143, 2695–2702.

Martin, G.R. & Wanless, S. 2015. The visual fields of Common Guillemots *Uria aalge* and Atlantic Puffins *Fratercula arctica*: foraging, vigilance and collision vulnerability. *Ibis*, 157, 798–807.

Martin, G.R., White, C.R., & Butler, P.J. 2008. Vision and the foraging technique of Great Cormorants *Phalacrocorax carbo*: pursuit or flush-foraging? *Ibis*, 150, 39–48.

Martin, G.R., Wilson, K.J., Wild, M.J., Parsons, S., Kubke, M.F., & Corfield, J. 2007c. Kiwi forego vision in the guidance of their nocturnal activities. *PLoS One*, 2(2), e198.

Martin, G.R. & Young, S.R. 1983. The retinal binocular field of the pigeon (*Columba livia*): English racing homer. *Vision Research*, 23, 911–915.

Martin, G.R. & Young, S.R. 1984. The eye of the Humboldt Penguin, *Spheniscus humboldti*: visual fields and schematic optics. *Proceedings of the Royal Society B: Biological Sciences*, 223, 197–222.

Martyniak, A., Wziatek, B., Szymanska, U., Hliwa, P., & Terlecki, J. 2003. Diet composition of great cormorants *Phalacrocorax carbo sinensis* at Katy Rybackie, NE Poland, as assessed by pellets and regurgitated prey. *Vogelwelt*, 124 (Suppl. 1), 217–225.

Mason, J.R. & Clark, L. 2000, 'The chemical senses in birds.' In: *Sturkie's Avian Physiology, 5th Edition*. G.C. Whittow, ed., San Diego, Academic Press, pp. 39–56.

Matson, K.D., Millam, J.R., & Klasing, K.C. 2001. Thresholds for sweet, salt, and sour taste stimuli in cockatiels (*Nymphicus hollandicus*). *Zoo Biology*, 20, 1–13.

Matsunami, H., Montmayeur, J.P., & Buck, L.B. 2000. A family of candidate taste receptors in human and mouse. *Nature*, 404, 601–602.

Maziarz, M. & Wesolowski, T. 2014. Does darkness limit the use of tree cavities for nesting by birds? *Journal of Ornithology*, 155, 793–799.

McGregor, P.K., Krebs, J.R., & Ratcliffe, L.M. 1983. The reaction of great tits (*Parus major*) to playback of degraded and undegraded songs: the effect of familiarity with the stimulus song type. *Auk*, 100, 898–906.

McIsaac, H.P. 2001. Raptor acuity and wind turbine blade conspicuity. *Proceedings of National Avian-Wind Power Planning Meeting IV*, 59–87.

McKenna, O.C. & Wallman, J. 1985. Accessory optic system and pretectum of birds: comparisons with those of other vertebrates. *Brain Behavior and Evolution*, 26, 91–116.

Medway, L. & Pye, D. 1977, 'Echolocation and systematic of swiftlets.' In: *Evolutionary Ecology*. B. Stonehouse & C.M. Perrins, eds, London, Macmillan, pp. 225–238.

Mench, J.A. & Andrew, R.J. 1986. Lateralization of a food search task in the domestic chick. *Behavioural and Neural Biology*, 46, 107–114.

Merton, D.V., Morris, R.B., & Atkinson, I.A.E. 1984. Lek behavior in a parrot—the Kakapo *Strigops habroptilus* of New Zealand. *Ibis*, 126, 277–283.

Meyer, D.B. 1977, 'The avian eye and its adaptations.' In: *Handbook of Sensory Physiology, Vol. VII/5*. F. Crescitelli, ed., Berlin, Springer-Verlag, pp. 549–611.

Middleton, W.E.K. 1958. *Vision Through the Atmosphere*. Toronto, University of Toronto Press.

Mikkola, H. 1983. *Owls of Europe*. Calton, T & A D Poyser.

Miller, W.H. 1979, 'Ocular optical filtering.' In: *Handbook of Sensory Physiology, Vol. VII/6A*. H. Autrum, ed., Berlin, Springer-Verlag, pp. 69–143.

Mitkus, M. 2015. *Spatial vision in birds: anatomical investigations of spatial resolving power*. Doctoral Dissertation. Lund, Lund University.

Mitkus, M., Chaib, S., Lind, O., & Kelber, A. 2014. Retinal ganglion cell topography and spatial resolution of two parrot species: budgerigar *Melopsittacus undulatus* and Bourke's parrot *Neopsephotus bourkii*. *Journal of Comparative Physiology A*, 200, 371–384.

Montgomerie, R. & Weatherhead, P.J. 1997. How do robins find worms? *Animal Behaviour*, 54, 143–151.

Moore, B.A., et al. 2012. Oblique color vision in an open-habitat bird: spectral sensitivity, photoreceptor distribution and behavioral implications. *Journal of Experimental Biology*, 215, 3442–3452.

Moore, B.A., Doppler, M., Young, J.E., & Fernandez-Juricic, E. 2013. Interspecific differences in the visual system and scanning behaviour of three forest passerines that form heterospecific flocks. *Journal of Comparative Physiology A*, 199, 263–277.

Moroney, M.K. & Pettigrew, J.D. 1987. Some observations on the visual optics of kingfishers (Aves, Coraciformes, Alcedinidae). *Journal of Comparative Physiology A*, 160, 137–149.

Morton, E.S. 1975. Ecological sources of selection on avian sounds. *American Naturalist*, 109, 17–34.

Morton, E.S., Howlett, J., Kopysh, N.C., & Chiver, I. 2006. Song ranging by incubating male Blue-headed Vireos: the importance of song representation in repertoires and implications for song delivery patterns and local/foreign dialect discrimination. *Journal of Field Ornithology*, 77, 291–301.

Müller, P. & Ahmad, M. 2011. Light activated cryptochrome reacts with molecular oxygen to form a flavin-superoxide radical pair consistent with magnetoreception. *Journal of Biological Chemistry*, 286, 21033–21040.

Munro, U., Luu, P., DeFilippis, L., & Freire, R. 2014. Ontogeny of magnetoreception in chickens (*Gallus gallus domesticus*). *Journal of Ethology*, 32, 69–74.

Muntz, W.R.A. 1972, 'Inert absorbing and reflecting pigments.' In: *Handbook of Sensory Physiology, Vol. VII/1*. H.J.A. Dartnall, ed., Berlin and New York, Springer-Verlag, pp. 72–92.

Myers, R.A. & Worm, B. 2003. Rapid worldwide depletion of predatory fish communities. *Nature*, 423, 280–283.

Naguib, M. 1997. Use of song amplitude for ranging in Carolina wrens, *Thryothorus ludovicianus*. *Ethology*, 103, 723–731.

Naguib, M. 1998. Perception of degradation in acoustic signals and its implications for ranging. *Behavioral Ecology and Sociobiology*, 42, 139–142.

Naguib, M., Klump, G., Hillmann, E., Grießmann, B., & Teige, T. 2000. Assessment of auditory distance in a territorial songbird: accurate feat or rule of thumb? *Animal Behaviour*, 59, 715–721.

Naguib, M. & Wiley, R.H. 2001. Estimating the distance to a source of sound: mechanisms and adaptations for long-range communication. *Animal Behaviour*, 62, 825–837.

Nelson, B.S. 2000. Avian dependence on sound-pressure level as an auditory distance cue. *Animal Behaviour*, 59, 57–67.

Nelson, B.S. & Stoddard, P.K. 1998. Accuracy of auditory distance and azimuth perception by a passerine bird in natural habitat. *Animal Behaviour*, 56, 467–477.

Nelson, B.S. & Suthers, R.A. 2004. Sound localization in a small passerine bird: discrimination of azimuth as a function of head orientation and sound frequency. *Journal of Experimental Biology*, 207, 4121–4133.

Nevitt, G.A. 2000. Olfactory foraging by Antarctic procellariiform seabirds: life at high Reynolds numbers. *Biology Bulletin (Woods Hole)*, 198, 245–253.

Nevitt, G.A. 2008. Sensory ecology on the high seas: the odor world of the procellariiform seabirds. *Journal of Experimental Biology*, 211, 1706–1713.

Nevitt, G.A. & Bonadonna, F. 2005. Sensitivity to dimethyl sulphide suggests a mechanism for olfactory navigation by seabirds. *Biology Letters*, 1, 303–305.

Nevitt, G.A., Veit, R.R., & Kareiva, P. 1995. Dimethyl sulphide as a foraging cue for Antarctic Procellariiform seabirds. *Nature*, 376, 680–682.

Newton, I. 2010. *Opticks; Or, A Treatise of the Reflections, Refractions, Inflections, and Colours of Light*. Project Gutenberg; http://www.gutenberg.org/ebooks/33504.

Nicol, J.A.C. & Arnott, H.J. 1974. Tapeta lucida in the eyes of goatsuckers (Caprimulgidae). *Proceedings of the Royal Society B: Biological Sciences*, 187, 349–352.

Nilsson, D.-E. 1988. A new type of imaging optics in compound eyes. *Nature*, 332, 76–78.

Nilsson, D.-E. 2009. The evolution of eyes and visually guided behaviour. *Philosophical Transactions of the Royal Society B: Biological Sciences*, 364, 2833–2847.

Nilsson, D.-E. & Pelger, R.F. 1994. A pessimistic estimate of the time required for an eye to evolve. *Proceedings of the Royal Society B: Biological Sciences*, 256, 53–58.

Nilsson, L. 1969. Food consumption of diving ducks wintering at the coast of South Sweden. *Oikos*, 20, 128–135.

Norberg, R.A. 1968. Physical factors in directional hearing in *Aegolius funereus* (Strigiformes), with special reference to the significance of the asymmetry of the external ears. *Arkive Zoology*, 20, 181–204.

Norberg, R.A. 1977. Occurrence and independent evolution of bilateral ear asymmetry in owls and implications on owl taxonomy. *Philosophical Transactions of the Royal Society B: Biological Sciences*, 280, 376–408.

Norberg, R.A. 1978. Skull asymmetry, ear structure and function and auditory localization in Tengmalm's Owl, *Aegolius funereus*. *Philosophical Transactions of the Royal Society B: Biological Sciences*, 282B, 325–410.

Norberg, U.M. 1990. *Vertebrate Flight: Mechanics, Physiology, Morphology, Ecology and Evolution. Zoophysiology Series, Vol. 27.* Berlin, Springer-Verlag.

Odeen, A. & Hastad, O. 2010. Pollinating birds differ in spectral sensitivity. *Journal of Comparative Physiology A*, 196, 91–96.

Odeen, A. & Hastad, O. 2013. The phylogenetic distribution of ultraviolet sensitivity in birds. *BMC Evolutionary Biology*, 13, doi:10.1186/1471-2148-13-36.

Odeen, A., Hastad, O., & Alstrom, P. 2010. Evolution of ultraviolet vision in shorebirds (Charadriiformes). *Biology Letters*, 6, 370–374.

Odeen, A., Hastad, O., & Alstrom, P. 2011. Evolution of ultraviolet vision in the largest avian radiation—the passerines. *BMC Evolutionary Biology*, 11, doi:10.1186/1471-2148-11-313.

Olsson, O. & North, A.W. 1997. Diet of the King Penguin *Aptenodytes patagonicus* during three summers at South Georgia. *Ibis*, 139, 504–512.

Orlowski, J., Harmening, W., & Wagner, H. 2012. Night vision in barn owls: visual acuity and contrast sensitivity under dark adaptation. *Journal of Vision*, 12, 1–8.

O'Rourke, C., Hall, M., Pitlik, T., & Fernandez-Juricic, E. 2010. Hawk eyes I: Diurnal raptors differ in visual fields and degree of eye movement. *PLoS One*, 5, doi: 10.1371/journal.pone.0012802.

Orta, J. 1992, 'Family Phalacrocoracidae (Cormorants).' In: *Handbook of the Birds of the World, Vol. 1. Ostrich to Ducks.* J. del Hoyo, A. Elliott, & J. Sargatal, eds, Barcelona, Lynx Edicions, pp. 326–353.

Osorio, D., Vorobyev, M., & Jones, C.D. 1999. Colour vision of domestic chicks. *Journal of Experimental Biology*, 202, 2951–2959.

Osterblom, H., Fransson, T., & Olsson, O. 2002. Bycatches of common guillemot (*Uria aalge*) in the Baltic Sea gillnet fishery. *Biological Conservation*, 105, 309–319.

Pakan, J.M.P. & Wylie, D.R.W. 2006. Two optic flow pathways from the pretectal nucleus lentiformis mesencephali to the cerebellum in pigeons (*Columba livia*). *Journal of Comparative Neurology*, 499, 732–744.

Park, T.J. & Dooling, R.J. 1991. Sound localization in small birds—absolute localization in azimuth. *Journal of Comparative Psychology*, 105, 125–133.

Parker, A. 2003. *In the Blink of an Eye.* Cambridge, MA, Persus Books.

Partridge, J.C., Douglas, R.H., Marshall, N.J., Chung, W.-S., Jordan, T.M., & Wagner, H.-J. 2014. Reflecting optics in the diverticular eye of a deep-sea barreleye fish (*Rhynchohyalus natalensis*). *Proceedings of the Royal Society B: Biological Sciences*, 281, doi: 10.1098/rspb.2013.3223.

Payne, R.S. 1971. Acoustic location of prey by barn owls. *Journal of Experimental Biology*, 54, 535–573.

Perrins, C.M. 2009. *The Encyclopedia of Birds*. Oxford University Press.

Petit, C., Hossaert-McKey, M., Perret, P., Blondel, J., & Lambrechts, M.M. 2002. Blue tits use selected plants and olfaction to maintain an aromatic environment for nestlings. *Ecology Letters*, 5, 585–589.

Philips, J.D., Nachtigall, P.E., Au, W.W.L., Pawloski, J.L., & Roitblat, H.L. 2003. Echolocation in the Risso's dolphin, *Grampus griseus*. *Journal of the Acoustical Society of America*, 113, 605–616.

Phillips, M.J., Gibb, G.C., Crimp, E.A., & Penny, D. 2010. Tinamous and Moa flock together: mitochondrial genome sequence analysis reveals independent losses of flight among ratites. *Systematic Biology*, 59, 90–107.

Piersma, T., Lindeboom, R., & Vaneerden, M.R. 1988. Foraging rhythm of great crested grebes *Podiceps cristatus* adjusted to diel variations in the vertical-distribution of their prey *Osmerus eperlanus* in a shallow eutrophic lake in the Netherlands. *Oecologia*, 76, 481–486.

Piersma, T., van Aelst, R., Kurk, K., Berkhoudt, H., & Maas, L.R.M. 1998. A new pressure sensory mechanisms for prey detection in birds: the use of principles of seabed dynamics? *Proceedings of the Royal Society B: Biological Sciences*, 265, 1377–1383.

Pirenne, M.H., Marriott, F.H.C., & O'Doherty, E.F. 1957. Individual differences in night vision efficiency. *Medical Research Council (GB) Special Report Series*, 294.

Podulka, S., Rohrbaugh, R.W., & Bonney, R. 2004. *Handbook of Bird Biology 2nd ed.* Ithaca, NY, Cornell Lab of Ornithology.

Pohl, N.U., Klump, G., & Langemann, U. 2015. Effects of signal features and background noise on distance cue discrimination by a songbird. *Journal of Experimental Biology*, 218, 1006–1015.

Popkin, R. 2003. *The History of Scepticism: From Savonarola to Bayle*. Oxford, Oxford University Press.

Portugal, S. J., Murn, C. P., & Martin, G. R. 2017. White-headed Vultures *Trigonoceps occipitalis* show visual field characteristics of hunting raptors. *Ibis*, in press.

Potier, S., et al. 2016. Visual abilities in two raptors with different ecology. *Journal of Experimental Biology*, 219, 2639–2649.

Poulter, T.C. 1969. Sonar of penguins and fur seals. *Proceedings of the California Academy of Sciences*, 36, 363–380.

Powesland, R.G., Merton, D.V., & Cockrem, J.F. 2006. A parrot apart: the natural history of the kakapo (*Strigops habroptilus*), and the context of its conservation. *Notornis*, 53, 3–26.

Pye, J.D. 1979. Why ultrasound? *Endeavour*, 3, 57–62.

Pye, J.D. 1985, 'Echolocation.' In: *A Dictionary of Birds*. B. Campbell & E. Lack, eds, Calton, Poyser, pp. 165–166.

Reeves, R.R., McClellan, K., & Werner, T.B. 2013. Marine mammal bycatch in gillnet and other entangling net fisheries, 1990 to 2011. *Endangered Species Research*, 20, 71–97.

Regular, P., Montevecchi, W., Hedd, A., Robertson, G., & Wilhelm, S. 2013. Canadian fishery closures provide a large-scale test of the impact of gillnet bycatch on seabird populations. *Biology Letters*, 9, 10.1098/rsbl.2013.0088.

Regular, P.M., Davoren, G.K., Hedd, A., & Montevecchi, W.A. 2010. Crepuscular foraging by a pursuit-diving seabird: tactics of common murres in response to the diel vertical migration of capelin. *Marine Ecology Progress Series*, 415, 295–304.

Regular, P.M., Hedd, A., & Montevecchi, W.A. 2011. Fishing in the dark: a pursuit-diving seabird modifies foraging behaviour in response to nocturnal light levels. *PLoS One*, 6, 10.1371/journal.pone.0026763.

Reiner, A., Yamamoto, K., & Karten, H.J. 2005. Organization and evolution of the avian forebrain. *Anatomical Record Part A*, 287A, 1080–1102.

Reymond, L. 1985. Spatial visual acuity of the eagle *Aquila audax*: a behavioural, optical and anatomical investigation. *Vision Research*, 25, 1477–1491.

Reymond, L. 1987. Spatial visual acuity of the falcon, *Falco berigora*: a behavioural, optical and anatomical investigation. *Vision Research*, 27, 1859–1974.

Reynolds, S.J. & Perrins, C.M. 2010, 'Dietary calcium availability and reproduction in birds.' In: *Current Ornithology, Vol. 17.* C. F. Thompson, ed., New York, Springer, pp. 31–74.

Robert, M., McNeil, R., & Le Duc, A. 1989. Conditions and significance of night feeding in shorebirds and other water birds in a tropical lagoon. *Auk*, 106, 94–101.

Robertson, H. A. 2013, 'Okarito brown kiwi.' In: *New Zealand Birds Online.* C. M. Miskelly, ed., Available at: www.nzbirdsonline.org.nz.

Roca, R.L. 1994. *Oilbirds of Venezuela: Ecology and Conservation.* Cambridge, MA, Nuttall Ornithological Club.

Rochon-Duvigneaud, A. 1943. *Les yeux et la vision des vertébrés.* Paris, Masson.

Rodieck, R.W. 1973. *The Vertebrate Retina: Principles of Structure and Function.* San Francisco, Freeman.

Rogers, L.J. 1991, 'Development of lateralization.' In: *Neural and Behavioural Plasticity: The Use of the Domestic Chicken and a Model.* R.J. Andrew, ed., Oxford, Oxford University Press, pp. 503–535.

Rogers, L.J. 2008. Development and function of lateralization in the avian brain. *Brain Research Bulletin*, 76, 235–244.

Rogers, L.J., Munro, U., Freire, R., Wiltschko, R., & Wiltschko, W. 2008. Lateralized response of chicks to magnetic cues. *Behavioural Brain Research*, 186, 66–71.

Rojas, L.M., Ramírez, Y., McNeil, R., Mitchell, M., & Marín, G. 2004. Retinal morphology and electrophysiology of two caprimulgiformes birds: the cave-living and nocturnal oilbird (*Steatornis caripensis*), and the crepuscularly and nocturnally foraging common pauraque (*Nyctidromus albicollis*). *Brain, Behavior and Evolution*, 64, 19–33.

Rossler, M., Nemeth, E., & Bruckner, A. 2015. Glass pane markings to prevent bird-window collisions: less can be more. *Biologia*, 70, 535–541.

Rothery, P., Newton, I., & Little, B. 2009. Observations of seabirds at offshore wind turbines near Blyth in northeast England. *Bird Study*, 56, 1–14.

Roura, E., Baldwin, M.W., & Klasing, K.C. 2013. The avian taste system: potential implications in poultry nutrition. *Animal Feed Science and Technology*, 180, 1–9.

Russell, B. 2001. *The Problems of Philosophy, 2nd Edition. Introduction by John Skorupski.* Oxford, Oxford University Press.

Sansom, A., Lind, J., & Cresswell, W. 2009. Individual behavior and survival: the roles of predator avoidance, foraging success, and vigilance. *Behavioural Ecology*, 20, 1168–1174.

Sarradj, E., Fritzsche, C., & Geyer, T. 2011. Silent owl flight: bird flyover noise measurements. *AIAA Journal*, 49, 769–779.

Sauer, E.G.F. & Sauer, E.M. 1955. Zur Frage der nachtlichen Zugorientierung von Grasmucken. *Revue Suisse de Zoologie*, 62, 250–259.

Schaub, M. & Pradel, R. 2004. Assessing the relative importance of different sources of mortality from recoveries of marked animals. *Ecology*, 85, 930–938.

Schuler, W. 1983. Responses to sugars and their behavioral mechanisms in the Starling (*Sturnus vulgaris* L). *Behavioral Ecology and Sociobiology*, 13, 243–251.

Semm, P. & Beason, R.C. 1990. Responses to small magnetic variations by the trigeminal system of the Bobolink. *Brain Research Bulletin*, 25, 735–740.

Serventy, D.L. 1936. Feeding methods of *Podargus*. *Emu*, 36, 74–90.

Shaw, J.M. 2009. *The end of the line for South Africa's national bird? Modelling power line collision risk for the Blue Crane.* University of Cape Town, Masters Thesis Percy Fitzpatrick Institute of African Ornithology.

Shaw, J.M., Jenkins, A.R., Smallie, J.J., & Ryan, P.G. 2010. Modelling power-line collision risk for the Blue Crane *Anthropoides paradiseus* in South Africa. *Ibis*, 152, 590–599.

Shi, P. & Zhang, J.Z. 2006. Contrasting modes of evolution between vertebrate sweet/umami receptor genes and bitter receptor genes. *Molecular Biology and Evolution*, 23, 292–300.

Shlaer, S. 1937. The relation between visual acuity and illumination. *Journal of General Physiology*, 21, 165–188.

Sitters, H.P. 2000. *The role of night-feeding in shorebirds in an estuarine environment with specific reference to mussel-feeding oystercatchers*. University of Oxford, Doctoral Thesis.

Sivak, J.G. 1976. The role of the flat cornea in the amphibious behaviour of the blackfoot penguin. *Canadian Journal of Zoology*, 54, 1341–1345.

Sivak, J.G. 1978, 'A survey of vertebrate strategies for vision in air and water.' In: *Sensory Ecology: Review and Perspectives*. M.A. Ali, ed., New York, Plenum Press, pp. 503–520.

Sivak, J.G. & Millodot, M. 1977. Optical performance of the penguin eye in air and water. *Journal of Comparative Physiology A*, 119, 241–247.

Sjoberg, S., et al. 2015. Weather and fuel reserves determine departure and flight decisions in passerines migrating across the Baltic Sea. *Animal Behaviour*, 104, 59–68.

Skelhorn, J. & Rowe, C. 2005. Frequency-dependent taste-rejection by avian predation may select for defence chemical polymorphisms in aposematic prey. *Biology Letters*, 1, 500–503.

Skiles, D. D. 1985, 'The geomagnetic field: its nature, history and biological relevance.' In: *Magnetite Biomineralization and Magnetoreception in Organisms*. J.L. Kirschvink, D.S. Jones, & B.J. MacFadden, eds, New York, Plenum Press, pp. 43–102.

Sliney, D.H. 1991, 'Measurement of light exposure and the geometry of exposure of the human eye.' In: *Vision and Visual Dysfunction*. J. Marshall, ed., London, Macmillan, pp. 23–29.

Smith, E.L., Greenwood, V.J., & Bennett, A.T.D. 2002. Ultraviolet colour perception in European starlings and Japanese quail. *Journal of Experimental Biology*, 205, 3299–3306.

Smythe, D.M. & Roberts, J.R. 1983. The sensitivity of echolocation by Grey Swiftlets *Aerodramus spodiopygius*. *Ibis*, 125, 339–345.

Snow, D.W. 1961. The natural history of the Oilbird, *Steatornis caripensis*, in Trinidad. 1. General behaviour and breeding habits. *Zoologica*, 46, 27–48.

Snyder, A.W., Laughlin, S.B., & Stavenga, D.G. 1977. Information capacity of eyes. *Vision Research*, 17, 1163–1175.

Sodhi, N.S. 2002. Competition in the air: birds versus aircraft. *Auk*, 119, 587–595.

Southern, H.N. 1970. The natural control of a population of Tawny Owls (*Strix aluco*). *Journal of Zoology*, 162, 197–285.

Srinivasan, M.V. 1996. Flies go with the flow. *Nature*, 384, 411.

Stahl, G. 2010. *Arthur Schopenhauer: On vision and Colors*. New York, Princeton Architectural Press.

Stapput, K., Gunturkun, O., Hoffmann, K.P., Wiltschko, R., & Wiltschko, W. 2010. Magnetoreception of directional information in birds requires nondegraded vision. *Current Biology*, 20, 1259–1262.

Steinbach, M.J. & Money, K.E. 1973. Eye movements of the owl. *Vision Research*, 13, 889–891.

Stevens, M. 2013. *Sensory Ecology, Behaviour, and Evolution*. Oxford, Oxford University Press.

Streby, H.M., Kramer, G.R., Peterson, S.M., Lehman, J.A., Buehler, D.A., & Andersen, D.E. 2015. Tornadic storm avoidance behavior in breeding songbirds. *Current Biology*, 25, 98–102.

Summers, R.W. & Dugan, D. 2001. An assessment of methods used to mark fences to reduce bird collisions in pinewoods. *Scottish Forestry*, 55, 23–29.

Suthers, R.A. & Hector, D.H. 1983. Mechanisms of sonar pulse production by Oilbirds. *Journal of the Acoustical Society of America*, 74 (Suppl. 1), 31.

Systad, G.H. & Bustnes, J.O. 2001. Coping with darkness and low temperatures: foraging strategies in Steller's eiders, Polysticta stelleri, wintering at high latitudes. *Canadian Journal of Zoology-Revue Canadienne de Zoologie*, 79, 402–406.

Taborsky, B. & Taborsky, M. 1992. Spatial organization of the North Island Brown Kiwi—sex, pairing status and territoriality. *Ibis*, 134, 1–10.

Taborsky, B. & Taborsky, M. 1995. Habitat use and selectivity by the brown kiwi (*Apteryx australis mantelli*) in a patchy environment. *Auk*, 112, 680–689.

The Avian Brain Nomenclature Consortium 2005. Avian brains and a new understanding of vertebrate evolution. *Nature Reviews Neuroscience*, 6, 151–159.

Theurich, M., Langner, G., & Scheich, H. 1984. Infrasound responses in the midbrain of the Guinea Fowl. *Neuroscience Letters*, 49, 81–86.

Thomas, B.T. 1999, 'Family Steatornithidae (Oilbird).' In: *Handbook of the Birds of the World, Vol. 5. Barn-owls to Hummingbirds*. J. del Hoyo, A. Elliott, & J. Sargatal, eds, Barcelona, Lynx Edicions, pp. 244–251.

Thoms, M. 2014. *Owls; A Natural History of British and Irish Species*. London, William Collins.

Thorn, F. & Boynton, R.M. 1974. Human binocular summation at absolute threshold. *Vision Research*, 14, 445–458.

Thorpe, J. 2003. Fatalities and destroyed civil aircraft due to bird strikes, 1912–2002. *IBSC26*, WPSA1.

Tinbergen, N. 1953. *The Herring Gull's World*. London, Collins.

Tennyson, A.J.D., Palma, R.L., Robertson, H.A., Worthy, T.H., & Gill, B.J. 2003. A new species of kiwi (Aves, Apterygiformes) from Okarito, New Zealand. *Records of the Auckland Museum*, 40, 55–64.

Tordoff, M.G., et al. 2008. Involvement of T1R3 in calcium-magnesium taste. *Physiological Genomics*, 34, 338–348.

Troscianko, J., von Bayern, A.M.P., Chappell, J., Rutz, C., & Martin, G.R. 2012. Extreme binocular vision and a straight bill facilitate tool use in New Caledonian crows. *Nature Communications*, 3, 1110.

Tucker, V.A. 2000. The deep fovea, sideways vision and spiral flight paths in raptors. *Journal of Experimental Biology*, 203, 3745–3754.

Tucker, V.A., Tucker, A.E., Akers, K., & Enderson, J.H. 2000. Curved flight paths and sideways vision in peregrine falcons *(Falco peregrinus)*. *Journal of Experimental Biology*, 203, 3755–3763.

Tyler, J.E. & Smith, R.C. 1970. *Measurements of Spectral Irradiance Under Water*. New York, Gordon and Breach.

Tyrell, L.P., Moore, B.A., Loftis, C., & Fernandez-Juricic, E. 2013. Looking above the prairie: localized and upward acute vision in a native grassland bird. *Scientific Reports*, 3231, doi:10.1038/srep03231.

Tyrrell, L.P., Moore, B.A., Loftis, C., & Fernandez-Juricic, E. 2013. Looking above the prairie: localized and upward acute vision in a native grassland bird. *Scientific Reports*, 3231, doi:10.1038/srep03231.

Vallortigara, G. 2000. Comparative neuropsychology of the dual brain: a stroll through animals' left and right perceptual worlds. *Brain and Language*, 73, 189–219.

van den Hout, P. & Martin, G.R. 2011. Extreme head-tilting in shorebirds: predator detection and sun avoidance. *Wader Study Group Bulletin*, 118, 18–21.

Veltman, C.J., Collier, K.J., Henderson, I.M., & Newton, L. 1995. Foraging ecology of blue ducks *Hymenolaimus malacorhynchos* on a New Zealand river: implications for conservation. *Biological Conservation*, 74, 187–194.

Veltman, C.J. & Williams, M. 1990. Diurnal use of time and space by breeding blue ducks. *Wildfowl*, 42, 65–86.

van Heezik, Y.M., Gerritsen, A.F.C., & Swennen, C. 1983. The influence of chemoreception on the foraging behaviour of two species of sandpipers, *Calidris alba* and *Calidris alpina*. *Netherlands Journal of Sea Research*, 17, 47–56.

Viitala, J., Korpimaki, E., Palokangas, P., & Koivula, M. 1995. Attraction of kestrels to vole scent marks visible in ultraviolet light. *Nature*, 373, 425–426.

Vorobyev, M. 2003. Coloured oil droplets enhance colour discrimination. *Proceedings of the Royal Society B: Biological Sciences*, 270, 1255–1261.

Vorobyev, M., Marshall, J., Osorio, D., & Hempel de Ibarra, N. 2001. Colourful objects through animal eyes. *Color Research Applications*, 26, S214–S217.

Vorobyev, M. & Osorio, D. 1998. Receptor noise as a determinant of colour thresholds. *Proceedings of the Royal Society B: Biological Sciences*, 265, 351–358.

Vorobyev, M., Osorio, D., Bennett, A.T.D., Marshall, N.J., & Cuthill, I.C. 1998. Tetrachromacy, oil droplets and bird plumage colours. *Journal of Comparative Physiology A*, 183, 621–633.

Voss, J. & Bischoff, H.J. 2009. Eye movements of laterally eyed birds are not independent. *Journal of Experimental Biology*, 212, 1568–1575.

Wade, N. 2000. *A Natural History of Vision*. Cambridge, MA, MIT Press.

Wallace, B.P., Kot, C.Y., DiMatteo, A.D., Lee, T., Crowder, L.B., & Lewison, R.L. 2013. Impacts of fisheries bycatch on marine turtle populations worldwide: towards conservation and research priorities. *Ecosphere*, 4, 1–49.

Wallman, J. & Pettigrew, J.D. 1985. Conjugate and disjunctive saccades in two avian species with contrasting oculomotor strategies. *Journal of Neuroscience*, 5, 1418–1428.

Walls, G.L. 1942. *The Vertebrate Eye and its Adaptive Radiation*. MI, Cranbrook Institute of Science.

Walsh, E.J., Whalen, C., Powell, L., Bomberger Brown, M., & McGee, J. 2015. Hearing sensitivity in the Greater Prairie Chicken *Tympanuchus cupido*. *Journal of the Acoustical Society of America*, 137, 2219.

Wang, J., Barkan, J., Fisler, S., Godinez-Reyes, C., & Swimmer, Y. 2013. Developing ultraviolet illumination of gillnets as a method to reduce sea turtle bycatch. *Biology Letters*, 9, doi: 10.1098/rsbl.2013.0383.

Wang, J.H., Fisler, S., & Swimmer, Y. 2010. Developing visual deterrents to reduce sea turtle bycatch in gill net fisheries. *Marine Ecology Progress Series*, 408, 241–250.

Wanless, S., Finney, S.K., Harris, M.P., & McCafferty, D.J. 1999. Effect of the diel light cycle on the diving behaviour of two bottom feeding marine birds: the blue-eyed shag *Phalacrocorax atriceps* and the European shag *P. aristotelis*. *Marine Ecology Progress Series*, 188, 219–224.

Warrant, E. 1999. Seeing better at night: life style, eye design and the optimum strategy of spatial and temporal summation. *Vision Research*, 39, 1611–1630.

Warrant, E. 2008, 'Nocturnal vision.' In: *The Senses: A Comprehensive Reference, Vol. 2. Vision II*. A.I. Basbaum et al., eds, Amsterdam, Elsevier, pp. 53–86.

Warren, J. 2009. *The Cambridge Companion to Epicureanism*. Cambridge, Cambridge University Press.

Watanuki, Y., et al. 2008. Microhabitat use and prey capture of a bottom-feeding top predator, the European shag, shown by camera loggers. *Marine Ecology Progress Series*, 356, 283–293.

Watanuki, Y., Takahashi, A., Daunt, F., Sato K, Miyazaki, N., & Wanless, S. 2007. Underwater images from bird-borne cameras provide clue to poor breeding success of European Shags in 2005. *British Birds*, 100, 466–470.

Weiner, J. 1994. *The Beak of the Finch: A Story of Evolution in Our Time*. New York, Alfred a Knopf.

Wendland, V. 1984. The influence of prey fluctuations on the breeding success of the Tawny Owl (*Strix aluco*). *Ibis*, 126, 284–295.

Wenzel, B. 1968. Olfactory prowess of the Kiwi. *Nature*, 220, 1133–1134.

Werner, S.J., Kimball, B.A., & Provenza, F.D. 2008. Food color, flavor, and conditioned avoidance among red-winged blackbirds. *Physiology & Behavior*, 93, 110–117.

Wesolowski, T. & Maziarz, M. 2012. Dark tree cavities—a challenge for hole nesting birds? *Journal of Avian Biology*, 43, 454–460.

Westheimer, G. 1972, 'Visual acuity and spatial modulation thresholds.' In: *Handbook of Sensory Physiology*. D. Jameson & L.M. Hurvich, eds, Berlin, Springer-Verlag, pp. 170–187.

White, C.R., Butler, P.J., Grémillet, D., & Martin, G.R. 2008. Behavioural strategies of cormorants (Phalacrocoracidae) foraging under challenging light conditions. *Ibis*, 150 (Suppl. 1), 231–239.

White, C.R., Day, N., Butler, P.J., & Martin, G.R. 2007. Vision and foraging in cormorants: more like herons than hawks? *PLoS One*, i2(7), e639.

Whittaker, D.J., Gerlach, N.M., Soini, H.A., Novotny, M.V., & Ketterson, E.D. 2013. Bird odour predicts reproductive success. *Animal Behaviour*, 86, 697–703.

Wilkie, S.E., Robinson, P.R., Cronin, T.W., Poopalasundaram, S., Bowmaker, J.K., & Hunt, D.M. 2000. Spectral tuning of avian violet- and ultraviolet-sensitive visual pigments. *Biochemistry*, 39, 7895–7901.

Williams, T.D. 1995. *The Penguins*. Oxford, Oxford University Press.

Wilson, K.-J. 2004. *Flight of the Huia: Ecology and Conservation of New Zealand's Frogs, Reptiles, Birds and Mammals*. Christchurch, Canterbury University Press.

Wilson, R.P., Puetz, K., Bost, C.A., Culik, B.M., Bannasch, R., Reins, T., & Adelung, D. 1993. Diel dive depth in penguins in relation to diel vertical migration of prey: whose dinner by candlelight? *Marine Ecology Progress Series*, 94, 101–104.

Wiltschko, R., Stapput, K., Thalau, P., & Wiltschko, W. 2010. Directional orientation of birds by the magnetic field under different light conditions. *Journal of the Royal Society Interface*, 7, S163–S177.

Wiltschko, R. & Wiltschko, W. 1995. *Magnetic Orientation in Animals*. Berlin, Heidelberg, New York, Springer-Verlag.

Wiltschko, R. & Wiltschko, W. 2006. Magnetoreception. *BioEssays*, 28, 157–168.

Wiltschko, R. & Wiltschko, W. 2013. The magnetite-based receptors in the beak of birds and their role in avian navigation. *Journal of Comparative Physiology A*, 199, 89–98.

Wiltschko, W., Munro, U., Ford, H., & Wiltschko, R. 2003. Lateralisation of magnetic compass orientation in silvereyes, Zosterops lateralis. *Australian Journal of Zoology*, 51, 597–602.

Wiltschko, W., Traudt, J., Gunturkun, O., Prior, H., & Wiltschko, R. 2002. Lateralization of magnetic compass orientation in a migratory bird. *Nature*, 419, 467–470.

Wittgenstein, L. 1978. *Remarks on Colour*. Berkeley, University of California Press.

Wood, C.A. 1917. *The Fundus Oculi of Birds Especially as Viewed by the Ophthalmoscope*. Chicago, Lakeside Press.

Wood-Gush, D.G.M. & Kare, M.R. 1966. The behaviour of calcium deficient chicken. *British Poultry Science*, 7, 285–290.

Woodhouse, J.M. & Campbell, F.W. 1975. The role of the pupil light reflex in aiding adaptation to the dark. *Vision Research*, 15, 649–653.

Worthy, T.H. & Holdaway, R.N. 2002. *The Lost World of the Moa: Prehistoric Life in New Zealand*. Christchurch, Canterbury University Press.

Wright, A.A. 1979, 'Color-vision psychophysics.' In: *Neural Mechanisms of Behavior in the Pigeon*. A.M. Granda & J.H. Maxwell, eds, New York, Plenum Press.

Wright, W.D. 1963. The rays are not coloured. *Nature*, 198, 1239–1244.

Wyatt, T.D. 2003. *Pheromones and Animal Behaviour: Communication by Smell and Taste*. Cambridge, Cambridge University Press.

Zeigler, H.P., Jager, R., & Palacios, A.G. 1993, 'Sensorimotor mechanisms and pecking in the pigeon.' In: *Vision, Brain and Behavior in Birds*. H.P. Zeigler & H.-J. Bischof, eds, Cambridge, MA, MIT Press, pp. 265–283.

Zhang, J.X., Wei, W., Zhang, J.H., & Yang, W.H. 2010. Uropygial gland-secreted alkanols contribute to olfactory sex signals in Budgerigars. *Chemical Senses*, 35, 375–382.

Zhang, Y.H., Du, Y.F., & Zhang, J.X. 2013. Uropygial gland volatiles facilitate species recognition between two sympatric sibling bird species. *Behavioral Ecology*, 24, 1271–1278.

Zheng, X., et al. 2013. Hind wings in basal birds and the evolution of leg feathers. *Science*, 339, 1309–1312.

Zweers, G.A. & Wouterlood, F.G. 1973. Functional anatomy of the feeding apparatus of the mallard (*Anas platyrhynchos*). *Proceedings of the 3rd European Anatomical Congress, Manchester*, 1, 88–89.

Zydelis, R., Small, C., & French, G. 2013. The incidental catch of seabirds in gillnet fisheries: a global review. *Biological Conservation*, 162, 76–88.

Index

Note: The Scientific and Common English names of species follows the International Ornithological Congress World Bird List. Gill, F and D Donsker (Eds). 2016. IOC World Bird List (v 6.2). http://www.worldbirdnames.org/ This list is under constant revision. Both the species English names and the scientific names used here may not be the same as those used in the sources cited in the list of References.

A

absolute visual threshold
  Rock Dove 122
  Tawny Owl 122
  Human 122
aerial photographs 1
Aerodramus spodiopygius White-rumped Swiflet 81
Accipiter nisus Eurasian Sparrowhawk 48
acuity 42, 58
  and distance 118
  comparative Tables 59, 247–52
Aethia cristatella Crested Auklet 90
African Penguin *Spheniscus demersus* 12
Agelaius phoeniceus Red-winged Blackbird 106
aircraft and collisions 217
albatrosses Diomedeidae
  foraging 162
Alcedo atthis Common Kingfisher 13
ambient light 3
  at night 120
  underwater 167
American Robin *Turdus migratorius* 51
American Kestrel *Falco sparverius*
  acuity 59
  contrast sensitivity 63
Anas clypeata Northern Shoveler 7
Anas penelope Eurasian Wigeon 7
Anas platyrhynchos Mallard 99
Antarctic Prion *Pachyptila desolata*
  olfactory individual recognition 90
Aptenodytes patagonicus King Penguin 169
Anthropoides paradiseus Blue Cranes 218
Apterygidae kiwi 27
Apteryx australis Southern Brown Kiwi 102
Apteryx mantelli North Island Brown Kiwi 136
Apus apus Common Swift 86
Aquila audax Wedge-tailed Eagle 59
Ardeotis kori Kori Bustard 55
Asio flammeus Short-eared Owl 128
Asio otus Long-eared Owl 76
Atlantic Canary *Serinus canaria*
  sound localization 74
Atlantic Puffin *Fratercula arctica*
  foraging 181
  visual field 193
audiogram 70
  birds 71
  human 71
audition, *see* hearing
auks Alcidae
  underwater foraging on evasive prey 162, 180
Australian Silver Eye *Zosterops lateralis*
  magnetoreception 111
Aythya marila Greater Scaup 162

B

Bar-tailed Godwit *Limosa lapponica*
  occasional nocturnal foraging 151
Balearica regulorum Grey Crowned Crane 218
beak, *see* bill
bird's eye view 2, 4, 159, 243
bill tip organs 98–103
  ibises 101
  kiwi 101
  parrots 99
  shorebirds 101
  waterfowl 98
bill shape 4
bill structure 6
bill, control of position 194, 196
bill, control of timing 194
binocular field 55, 192
  abolition 205

binocular vision, function 198
  and bill control 196
  and nocturnality 204, 206
  and outer ears in owls 204, 206
  see also contralateral vision
bird defined
  a bill guided by and eye 215
  a wing guided by an eye 185
bird's eye view 2, 4, 159, 243
Black-crowned Night Heron *Nycticorax nycticorax*
  binocularity and nocturnality 206
Black Skimmer *Rynchops niger*
  binocularity and nocturnality 206
Black Vulture *Coragyps atratus*
  ears 69
Blind foraging in Common Guillemots 181
Blind Mole Rat *Spalax ehrenbergi* 67
Blue Cranes *Anthropoides paradiseus*
  collisions with power lines 218
  not looking ahead 227–8
  visual fields 228
Blue Duck *Hymenolaimus malacorhynchos*
  tactile foraging underwater 171
  visual fields 171
Blue Petrel *Halobaena caerulea*
  olfactory individual recognition 90
body odours 89
Bobolink *Dolichonyx oryzivorus*
  magnetoreception 111
Bourke's Parrot *Neopsephotus bourkii* 46
Branta canadensis Canada Goose 47
brow ridges 208
Brown Falcon *Falco berigora*
  acuity 61
Brown-headed Cowbird *Molothrus ater*
  and collisions 234
Bubo scandiacus Snowy Owl 128
Bubo virginianus Great Horned Owl 59
Bubulcus ibis Western Cattle Egret 193
Budgerigar *Melopsittacus undulatus*
  contrast sensitivity 63
  ears 69
  ganglion cell distribution 46
  olfactory species recognition 89
  sound localization 74
Buff-necked Ibis *Theristicus caudatus*
  bill tip organ 101
Bustard, *see* Denham's Bustard; Kori Bustard; Ludwig's Bustard
Buzzard, *see* Common Buzzard
ducks 6
Buteo buteo Common Buzzard 49

C

Calidris alba Sanderling 107
Calidris alpina Dunlin 107
Calidris canutus Red Knot 107
Calidris maritima Purple Sandpiper 107
California Quail *Callipepla californica*
  retina 13
Callipepla californica California Quail 13
Calonectris borealis Cory's Shearwater 86
camera eye 19, 28, 31
Cambrian Explosion 19
Canada Goose *Branta canadensis* 47
  collisions with aircraft 233
Caprimulgus europaeus European Nightjar 146
car driving and collisions 224
Carpodacus mexicanus House Finch 91
Cathartes aura Turkey Vulture 87
Cattle Egret *Bubulcus ibis*
  visual field 193
cave dwelling birds 79
cave swiftlets Apodidae 79
cetaceans, hearing 67
Chicken (Red Jungle Fowl) *Gallus gallus*
  acuity 61
  hearing 72
  magnetoreception 110
  taste, bitter 106
  taste, calcium 106
  taste salt 106
  taste sour, 106
  taste, sweet 105
  taste, umami 106
Chordeiles minor Common Nighthawk 145
Ciconia ciconia White Stork 55
Circaetus gallicus Short-toed Snake Eagle 208
Clangula hyemalis Long-tailed Duck 162
cloud cover, and light levels 122
Cockatiel *Nymphicus hollandicus*
  taste, sweet 105
cognition and nocturnality 130, 137
collisions 217–235
  and car driving 224–5
  and predatory escape response 234
  with gillnets 235–41
  with glass panes 241
  with power lines 217–9
  with wind turbines 217–9
  with vehicles 233–4
colour discrimination 23
colour vision 21, 23, 34, 36
  evolutionary drivers 213
  tetrachromatic 40
Columba livia Rock Dove 35
Columba palumbus Common Wood Pigeon 12

Common Buzzard *Buteo buteo*
  fovea 49
Common Chaffinch *Fringilla coelebs*
  sound ranging 78
Common Guillemot *Uria aalge*
  foraging 181
  visual field 193
Common Kingfisher *Alcedo atthis*
  retina 13
Common Mure *see* Common Guillemot
Common Nighthawk *Chordeiles minor* 145
Common Nightingale *Luscinia megarhynchos*
  nocturnal behaviour 127
Common Ostrich *Struthio camelus* 13, 33, 58,
  precision pecking 200
Common Redshank *Tringa totanus*
  occasional nocturnal foraging 151
Common Starling *Sturnus vulgaris*
  olfaction and nest material 92
  schematic eye 54
  sound localization 74
  taste, bitter 106
  taste, sweet 105
  taste, umami 106
Common Swift *Apus apus*
  olfactory navigation 86
Common Wood Pigeon *Columba palumbus*
  retina 12
cones 34, 36, 38
contralateral vision 210–212
contrast sensitivity 61
control of bill position 194
Coragyps atratus Black Vulture 69
cormorant, *see* Great Cormorant
cormorants Phalacrocoracidae
  underwater foraging on evasive prey 162
cornea 31, 164
Corvidae 34
Corvus corax, Northern Raven 12
Corvus moneduloides New Caledonian
    Crow 201–2
Cory's Shearwater *Calonectris borealis*
  olfactory navigation 86
  olfactory species recognition 89
Crested Auklet *Aethia cristatella*
  olfactory recognition of individual
    quality 90
crytochromes 111
cutaneous nociceptors 98
Cyanocitta stelleri Steller's Jay 13

D

dark adaptation 25
  and diving 179
  and gillnet by catch 238
Dark-eyed Junco *Junco hyemalis*
  olfactory species recognition 89
daylight 26, 124
day length 121
Denhams' Bustard *Neotis denhami*
  collisions with power lines 218
dietary change and vision 197
dimethyl sulphide (DMS) 87
display behaviours 23, 35
divers Gaviidae
  underwater foraging on evasive
    prey 162
diving petrels Pelecanoididae
  underwater foraging on evasive
    prey 162
double cones 38, 40
dove, *see* Rock Dove
DMS (dimethyl sulphide)
  and foraging in seabirds 87
Dolichonyx oryzivorus Bobolink 111
Dromaius novaehollandiae Emu 85
Dumetella carolinensis Grey Catbird 86
Dunlin *Calidris alpina*
  taste and foraging 107,154
Dusenbery, David 11

E

eagle, *see* Wedge-tailed Eagle
ears 8
  Western Barn Owl 76
  Black Vulture 69
  Budgerigar 69
  Guinea-fowl 69
  Long-eared Owl 76
  and binocularity 204
Eastern Meadowlark *Sturnella magna* 47
Eastern Towhee *Pipilo erythrophalmus*
  sound localization 74
Echolocation (active SONAR) 78
  Cave Swiftlets 80
  object detection thresholds 81
  Oilbirds 80
Ecology of Vision 16
eider ducks *Somateria spp.*
  under water foraging on sessile prey 162
elephants, hearing 67
Emu *Dromaius novaehollandiae*
  olfactory bulb size 85
Epicurus 9
Erithacus rubecula European Robin 111
Eudyptula minor Little Penguin 158
Eurasian Blackcap *Sylvia atricapilla*
  star compass 156

Eurasian Curlew *Numenius arquata*
  foraging technique 153
  occasional nocturnal feeding 153
Eurasian Oystercatcher *Haematopus ostralegus*
  occasional night foraging 151
Eurasian Sparrowhawk *Accipiter nisus* 48
Eurasian Spoonbill *Platalea leucorodia*
  foraging 203
  visual field 203
Eurasian Wigeon *Anas Penelope* 7
  visual field 172
Eurasian Woodcock *Scolopax rusticola*
  occasional night time feeding 152
  visual fields 155
European Nightjar *Caprimulgus europaeus*
  diet 146
  foraging technique 146
  jaw apparatus 147
  nocturnal foraging 146
  rictal bristles 146
  touch sensitivity 147
  visual field 148
European Robin *Erithacus rubecula*
  magnetoreception 111
European Storm Petrel *Hydrobates pelagicus*
  olfactory individual recognition 90
evolution of eyes 19
eye lashes 208
eye regressive evolution 139
eye size and imaging the sun 207
eye types 22

F

Falco berigora Brown Falcon 61
Falco peregrinus Peregrine Falcon 49
Falco sparverius American Kestrel 59
Falcon, *see* American Kestrel; Brown Falcon; Peregrine Falcon
feet, control of position in prey capture 209
field of view 21
fishing nets *see* gillnets
flamingo, *see* Greater Flamingo; Lesser Flamingo
Flappet Lark *Mirafra rufocinnamomea* 66
flight, perceptual challenges 187
f-number 52, 125, 133, 180
focal length 52
foraging
  at night 127–156
  control of bill position 194
  perceptual challenges 188
  underwater 161–82
  and olfaction 88
  and taste 107
  and tactile information 98–103
  and visual fields 209
foraging underwater
  and light level challenges 166
  and optical challenges 163
  and spectral challenges 166
  and visual field changes 165
fovea 47, 48
Fratercula arctica Atlantic Puffin 181
Fringilla coelebs Common Chaffinch 78
frogmouths (Podargidae) 145–151
Fulmarus glacialis Northern Fulmar 46
fundus oculi 12

G

galliforms 45, 50
Gallus gallus Chicken (Red Jungle Fowl) 61
ganglion cells 33
  see also image analysis
ganglion cell distribution patterns 44, 53, 212
gannet, *see* Northern Gannet
gannets Sulidae
  underwater foraging on evasive prey 162
Garden Warbler *Sylvia borin*
  magnetoreception 111
Gelochelidon nilotica Gull-billed Tern 224
geomagnetic field 108
Geronicus eremita Northern Bald Ibis 196
gillnets and entrapment of diving birds 235–41
  bycatch species 235–6
  lighting nets 238
  marking nets 237–41
  mitigations 239–40
  vision and bycatch 236
Glossy Ibis *Plegadis falcinellus*
  bill tip organ 101
Golden Plover *Pluvialis apricaria*
  hearing and foraging 154
  binocularity and nocturnality 206
Golden-winged Warbler *Vermivora chrysoptera* 72
Grandry corpuscles 97
Great Cormorant *Phalacrocorax carbo* 18
  acuity and contrast 176
  acuity underwater 175
  eye movements 205
  flush-foraging close-capture 174
  foraging technique 174
  modelled vision 177
  nostrils 84
  visual fields 178, 205
Great Horned Owl *Bubo virginianus*
  acuity 59, 123
Great Tit *Parus major*
  olfactory foraging 88
  sound localization 74
  sound ranging 78
Greater Flamingo *Phoenicopterus roseus* 13

Greater Scaup *Aythya marila*
underwater foraging on sessile prey 162
grebes Podicipedidae
underwater foraging on evasive prey 162
Grey Catbird *Dumetella carolinensis*
olfactory navigation 86
Grey Crowned Crane *Balearica regulorum*
collisions with power lines 218
Grey Plover *Pluvialis squatarola*
occasional nocturnal foraging 151
Griffon Vulture *Gyps fulvus* 18
brow ridges 208
not looking ahead 227–8, 230–1
visual field 193
Gull-billed Tern *Gelochelidon nilotica*
lateral vision and foraging 224
gulls Laridae 34
Gyps fulvus Griffon Vulture 18

H

hair cells 70
Haematopus ostralegus Eurasian Oystercatcher 151
Halobaena caerulea Blue Petrel 90
hearing 27, 66
absolute threshold 72
hearing frequency range 69
Helmeted Guinea-fowl *Numida meleagris*
hearing 72
Herbst corpuscles 96
House Finch *Carpodacus mexicanus*
olfactory recognition of individual quality 91
House Sparrow *Passer domesticus*
acuity, 59
human, absolute visual threshold 122
acuity 59
f-number 126
visual field 56
Humboldt Penguin *Sphensicus humboldti* 37
Hummingbirds (Trochilidae)
and optic flow-fields 211
taste, sweet 105
Hydrobates pelagicus European Storm Petrel 90
Hymenolaimus malacorhynchos Blue Duck 171

I

Ibis
Buff-necked 101
Glossy 101
Madagascar Crested 102
Northern Bald 196
Puna 196
illumination, natural 120–122
infra-sound 72
Indian Peafowl *Pavo cristatus* 37
inner ear 70
image analysis 28, 29, 33

J

jay, *see* Steller's Jay
Junco hyemalis Dark-eyed Junco 89

K

Kakapo *Strigops habroptilus*
diet 141
eye size 143
longevity 143
nocturnal behaviour 141
olfaction 143
sedentary habit 143
kingfisher, *see* Common Kingfisher
King Penguin *Aptenodytes patagonicus*
rapid light levels changes and foraging 169
eye optics 179
eye pupil 179
f-number 180
Kite, *see* RedKite
kiwi Apterygidae 27, 36, 101
bill tip organ 101–2
nocturnality 134
nostrils 84, 102
see also North Island Brown Kiwi; Southern Brown Kiwi
Kori Bustard *Ardeotis kori*
collisions with power lines 218
not looking ahead 227–8
visual field 56

L

lateral vision 223
lateralized brain function 223
Leach's Storm-petrel *Oceanodroma leucorhoa* 46
olfactory recognition of nest 92
lens 31
Lesser Flamingo *Phoeniconaias minor*
chick provisioning 199
visual field 199
linear area 43, 47
Limosa lapponica Bar-tailed Godwit 151
Little Penguin *Eudyptula minor*
nocturnal nest attendance 158
Long-eared Owl *Asio otus*
outer ears 76
Long-tailed Duck *Clangula hyemalis*
underwater foraging on sessile prey 162
looking but failing to see 224
loons, *see* divers
Lophotibis cristata Madagascar Crested Ibis 102
Ludwing's Bustard *Neotis ludwigii*
collisions with power lines 218

Luscinia luscinia Thrush Nightingale 127
Luscinia megarhynchos Common Nightingale 127
Lythgoe, John 14, 16

M

Madagascar Crested Ibis *Lophotibis cristata*
  bill tip organ 102
magnetite 111
magnetoreception
  and homing 108
  and navigation 108
  crytochromes 111
  magnetite 110
  mechanisms 110
  radical pair model 111
Malacorhynchus membranaceus Pink-eared Duck 173
Mallard *Anas platyrhynchos*
  bill tip organ 99
  nocturnal foraging 154
  taste buds 104
Manx Shearwater *Puffinus puffinus* 49
  nocturnal nest attendance 158
  retina 53
  schematic eye 52
mechanoreception 96–98
Melopsittacus undulatus Budgerigar 46
middle ear 70
microbats, hearing 67
migration at night
  see nocturnal migration
Milvus milvus Red Kite 48
Mirafra rufocinnamomea Flappet Lark 66
Mimus polyglottos Northern Mockingbird 128
Molothrus ater Brown-headed Cowbird 234
moonlight 26, 120
Morus Bassanus Northern Gannet 18

N

natural light environment 3, 120–122
Neopsephotus bourkii Bourke's Parrot 46
Neotis denhami Denhams' Bustard 218
Neotis ludwigii Ludwing's Bustard 218
nest attendance at night 158–159
nest construction and vision 197
New Caledonian Crow *Corvus moneduloides*
  tool use 201
New World Vultures 45
Newton Issac, *Opticks* 22, 42
nightingale, *see* Common Nightingale; Thrush Nightingale
nightjars 145–151
night length 121
Night Parrot *Pezoporus occidentalis*
  nocturnal behaviour 141
night-time 119–125
  light levels 120, 122
nocturnal birds 127
nocturnal migration 156–157
  migratory restlessness, Zugunruhe 156
  star compass 156
nocturnal nest attendance 158–159
  and hearing 158
  and olfaction 158
  and vision 158
nocturnality
  and binocularity 204
  and cognition 130, 137, 14
  and diet 131, 137
  and sedentary life 132
  and territoriality 130, 137
North Island Brown Kiwi *Apteryx mantelli* 136
  see also Southern Brown Kiwi
Northern Bald Ibis *Geronicus eremita*
  visual field and foraging 196
Northern Fulmar *Fulmarus glacialis* 46
Northern Gannet *Morus Bassanus* 18
  plunge diving and optic-flow fields 211
Northern Lapwing *Vanellus vanellus*
  hearing and foraging 154
Northern Mockingbird *Mimus polyglottos*
  nocturnal behaviour 128
Northern Raven *Corvus corax 12*
Northern Shoveler *Anas clypeata* 7
  occasional nocturnal foraging 154
nostrils
  in auks 84
  in kiwi 84
  in gannets and cormorants 84
Numenius arquata Eurasian Curlew 153
Numida meleagris Helmeted Guinea-fowl 72
Nycticorax nycticorax Black-crowned Night Heron 206
Nymphicus hollandicus Cockatiel 105

O

Oceanites oceanicus Wilson's Storm Petrel 90
Oceanodroma leucorhoa Leach's Storm-petrel 46
Oilbird *Steatornis caripensis*
  binocularity and nocturnality 206
  echolocation 79, 133
  f-number 133
  nocturnality 132
  olfaction 133
  retina 132
oil droplets 37, 38, 44
Olfaction 83–93
  and foraging 86
  and individual quality recognition 90

and individual recognition 89
and predator detection 86
and navigation 86
and nest material 91
and nest recognition 92
and species recognition 89
olfactory bulb (brain) 85
olfactory system 84
optic flow-field 194, 211
and binocular vision 211
and contralateral vision 211
optics, *see* optical system
optical adnexa 207
optical axis 30
optical system 28, 29, 31, 52
optimal eyes 22
ostriches, Struthioniformes 34
see also Common Ostrich
owls Strigidae 27, 31, 33, 34, 36, 45
owls
and binocularity 204
wisdom of owls 118
see Great Horned Owl; Short-eared Owl; Snowy Owl; Tawny Owl; Western Barn Owl

P

Pachyptila desolata Antarctic Prion 90
panoramic vision 195
Paraque *Nyctidromus albicollis*
binocularity and nocturnality 206
Pardalotus punctatus Spotted Pardalotes 85
parrots 33, 45
bill tip organs 99
see Kakapo; Night Parrot; Senegal Parrot
Parus major Great Tit 74
Passer domesticus House Sparrow 59
passerines 31, 45
Pavo cristatus Indian Peafowl 37
pecten 29
pelicans Pelecanidae
underwater foraging on evasive prey 162
penguin, *see* African Penguin; Humboldt Penguin; King Penguin
penguins Spheniscidae
underwater foraging on evasive prey 162
perch-and-pounce hunting 129
perceptual challenges
flight 187
foraging 188
predator detection 189
reproduction 189
perceptual limit and collisions 224
Peregrine Falcon *Falco peregrinus* 49
prey approach and capture 210
petrels Procellariidae 27
underwater foraging on evasive prey 162
and olfactory foraging 88
see Leach's Storm-petrel
Pezoporus occidentalis Night Parrot 141
Phalacrocorax carbo Great Cormorant 18
Phoeniconaias minor Lesser Flamingo 199
Phoenicopterus roseus Greater Flamingo 13
photoreceptors 33, 38
photopigments 36
pigeon, *see* Common Wood Pigeon; Rock Dove
Pink-eared Duck *Malacorhynchus membranaceus*
visual field 173, 193
Pinnipeds
tactile foraging 182
vibrissae 182
Pipilo erythrophalmus Eastern Towhee 74
Platalea leucorodia Eurasian Spoonbill 203
Plegadis falcinellus Glossy Ibis 101
Plegadis ridwayi Puna Ibis 196
plumage patterns 23, 35
Pluvialis apricaria Golden Plover 154
Pluvialis squatarola Grey Plover 151
Podargus strigoides Tawny Frogmouth 150
Poicephalus senegalus Senegal Parrot 100
power lines and collisions 217
potoos (Nyctibiidae) 145–151
predator detection perceptual challenges 189, 196
and foraging 209
preen oil 89
Procellariidae petrels 27
provisioning of young and vision 197–8
Puffinus pacificus Wedge-tailed Shearwater 37
Puffinus puffinus Manx Shearwater 49
Puna Ibis *Plegadis ridwayi*
visual field and foraging 196
Purple Sandpiper *Calidris maritima*
taste and foraging 107, 154

Q

quail, *see* California Quail

R

radical pair model of magnetoreception 111
rate of gain of information 225
raven, *see* Northern Raven
red field 43
Red Jungle Fowl *Gallus gallus*, *see* Chicken
Red Kite *Milvus milvus* 48
Red Knot *Calidris canutus*
bill tip organ 101
dietary change 197
taste and foraging 107, 154

Red-winged Blackbird *Agelaius phoeniceus*
  taste, umami 106
  taste, bitter 106
remote touch, *see* bill tip organ
reproduction perceptual challenges 189
retina, *see* image analysis
rhodopsin 37
rictal bristles 134, 137
Robin, *see* American Robin; European Robin
Rock Dove *Columba livia* 35, 43, 49,
  absolute visual threshold 122
  acuity 59
  contrast sensitivity 63
  f-number 126
  magnetoreception 110
  precision pecking 200
  retina 53
  schematic eye 52
Rochon-Duvigneaud, André 15
rods 34, 36, 38
Russell, Bertrand 9
Rynchops niger Black Skimmer 206

S

Sanderling *Calidris alba*
  taste and foraging 107, 154
Scepticism 10
shags Phalacrocoracidae
  underwater foraging on evasive prey 162
schematic eye models 52, 54
Scolopacidae shorebirds 27
Scolopax rusticola Eurasian Woodcock 152
semiochemicals 89
Senegal Parrot *Poicephalus senegalus*
  bill tip organ 100
sensory ecology, definition 11
Serinus canaria Atlantic Canary 74
Sextus Empiricus 9
Shearwater, *see* Cory's Shearwater; Manx Shearwater; Wedge-tailed Shearwater
shearwaters Procellaridae
  underwater foraging on evasive prey 162
shorebirds Scolopacidae 27
  occasional nocturnal feeding 153
Short-eared Owl *Asio flammeus*
  nocturnal behaviour 128
Short-toed Snake Eagle *Circaetus gallicus*
  not looking ahead 227
  sunshades 208
signal-to-noise 24
simple eye, *see* camera eye
smell, *see* olfaction, odour
Snowy Owl *Bubo scandiacus*
  nocturnal behaviour 128
somatic sensitivity (touch), 27, 94–103
SONAR, *see* echolocation
sound 68
sound localization 73–8, 130
  direction 73
  distance 77
sound ranging 77
Southern Brown Kiwi *Apteryx australis*
  bill tip organ 102, 137
  diet 137
  eye size 135
  nocturnality 134
  nostrils 102
  olfaction 135
  olfactory bulbs 136
  regressive evolution of eyes
  tactile foraging 135
  visual fields 140
Southern Ground Hornbill *Bucorvus leadbeateri*
  eye lashes 208
  precision grasping 201
  sunshades 208
  visual field
Spalax ehrenbergi Blind Mole Rat 67
Sparrowhawk, *see* Eurasian Sparrowhawk
spatial resolution 23, 25, 58
  see also acuity
Sphensicus humboldti Humboldt Penguin 37
Spotless Starling *Sturnus unicolor*
  olfactory species recognition 89
Spotted Pardalotes *Pardalotus punctatus*
  olfactory bulb size 85
star compass 156
starlight 26, 120
Starling, *see* Common Starling
Steatornis caripensis Oilbird 79
Steller's Jay *Cyanocitta stelleri* 13
Stork, *see* White Stork
storm petrels Hydrobatidae
  underwater foraging on evasive prey 162
Strigidae owls 27
Strigops habroptilus Kakapo 141
Strix aluco Tawny Owl *12*
Struthio camelus Common Ostrich 13
Sturnella magna Eastern Meadowlark 47
Sturnus unicolor Spotless Starling 89
Sturnus vulgaris Common Starling 54
sunlight 120
sunshades 207
  and not looking ahead 227
Sylvia atricapilla Eurasian Blackcap 156
Sylvia borin Garden Warbler 111

T

tactile receptors 8, 94–103
Taeniopygia guttata Zebra Finch 74

tapetum, in nightjars 148
taste 103–108
taste and foraging 107, 154
taste categories
  bitter 106
  calcium 106
  fat 107
  salt 106
  sour 107
  sweet 105
  umami 106
  taste buds 104
  taste genes 104
  taste receptors 104
Tawny Frogmouth *Podargus strigoides*
  eye size 150
  nocturnal behaviour 150
Tawny Owl *Strix aluco* 12,
  absolute visual threshold 122
  acuity 61
  f-number 126
  nocturnal behaviour 129
  visual field 32
  schematic eye 54
  sound localization 76
terrain hypothesis 47, 212
territoriality and nocturnality 130
Theristicus caudatus Buff-necked Ibis 101
thermo-sensitive receptors 97
Thrush Nightingale *Luscinia luscinia*
  nocturnal behaviour 127
time-to-contact and collisions 221
Tinbergen, Niko 16
touch (somatic sensitivity), 27, 94–103
  see also tactile receptors
tree canopy 122
tropic birds Phaethontidae
  underwater foraging on evasive prey 162
Tringa totanus Common Redshank 151
Turdus migratorius American Robin 51
Turkey Vulture *Cathartes aura*
  olfactory guided foraging 87
tympanic membrane 70
Tyto alba Western Barn Owl 13

U

ultrasound 67
ultraviolet light (UV) 34
  and acuity 60
  and collisions with glass panes 242
  and foraging of raptors 214
  and detection of fruits 41
  and plumage patterns 214
Uria aalge Common Guillemot 181
uropygial gland 89

V

Vanellus vanellus Northern Lapwing 154
Vermivora chrysoptera Golden-winged Warbler 72
visual fields 32, 55
  comparative Table 253–8
  general characteristics 191, 193
visual pigments 36
visual streak 43
Vulture, *see* Black Vulture
  Griffon Vulture
  Turkey Vulture
vultures 33
  see also New World vultures

W

Walls, Gordon Llyn 14
waterfowl 45
  bill tip organ 98
Wedge-tailed Eagle *Aquila audax*
  acuity 59
Wedge-tailed Shearwater *Puffinus pacificus* 37
Western Barn Owl *Tyto alba* 13
  acuity 61, 123
  contrast sensitivity 63
  nocturnal behaviour 129
  outer ears 76
  sound localization 75
Western Cattle Egret *Bubulcus ibis*
  visual field 203
  foraging 203
White-rumped Swiflet *Aerodramus spodiopygius*
  echolocation 81
White-tailed Eagle 2
White Stork *Ciconia ciconia*
  visual field 56
wildfowl 33
  occasional nocturnal feeding 153
  see also waterfowl
Wilson's Storm Petrel *Oceanites oceanicus*
  olfactory individual recognition 90
wind turbines and collisions 217
wing shape 5
Wood, Casey Albert 11, 14
woodland canopy 122, 127

Y

yellow field 43

Z

Zebra Finch *Taeniopygia guttata*
  sound localization 74
Zosterops lateralis Australian Silver Eye 111
Zugunruhe 156

The manufacturer's authorised representative in the EU for product safety is Oxford University Press España S.A. of el Parque Empresarial San Fernando de Henares, Avenida de Castilla, 2 – 28830 Madrid (www.oup.es/en or product.safety@oup.com). OUP España S.A. also acts as importer into Spain of products made by the manufacturer.

www.ingramcontent.com/pod-product-compliance
Lightning Source LLC
LaVergne TN
LVHW011915200926
848484LV00002B/634
9780199694532